Geology and Geochemistry of Host Rocks

Geology and Geochemistry of Host Rocks

Samantha Bennett

Larsen & Keller
www.larsen-keller.com

Geology and Geochemistry of Host Rocks
Samantha Bennett
ISBN: 979-8-88836-058-3 (Hardback)

Larsen & Keller
Published by Larsen and Keller Education,
5 Penn Plaza,
19th Floor,
New York, NY 10001, USA

Cataloging-in-Publication Data

Geology and geochemistry of host rocks / Samantha Bennett.
p. cm.
Includes bibliographical references and index.
ISBN 979-8-88836-058-3
1. Rock-forming minerals. 2. Minerals. 3. Geology. 4. Geochemistry.
5. Mineralogy. I. Bennett, Samantha.

QE397 .G46 2023
549--dc23

For more information regarding Larsen and Keller Education and its products, please visit the publisher's website www.larsen-keller.com

Table of Contents

Preface

The host rocks are sedimentary rocks that include quartzites, slates, older cherty carbonate and iron formations. There are three types of host rocks, namely, crystalline rocks, clay rocks and salt rocks. In the context of radioactive ore deposits, safe enclosure of the host rocks is affected by technical, geo-technical and geological barriers. In the context of mining, majority of mining wastes are connected to the acquisition of host rock along with other processes that result in the production of metals including copper, aluminum, gold and iron, or nonmetals like gravel, asbestos, coal, gypsum, etc. Host rocks in the vicinity of hydrothermal mineral deposits frequently exchange elements with hydrothermal fluids and thus lead to the formation of alteration minerals. The density and volume of the host rock is affected by mineralogical alterations. This book is a compilation of chapters that discuss the most vital concepts on the geology and geochemistry of host rocks. It will provide comprehensive knowledge to the readers.

This book is a result of research of several months to collate the most relevant data in the field.

When I was approached with the idea of this book and the proposal to edit it, I was overwhelmed. It gave me an opportunity to reach out to all those who share a common interest with me in this field. I had 3 main parameters for editing this text:

1. Accuracy – The data and information provided in this book should be up-to-date and valuable to the readers.

2. Structure – The data must be presented in a structured format for easy understanding and better grasping of the readers.

3. Universal Approach – This book not only targets students but also experts and innovators in the field, thus my aim was to present topics which are of use to all.

Thus, it took me a couple of months to finish the editing of this book.

I would like to make a special mention of my publisher who considered me worthy of this opportunity and also supported me throughout the editing process. I would also like to thank the editing team at the back-end who extended their help whenever required.

Samantha Bennett

1

GeomInt: An Overview

Thomas Nagel, Uwe-Jens Görke, Heinz Konietzky, Jobst Maßmann, Mathias Nest, Holger Steeb, Frank Wuttke, and Olaf Kolditz

1.1 Background

The use of the subsurface as a source of resources, a storage space and for installing underground municipal or traffic infrastructure has become much more intensive and diverse in recent years. In addition to classical anthropogenic interventions such as mining, oil and gas production or tunnel construction, other forms of underground use have come into the focus of economic, political and scientific research, particularly in connection with the transformation of energy systems. These include, for example, the extraction of energy (e.g. geothermal energy) and energy sources using new technologies (e.g. unconventional gas extraction) as well as the geological short-term and long-term storage of energy carriers (e.g. compressed air, hydrogen, methane) and the safe storage of waste generated during energy production or in other industrial sectors (e.g. carbon dioxide, radioactive waste).

T. Nagel (✉) · H. Konietzky
TUBAF, Technische Universität Bergakademie Freiberg, Freiberg, Germany
e-mail: thomas.nagel@ifgt.tu-freiberg.de

U.-J. Görke
UFZ, Helmholtz Centre for Environmental Research, Leipzig, Germany

J. Maßmann
BGR, Federal Institute for Geosciences and Natural Resources, Hannover, Germany

M. Nest
IfG, Institut für Gebirgsmechanik Leipzig, Leipzig, Germany

H. Steeb
UoS, University of Stuttgart, Stuttgart, Germany

F. Wuttke
CAU, Christian-Albrechts-Universität zu Kiel, Kiel, Germany

O. Kolditz
UFZ, Helmholtz Centre for Environmental Research, Technische Universität Dresden, Dresden, Germany

Increasing utilization of the geological environment requires careful analyses of the rock-fluid systems as well as assessments of the feasibility, efficiency and environmental impacts of the technologies under consideration. The establishment of safe, economic and ecological operation of underground geosystems requires a comprehensive understanding of the physical, (geo)chemical and microbiological processes on all relevant time and length scales. This understanding can only be deepened on the basis of intensive laboratory and in-situ experiments in conjunction with reliable studies on the modelling and simulation (numerical experiments) of the corresponding multi-physical/chemical processes.

The rocks and rock mass themselves are characterized by complex material behaviour. Irreversible deformation, rate-dependent strength behaviour as well as creep, swelling and shrinkage effects occur under real loading conditions. Damping aspects, for example in the range of low-frequency seismic waves, damage and physical-chemical aging (e.g. solution and/or precipitation reactions) play an important role in many geotechnical applications and have roots in the multiphase nature of geomaterials. In particular, the topics of damage, crack formation and propagation as well as interface problems are aspects that require further clarification. Because of prevailing knowledge gaps in these areas, many such processes can currently not be adequately modelled with numerical simulation systems commonly used in geotechnical practice. They therefore represent topics in urgent need of research. The corresponding processes manifest themselves in diverse micro- and macromechanical behavioral structures of the considered materials and are here summarized under the umbrella term **discontinuities** (Fig. 1.1).

1.2 The GeomInt Project

GeomInt (Geomechanical Integrity of Host and Barrier Rocks—Experiment, Modeling and Analysis of Discontinuities) contributes to the realistic and application-oriented experimental-numerical analysis of the formation and development of discontinuities in underground rock salt, clay rocks and crystalline rocks. In the following, these rocks will also be briefly referenced as salt, clay and crystalline. The understanding and quantification of interactions with dynamically developing geological rock properties (e.g. permeability), which determine the geomechanical integrity and tightness of geological reservoir-barrier systems, are at the centre of this work. Included in the investigations are discontinuities of volumetrically distributed damage types as they occur in the damage zone of solid rocks, discontinuities that can form uncontrolled or controlled at phase boundary surfaces as well as discrete crack and fracture networks. The pathways created or extended by these discontinuities for fluids in host and barrier rocks harbour the risk that, for example by migration of fluid phases from deep to near-surface geological strata and ultimately into the biosphere, vital ecosystems can be impaired significantly. A range of thermo-hydro-mechanical-chemical (THMC) processes can cause an evolution of discontinuities in the near field of a geotechnical infrastructure. This in turn may lead to the estab-

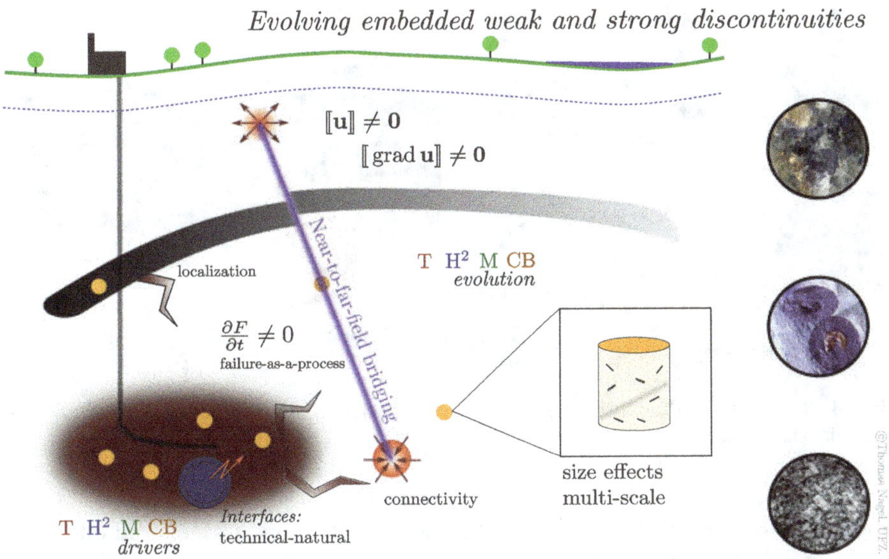

Fig. 1.1 Graphical abstract of the GeomInt project: geological reservoir-barrier systems and geomechanical integrity

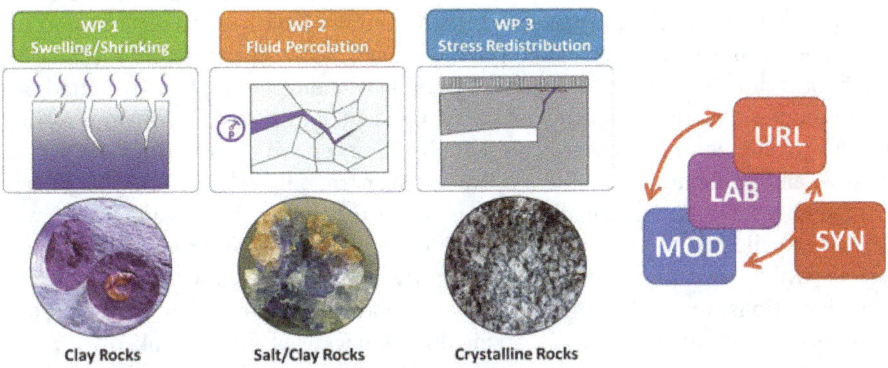

Fig. 1.2 Work package structure of GeomInt according to process and rock types (left) and combining experimental as well as modelling works in a synergistic workflow (right)

lishment of previously not existent connectivity. When connectivity is established to conductive fault zones or fracture networks with a certain range, transport into the far field can become possible on a much shorter time scale than before (Fig. 1.1).

GeomInt is conducted by an interdisciplinary consortium of partners from universities, governmental and private research institutions with complementary, long-term experience in the analysis of underground geosystems. Three typical effects leading to the emergence and development of specific discontinuities are considered as main research areas: Swelling and shrinkage processes, pressure-driven fluid perco-

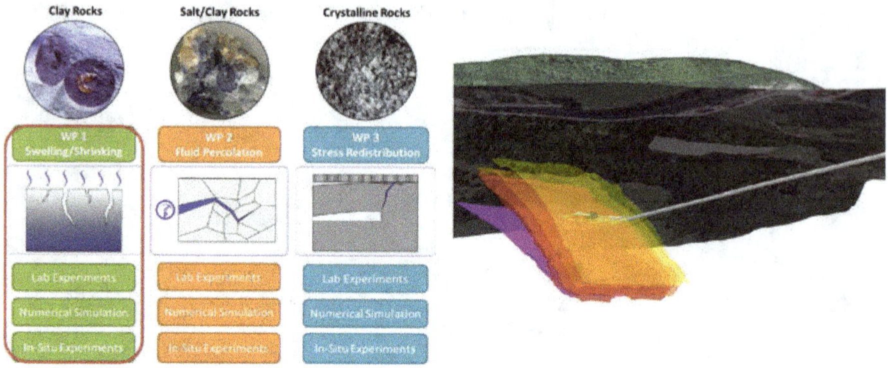

Fig. 1.3 Synthesis: Interlinking process-oriented work packages with VR methods (VR Mt. Terri—visual data and model integration)

lation and stress redistribution (Fig. 1.2). The research work is structured into laboratory experiments, numerical simulation and in-situ experiments (in underground research laboratories, URLs). New insights into process understanding have been gained from the laboratory experiments. In particular, contributions of coupled thermal, hydraulic, and mechanical (THM) processes for the formation and development of discontinuities have been considered. These findings have been used to support the further development of different continuum mechanical, discontinuum mechanical and hybrid numerical approaches and to compare their potentials and limitations. Novel models and algorithms have been implemented in scientific software mainly which is available to the research community. The suitability of the models for the analysis and prediction of realistic operating scenarios of underground geosystems (e.g. geothermal reservoirs, geological reservoirs for energy sources and waste) will be further validated on the basis of test field simulations of different in-situ experiments, which have been carried out primarily using synergies with other national and international research projects in underground laboratories accessible to project partners of GeomInt. The process-oriented work packages are interlinked with synthesis activities such data and model integration using virtual reality (VR) methods (Fig. 1.3). A corresponding pilot demonstrator is being implemented in the Mont Terri project.

The project results allow an improved process understanding. The applied methods and the application-oriented systems for relevant time and length scales will support safer, more reliable and more efficient planning and realization of geotechnical applications. An important advantage of GeomInt is the principle transferability of experimental-numerical concepts, models and methods to a multitude of urgent scientific-technical questions in the geosciences. This allows the exploitation of project results for different politically and socially relevant geotechnological uses (e.g. deep geothermal energy, energy storage, repository problems, methods for hydraulic stimulation, conventional and unconventional resource extraction or tunnel construction). This transferability has been particularly taken into account in the methodological design of the project as well as in the documentation of methods, approaches and results and can at the same time serve as a basis for potential follow-up work to GeomInt.

1.3 GeomInt Approach: Lab, In-situ, In-silico, Virtual Reality

GeomInt relies on a very close link between experimental and modelling work. Three work packages focus on different mechanisms affecting the barrier integrity of potential host rocks: (WP1) swelling/shrinking of clays, (WP2) pressure driven fluid percolation in salt and clay rocks as well as (WP3) stress redistribution in crystalline rocks. Figure 1.4 illustrates the geographical WP workflows from in-situ sampling to geomechanical laboratories and modelling. The main sources for rock samples are (i) Mt. Terri for clay (ii) Springen for salt and (iii) Freiberg/Kirchberg for crystalline rock specimen. Moreover, there is collaboration with other URLs (Bure, Grimsel) concerning experimental and modelling work—mainly for testing transferability of the methodology to other rock types (e.g. Callovo-Oxfordian Clay—COx).

In the following we briefly introduce the "geographical workflow" for the individual work packages. Please not that the experimental and modelling platforms for the analysis of discontinuities are being established independent of specific rock types.

Output from the following research infrastructures will be combined:

- **Rock-mechanical laboratories**
  - THM-coupled testing under controlled boundary conditions
  - Material and process characterization
- **Numerical methods and software**
  - OpenGeoSys (XFEM, PFM)
  - md-LEM (LEM)
  - pythonSPH (SPH)
  - UDEC, 3DEC (DEM)
  - FLAC3D (FDM)
- **Underground laboratories**
  - Springen (rock salt, potash)
  - Mont Terri (clay rock)
  - Reiche Zeche (crystalline rock)

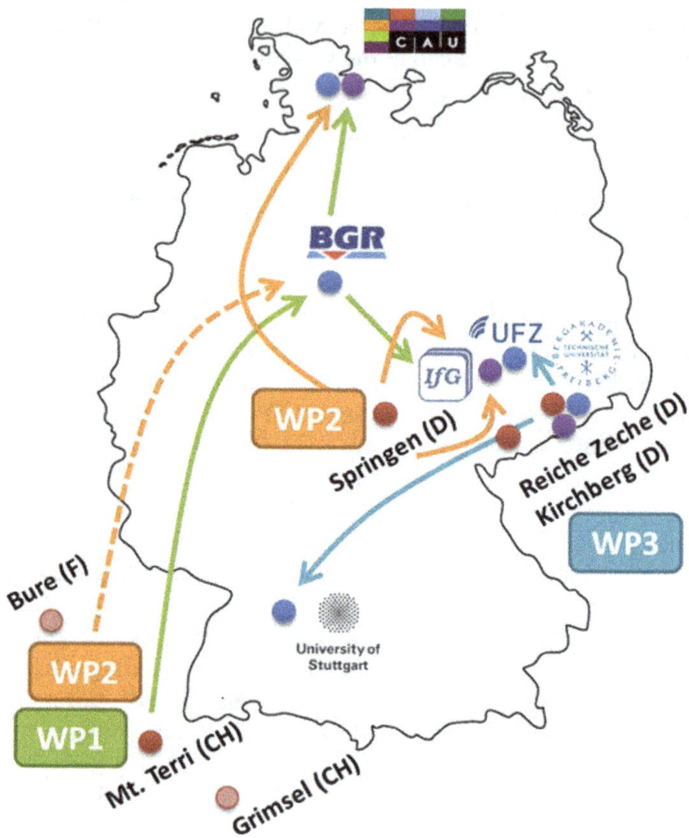

Fig. 1.4 Geographical work flow including interlinked experimental and modeling works in GeomInt

- **Simulation- and development infrastructure**
  - HPC Cluster
  - Version management
  - VISLAB (www.ufz.de/vislab).

In addition to experimental (lab and in-situ) and modelling work, we use Virtual Reality (VR) methods for data and model integration as well as visualization. Figure 1.8 shows a snapshot of the Virtual URL project for Mont Terri. The basic idea is to combine all information from in-situ (and even lab) experiments as well as the corresponding models in a Virtual Reality context—like a visual data base. VR embedded data can be accessed in an interactive and interoperable manner [1].

WP1 is dealing with swelling and shrinking processes in clay stone. BGR obtains **Opalinus Clay** samples from the Mont Terri Underground Research Lab (URL) in Switzerland and providing them for laboratory testing to CAU and IfG partners in Kiel and Leipzig, respectively, cf. Fig. 1.5. Samples come from various clay and

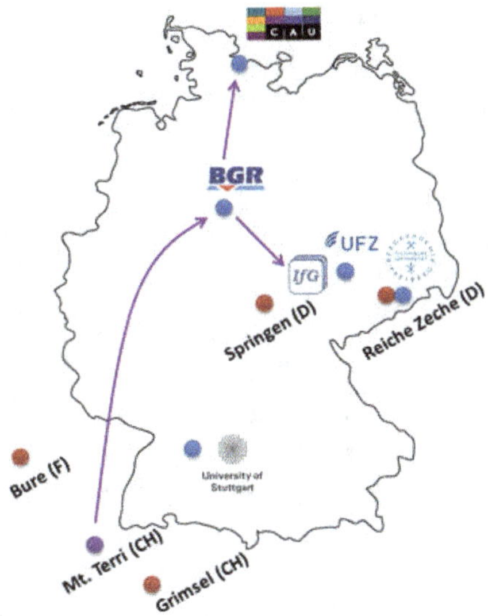

Fig. 1.5 WP1: Swelling/shrinking

sandy facies. Experimental work with specimen from the sandy facies is particularly challenging due to the large heterogeneity of the material. WP1 is closely linked with the Mont Terri Project in cooperation with swisstopo.

Fluid percolation processes are studied in both **clay and salt rocks**, i.e. ductile materials. Samples of Opalinus Clay come from the Mt. Terri URL (see above). Salt rock samples are mainly obtained from the Springen site in Thuringia. Experimental works concerning percolation are conducted in the IfG and CAU labs in Leipzig and Kiel. WP2 investigates mechanisms of percolation threshold for both rock types depending on hydro-mechanical (HM) processes (i.e. fluid pressure and mechanical stress field) (Fig. 1.6).

WP3 is investigating discontinuities formed by stress redistribution in brittle materials. **Granite rock** samples are obtained from locations in the Ore Mountains, i.e. from Kirchberg and Freiberg (URL Reiche Zeche). Experimental investigations are conducted in the Freiberg (TU Freiberg) and Stuttgart labs (University of Stuttgart). Constant Normal Load (CNL) and Constant Normal Stiffness (CNS) experiments are conducted to study fluid flow in rough fractures under confining stresses. Rock samples from Freiberg will be also used with in the "Crystalline Task" of the new DECOVALEX-2023 phase (Fig. 1.7).

The book is organized in the following way: Experimental and modelling platforms are described in detail in Chaps. 2 and 3, respectively. It follows the "Model-Experiment-Exercises" (MEX) combining experimental and modeling work for a "Proof-of-Concept", i.e. are experiments and models adequate to better understand fracturing processes in various rock types (Chap. 4). The Data-Management concept is introduced in Chap. 5 accompanied by examples for data and model storage. The book end with a synthesis and outlook Chap. 6. Lists of symbols, literature references, keywords (index) can be found in the appendices.

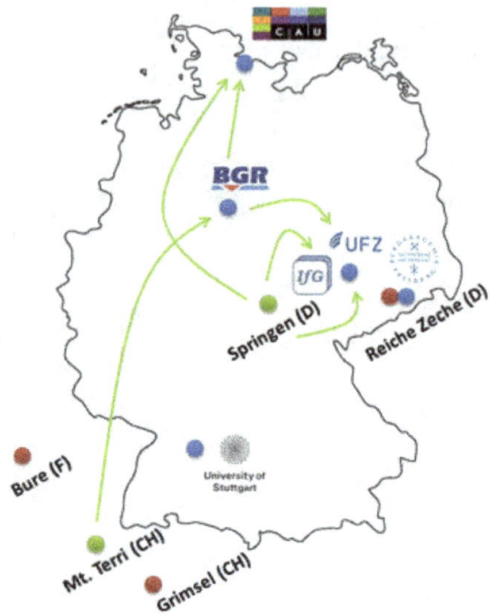

Fig. 1.6 WP2: Fluid percolation

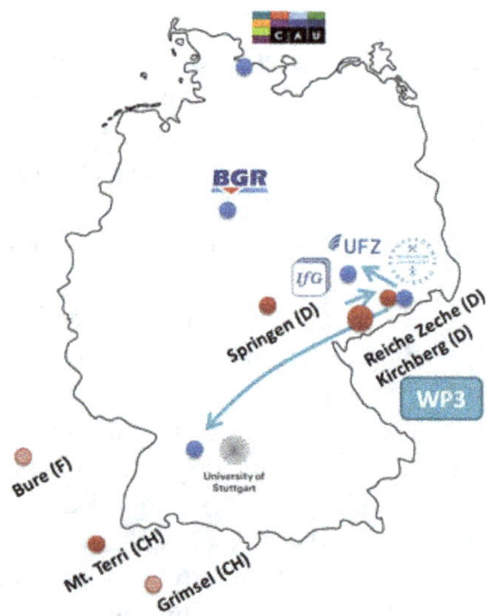

Fig. 1.7 WP3: Stress redistribution

1.4 GeomInt Team

The GeomInt consortium is made up of the partners Federal Institute for Geosciences and Natural Resources, Christian-Albrechts-University of Kiel, Helmholtz Centre for Environmental Research—UFZ, Institut für Gebirgsmechanik GmbH, TU Bergakademie Freiberg and University of Stuttgart. The consortium is based on the complementary expertise and resources of the partners, partly unique laboratory equipment, many years of experience with different numerical approaches

Fig. 1.8 Virtual URL Mont Terri: Virtual data and model integration in a precise geo-referenced context

and access to underground laboratories for the rocks under consideration. Detailed information on the partners and their previous project-relevant work is presented in the following.

1.4.1 BGR

The Federal Institute for Geosciences and Natural Resources (BGR, Germany) is the central geoscientific authority providing advice to the German Federal Government in all geo-relevant questions inter alia the safe final disposal of radioactive waste and the related geological-geotechnical safety analyses. As such, the BGR with its wide-ranging experience has an interest in investigating the host rocks salt, claystone and crystalline rocks to guarantee an independent recommendation for the site selection process. This includes the use of state-of-the-art modelling and experimental techniques. Within the GeomInt project, the BGR is focusing on the integrity of clay rock formations as they are currently investigated in the Mont Terri Underground Rock Laboratory (URL) in Switzerland. The BGR is currently coordinating and participating in several experimental campaigns that are expected to enhance the fundamental understanding of clay rock exposed to stress redistribution, seasonally varying boundary conditions and pressure driven percolation. For the simulation of the hydraulic conditions of unsaturated claystone coupled with unsaturated linear elastic mechanical deformation, a strong collaboration with the Helmholtz Centre for Environmental Research (UFZ) has been pursued to employ the open source code OpenGeoSys for in-situ scale problems of the Mont Terri URL.

1.4.2 CAU

The geomechanical and geotechnical group of Christian-Albrechts-U (CAU) of Kiel has vast experience in the experimental and numerical analysis of coupled thermo-hydro-mechanical (THM) processes in geomaterials. Some research fields which are under investigation and development at CAU Kiel are: material characterization by static, cyclic and dynamic loaded soil-structure-interaction, development of sustainable geomaterial and earth structure, material characterization by geophysical and advanced geotechnical methods, energy geotechnics, energy geo-storages, and cyclic thermo-hydro-mechanical loaded geomaterial. In 2016, the first international conference on energy geotechnics (ICEGT-2016) was organized and hosted in Kiel University. The numerical simulations such as Coupled hybrid models, boundary element modeling (BEM), finite element modeling (FEM), static and dynamic soil-structure-interface and constitutive modeling, cyclic system/macro modeling for complex soil-structure-conditions and loads as well as fracture simulation under coupled THM processes using lattice element method (LEM) have all been developed here and have been applied in different research fields and projects, such as ANGUS, DuoFill, FIBERSLAG, BioSolidEncap and GeomInt.

1.4.3 IfG

The Institut für Gebirgsmechanik GmbH—IfG is one of the leading companies for all questions regarding the use of salt formations, e.g. salt mining, storage, and waste disposal projects. We adopt an interdisciplinary approach, consisting of numerical and experimental geomechanics, as well as geotechnical in-situ field measurements. The company provides fundamental geomechanical/geotechnical safety concepts, and produces evidence reports on rock-mechanical stability and long-term safety in the form of design concepts, site planning procedures and public acceptance procedures. Supported by combinations of in situ observations, geomechanical laboratory testing and numerical modelling we provide comprehensive expertise in rock mechanics, necessary to solve practical mining or repository problems and to reduce project risks. Our research includes such diverse phenomena like creep, strain softening and post-failure behavior, rock burst mechanisms, and pressure-driven percolation in polycrystalline salt rocks (both numerically and experimentally). Overall, the IfG can draw on more than five decades of continuous application of research in the field of rock mechanics.

1.4.4 TUBAF

The Technical University of Freiberg as so called "University of Resources" focuses on energy, resources and materials. The Geotechnical Institute works in the field of utilizing the earth's crust in a safe and sustainable manner. This includes topics like radioactive waste repositories in different host rocks (salt, crystalline, clay), geothermal energy production, gas storage or tunneling. To address these challenges the facilities of the rock mechanical laboratory, the "Reiche Zeche" (the local research mine) and numerical simulations are used. Involving practical and numerical results on different scales helps to give a useful contribution to open questions in the related research topics. The staff of the institute works in different national and international research projects. The topic of deep geothermal energy has been on the agenda since 2008. Particularly the genesis and development of cracks and fracture formations was intensively studied. E.g. the hydraulic fracturing, the induced seismicity or the response of existing cracks to static and dynamic loads have been experimentally and numerically investigated.

1.4.5 UFZ

The Helmholtz Centre for Environmental Research—UFZ deals with a variety of tasks in environmental, climate and energy research, the results of which are internationally recognized. In the energy sector, the spectrum of topics covered ranges from process modelling and simulation to the development of innovative monitoring strategies and the investigation of socio-economic aspects. Since its foundation in 2007, the Department of Environmental Informatics (ENVINF) has been concerned with the development of numerical methods and software components for the simulation of coupled processes in porous media based on the finite element method. In addition, the development of simulation platforms for the treatment of these problems as well as benchmarking for model and software validation will be discussed. Workflows and system components for the 3D visualization of complex, heterogeneous data from different sources are an integral part of these platforms. In this context, the UFZ acts as main developer and coordinator of the international scientific open source software project OpenGeoSys. In addition to method and software development, there is a strong connection to applications in hydrology, geotechnics and energy storage research. Aspects of modelling and numerical simulation of discontinuities in rocks were considered in phase field, non-local deformation and modified XFEM approaches, especially in connection with process analysis in Enhanced Geothermal Systems. In order to increase the efficiency of numerical simulations and for the clear evaluation of results, the UFZ has capacities for high-performance computing and scientific 3D visualization.

1.4.6 UoS

University of Stuttgart

The Institute of Applied Mechanics (Chair of Continuum Mechanics) at the University of Stuttgart (UoS) contributes with a profound knowledge of coupled geomechanical problems in different scales. Investigations in the field of subsurface flow in fractured and unfractured porous media to answer open questions regarding underground storage, geomaterial characterization and geotechnical energy and safety concerns is just a small selection of the overall research fields of interest. Experimental state of the art infrastructure including a custom designed μXRCT-Scanner with for image-based characterization combined with hydro-mechanical in-situ experiments, imaging and PIV of complex single and multi-phase flow processes in micro-fluidic setups with sub-micrometer pore-scale resolution and various experimental devices for the harmonic characterization of complex fluids, solids, and especially rock (and granular media) samples form the basis of the so called Porous Media Lab(oratory) at UoS. The tight interplay of experimental and theoretical work is closed by a number of advanced numerical in house codes such as the solver package HOOSPH (based on the HOOMD-blue library), a massive parallel CPU/GPU Smoothed Particle Hydrodynamics (SPH) framework, PANDAS a general Finite Element (FE) package for strongly coupled multiphasic porous media problems, and a newly developed extension of the DUNE PDELab framework for coupled problems in fractured porous media using hybrid-dimensional element formulations. Important contribution of the experimental and numerical work in numerous projects within the CRC 1313 at University of Stuttgart (https://www.sfb1313.uni-stuttgart.de), GeomInt (https://www.ufz.de/geomint/), SHynergie and the Cluster of Excellence SimTech (https://www.simtech.uni-stuttgart.de) help to prove the deep understanding of geomechanical processes.

Reference

1. K. Rink, E. Nixdorf, C. Zhou, M. Hillmann, and L. Bilke. A virtual geographic environment for multi-compartment water and solute dynamics in large catchments. *Journal of Hydrology*, 582, 2020.

2

Experimental Basis: Thermo-Hydro-Mechanical Laboratory Tests

Amir Shoarian Sattari, Thomas Frühwirt, Jobst Maßmann, Mathias Nest, Dirk Naumann, Daniel Pötschke, Matthias Ruf, Tilo Kneuker, Bernhard Vowinckel, Markus Furche, and Gesa Ziefle

In order to investigate the barrier rocks, such as saltstone, claystone and crystalline, response under the coupled thermo-hydro-mechanical (THM) processes, a series of laboratory and field tests in the scope of the GeomInt project are carried out. A graphical abstract of experimental activities is depicted in Fig. 2.4.

Initially, an introduction to the tested barrier rocks specifications and characteristics is provided (Sect. 2.1). Next, the Opalinus Clay and saltstone responses under the coupled THM processes are investigated. Three-point bending, Brazilian disk and true triaxial tests are used to determine fracture toughness, splitting strength as well as changes of the mechanical and thermal properties under the coupled loadings, respectively. It is shown that the embedded layering orientation in the claystone samples has a significant effect on materials strength, deformation and frack paths (Sect. 2.2). After determining the general material properties, distinct laboratory tests, which are chosen based on the unique purpose of the GeomInt project, are performed to analyze the shrinkage and swelling behavior of the claystone samples under free or in-situ loading conditions. The results again indicate the effect of the anisotropy on the swelling and shrinkage direction of the claystone material (Sect. 2.3). The fluid

A. S. Sattari (✉)
CAU, Christian-Albrechts-Universität zu Kiel, Kiel, Germany
e-mail: amir.shoarian-sattari@ifg.uni-kiel.de

T. Frühwirt · D. Pötschke
TUBAF, Technische Universität Bergakademie Freiberg, Freiberg, Germany

J. Maßmann · T. Kneuker · B. Vowinckel · M. Furche · G. Ziefle
BGR, Federal Institute for Geosciences and Natural Resources, Hannover, Germany

M. Nest · D. Naumann
IfG, Institut für Gebirgsmechanik Leipzig, Leipzig, Germany

M. Ruf
UoS, University of Stuttgart, Stuttgart, Germany

or gas driven percolation tests on Opalinus Clay and saltstone samples explore the change of the permeability, the required fracking pressure and the principle stress dependent frack paths. Additionally, the healing and frack closure characteristic of the barrier rocks are investigated. Moreover, the effect of the compressibility on the fluid flux and drop of the reservoirs pressure when using brine or gas as fracking medium is presented (Sect. 2.4). The direct shear test on the crystalline gives an insight of the shear characteristics of the joints under the constant normal load (CNL) boundary condition. Additionally, the constant normal stiffness (CNS) test with the additional of extra load to the CNL procedure is performed. Specifically, for the crystalline samples, the diffusion and time-dependence states of the fractures under the cyclic loading conditions are characterized. Eventually, the flow and pressure dependency on the deformations of the frack volumes and paths are studied (Sect. 2.5) (Fig. 2.1).

2.1 Rock Material Properties

2.1.1 Opalinus Clay from Mont Terri, Switzerland

Tilo Kneuker, Bernhard Vowinckel, Markus Furche, Gesa Ziefle, Jobst Maßmann (BGR)

2.1.1.1 The Mont Terri Rock Laboratory

Opalinus claystone is a very promising host rock for the safe disposal of heat emitting nuclear waste. This type of host rock has been investigated in the Mont Terri Rock laboratory for more than 24 years. The Mont Terri rock laboratory is a facility to conduct research in the deep geological underground at in-situ scale, such as the safe deposition of radioactive waste, where the local host rock is Opalinus Clay. The rock laboratory is located within the Jura Mountain fold belt. The development of the Jura fold belt began in the Middle Miocene around 12 million years ago, which was constrained by the first occurrence of overthrusted and folded molasse sediments [1]. The overthrust of the frontal fold and thrust belt over the allochthonous foreland (Tabular Jura) occurred ca. 10.5 million years ago [2]. More specifically, the Mont Terri rock laboratory is located in the southeast dipping fold limb of the NW-vergent Mont Terri anticline [3]. The total amount of shortening of the anticlinal structure is approximately 2.1 km [4]. During the folding process, the northwestern fold limb of the Mont Terri anticlinal structure was sheared-off and now lies on top of the Tabular Jura (Fig. 2.2). The Mont Terri anticlinal structure developed in a special structural setting at the intersection between the frontal part of the Jura fold belt (main shortening direction NW-SE) and the roughly N-S oriented structural elements of the Rhine-Bresse-graben transfer zone [3].

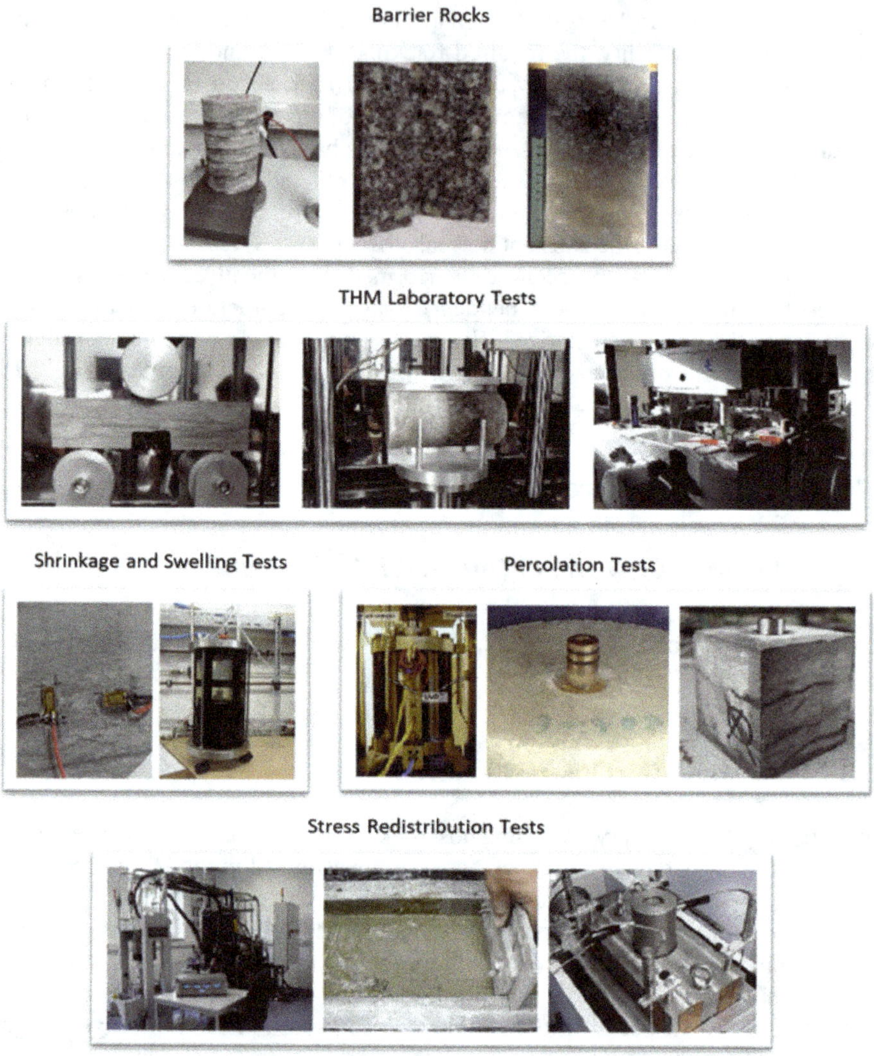

Fig. 2.1 The conducted laboratory tests in the scope of GeomInt research project

The Mont Terri rock laboratory branches off from the security gallery of the motorway tunnel near the town of St. Ursanne (NW Switzerland). The rock laboratory is located mainly within the Middle Jurassic Opalinus Clay formation. The thickness of the Opalinus Clay in the rock laboratory is around 130 m [6] and the layers are dipping with ca. 40 towards SE. The depth below ground varies between 230 and 320 m, depending on the topography [7]. Since 1996, a total of 1400 m of galleries and niches have been excavated in the Mont Terri rock laboratory (Fig. 2.3). The

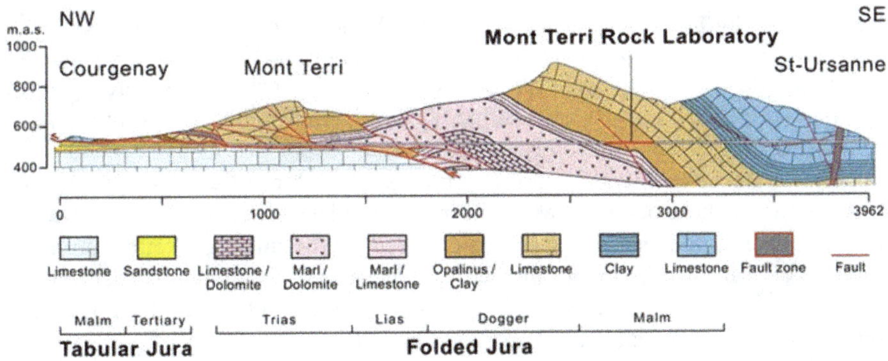

Fig. 2.2 Geological cross section along the motorway tunnel through the Mont Terri anticline. From: Kaufhold et al. [5], based on Freivogel and Huggenberger [4]

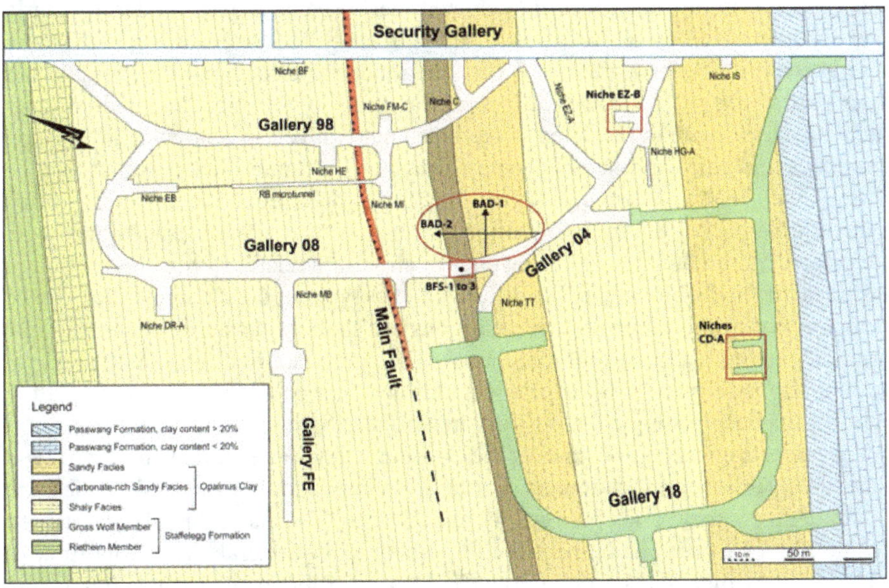

Fig. 2.3 Geological map of the rock laboratory with all in-situ experiments relevant for the GeomInt project including the locations of the two AD-boreholes (BAD-1, BAD-2), the drilling for the FS experiment (BFS-1 to 3), the EZ-B niche for the CD/LP experiment and its successor experiment CD-A in Gallery 18. The different facies types of the Opalinus Clay can be recognized by the different shades of brown and yellow, the new Gallery 18 and the new experiment niches are highlighted in green on the map. (map modified from Mont Terri Consortium, swisstopo)

Mont Terri rock laboratory is a generic scientific research laboratory. At Mont Terri, there will be no storage of radioactive waste.

The Opalinus Clay in the rock laboratory is composed of a dark gray claystone that was named after the ammonite species *Leioceras opalinum*. This claystone formation was deposited during the period of the Toarcian/Aalenian, at an age of approximately 174 million years. The Opalinus Clay is exposed along the rim of the Swabian and Franconian Alp in Germany and stretches into northern Switzerland [8]. The Opalinus Clay was deposited in a shallow-marine, epicontinental milieu in the area of the storm wave base at approximately 20–50 m water depth [9]. Coarser siliciclastic components are of detritical origin. Potential sources for the detritic components are the areas of the Bohemian Massif and the Vindelician Landmass [9]. During Cretaceous burial, the Opalinus Clay experienced maximum palaeo temperatures of 75–90 °C [10] at a maximum burial depth of 1.35 km. The Opalinus Clay at the Mont Terri site is underlain by marls of Upper Toarcian age and overlain by limestones (Bajocian), some of which act as karst aquifer [11].

The Opalinus Clay at the Mont Terri rock laboratory can be subdivided into three main facies types [10]. First, the lower shaly facies occupies the largest area of the rock laboratory (Fig. 2.3). It dominates the lower part of the Opalinus Clay formation. It consists of mica-bearing clay and marly shales as well as flasery, marly layers characterized by bioturbation. The upper shaly facies contains a higher volumetric content of quartz grains. Second, the sandy facies occurs in the middle and upper part of the profile (lower and upper sandy facies). It includes medium gray marly claystones with intercalated, bioturbated marly layers or lenticular, gray sandy limestones and pale sand layers of approximately 1–10 mm thickness that include pyrite as well. Third, a carbonate-rich sandy facies of approx. 5 m thickness occurs in the middle part of the rock formation. It consists of calcareous sandstones with intercalated bioturbated limestone layers, which show a relatively high proportion of detritic quartz and white mica. The different facies types of the Opalinus Clay can be attributed to varying sedimentation conditions in a shallow marine environment (like variations in depth and current directions). The carbonate-rich facies is typical for the Jura region in western Switzerland and it does not occur in the proposed siting regions for a deep geological repository in Northern Switzerland.

The mineralogical composition of the Opalinus Clay was examined by Traber and Blaser [12] for several locations in Northern Switzerland. For the shaly facies, the clay mineral content varies between 40 and 75 wt%. The clay minerals determined include illite, kaolinite and smectite-illite mixed layer minerals, the proportion of swellable clay minerals is around 10 wt%. Detritic components such as quartz and feldspars typically make up to 20 wt% of the investigated samples. The carbonate content (calcite and dolomite) is around 20 wt%. The sandy facies of the Opalinus Clay is composed of up to 40 wt% clay minerals and ca. 30 wt% quartz; it shows a lower amount of clay minerals in favor of a higher quartz content, compared to the shaly facies [7].

The BGR has been involved in a number of campaigns to study in-situ conditions of clay rock, of which the following four are particularly noteworthy within the context of the GeomInt-Project. First, in the CD experiment the long-term cyclic deformation (CD) due to seasonally induced cyclic swelling and shrinkage is investigated in a niche of the rock laboratory. These measurements are continued in the LP (long-term monitoring of pore parameters) experiment. In addition, a follow-up project, the CD-A experiment, has been prepared in recent years, to distinguish between deformation processes due to stress redistribution and seasonal variations in air humidity that cause saturation (swelling) and desaturation (shrinkage) of the rock and stress redistribution alone. To this end, two identical niches were excavated, one sealed towards the gallery and with a high humidity inside to minimize desaturation and one open to the general air circulation of the rock laboratory. The measurement campaign was started in October 2019 [13]. Third, the AD experiment (experimental-numerical Analysis of Discontinuities) intends to provide an improved process understanding for the experimental-numerical analysis of discontinuities. Finally, the Fault Slip (FS) experiment addresses the fault reactivation due to pressure-induced percolation in a low-permeability, large-scale discontinuity in the Mont Terri rock laboratory. The AD experiment is directly relevant to the numerical and experimental investigations presented in Sects. 4.4–4.6, because the rock material used in these experiments were drilling samples from the AD experiment. Hence, we provide a brief overview for the campaign in the following.

2.1.1.2 The CD/LP Experiment in the Mont Terri Rock Laboratory

The Mont Terri Rock laboratory in Switzerland hosts a multitude of in-situ experiments that investigate the response of Opalinus Clay to various geotechnical applications. An overview of the rock laboratory is given in Sect. 2.1.1.1. In particular, the CD (Cyclic Deformation) experiment has been a valuable site to gather experimental data at the in-situ scale to investigate the hydraulic-mechanial coupling induced by swelling and shrinking of Opalinus Clay due to cyclic variations of air humidity. Sect. 4.7 focuses on the numerical investigation of these processes. Here, we provide a brief overview of the experimental CD campaign at the Mont Terri Rock Laboratory.

The experiment itself is located in the EZ-B niche (Fig. 2.3). The experiment has been conducted for more than 13 years to provide information on the swelling and shrinkage behavior of Opalinus Clay in the Mont Terri rock laboratory. The idea was to analyze a niche that is not covered by shotcrete. Instead, the clay rock remains in direct contact with the atmospheric conditions of the main gallery for the entire time. Consequently, the swelling and shrinkage is induced by changes in temperature and relative humidity, which can decrease to values as low as 65% in the winter and reaches values of up to 100% in the summer.

A special focus of this experiment was to investigate the long-term impact of these seasonal variations on the temporal evolution of the cracks that occur during the excavation process and make up the Excavation Damaged Zone (EDZ). To this end, the EZ-B niche was excavated in the years 2004/2005 (Fig. 2.4). Subsequently, the niche was equipped with a comprehensive set of measurement devices to record the evolution of temperature, water content, convergence of the niche and crack development at the tunnel walls over time. This measurement campaign was started in 2006 and has been continued until today to investigate long-term effects. Note that the experiment was transferred into the LP-A experiment to explicitly focus on the long-term monitoring of pore pressure. The CD/LP experiment under in-situ

Fig. 2.4 The EZ-B niche in the Mont Terri rock laboratory, where the CD/LP experiment has been conducted since 2006 (photo: Mont Terri consortium, swisstopo)

conditions was supplemented by laboratory experiments with drill cores to determine hydraulic-mechanical properties of the clay rock, such as porosity, grain density, etc. [14].

Characteristic macroscopic cracks on the tunnel walls have been monitored and the field data of the crack opening show a good correlation with the seasonal variation of temperature and humidity. The cyclic deformation of the crack opening yields a re-occurring compression perpendicular to the crack during summer, which typically is a time of high relative humidity and, hence, the swelling causes an increase of rock volume [15]. This characteristic behavior of swelling and shrinking was successfully reproduced by means of hydraulic-mechanically coupled numerical simulations [16], which provides valuable benchmark data for future investigations of the cyclic deformation of clay rock.

2.1.1.3 The AD Experiment

The aim of the experiment is to provide core samples from the sandy facies of the Opalinus Clay as a typical example of an argillaceous host rock for the safe disposal of nuclear waste. These samples were used for experimental-numerical analysis in the framework of the GeomInt project. Additionally, a geological characterization of the cores and seismic (Interval Velocity Measurements—IVM) and geolelectrical measurements (Electrical Resistivity Tomography—ERT) in the boreholes were performed. The results of the experimental campaign yield a valuable description of the sandy facies in addition to the well characterized shaly facies of the Opalinus Clay [10, 17].

From a geological perspective, the AD experiment gave opportunity to study the lower sandy facies of the Opalinus Clay at the Mont Terri rock laboratory in detail. The two fully cored boreholes BAD-1 and BAD-2 with a diameter of 131 mm (yielding samples of 101 mm diameter) were drilled parallel and perpendicular to the sedimentary bedding, respectively. The 15.35 m long horizontal borehole BAD-1 was drilled from 7th–10th of July 2018 by the BGR. It is located entirely in the lower sandy facies. The geological mapping was performed by swisstopo [18]. The core material of BAD-1 was entirely sampled for laboratory experiments performed by the Christian-Albrechts University of Kiel (Germany) and the Institute of Geomechanics (IfG) Leipzig (Germany). The core samples were conditioned in aluminum foil and pressurized in special nitrogen-filled BGR-liners (autoclaves) to avoid further alteration.

The BAD-2 borehole has a length of 27.0 m. It was drilled by the BGR team from 9th–17th of May 2018. The borehole is oriented perpendicular to the bedding (with a dip of 43 deg), thus crossing several facies types of the Opalinus Clay. The geological mapping by Swisstopo reported by Galletti and Jaeggi [18] revealed the following sequence with varying quantities of quartz, carbonates (cements and fossils) and clay minerals:

- 0.0–4.75 m depth: upper shaly facies,
- 4.75–19.4 m depth: lower sandy facies,
- 19.4–24.57 m depth: carbonate-rich sandy-facies,
- 24.57–27.0 m depth: lower shaly facies.

This subdivision is confirmed by petrographic-structural studies and geoelectrical resistivity measurements (ERT) performed by the BGR. The BAD-2 drillcores were sampled from 4.0 to 14.8 m (lower sandy facies) for laboratory experiments by the Universities of Kiel and IFG Leipzig. Following the procedure employed for the BAD-1 drilling, the core material was conditioned in aluminum foil and pressurized in special nitrogen-filled BGR-liners (autoclaves). The drillcore material from the intervals between 0.0–4.0 m and 14.8–27 m, including the transition towards the underlying carbonate-rich facies, are stored at the BGR facility in Hannover (Germany) for further geological characterization. The first results revealed a good core quality and confirm a rather uniform appearance of the sampled profile inside the lower sandy facies, the drillcore material is thus suitable for the planned experiments (cf. Fig. 2.5).

High resolution ERT measurements in borehole BAD-2

The task of geophysics was to characterise the rock unit that had been drilled through, to precisely determine the locations of the facies boundaries and to describe the variability, especially of the sandy facies. The electrical resistivity tomography (ERT) method was used for this purpose.

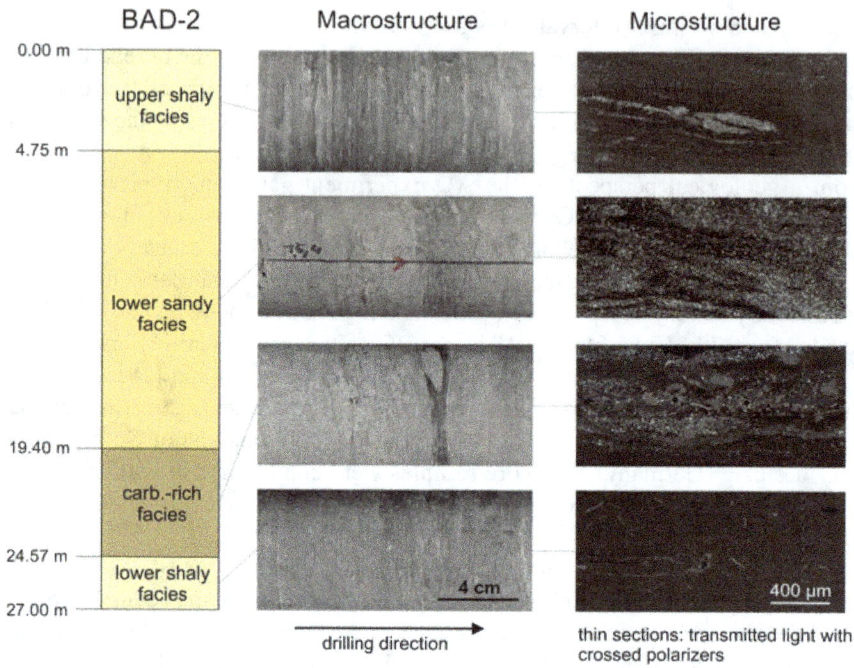

Fig. 2.5 Schematic profile of borehole BAD-2 as marked in Fig. 2.3 (left), macrostructural (on drillcore scale) and microstructural features (on thin section scale) of the different facies types of Opalinus Clay encountered in the BAD-2 borehole

Principle

To determine the spatial electrical resistivity distribution (or its reciprocal – the electrical conductivity) in the ground, a direct current (DC) is injected into the ground through two point electrodes (A, B), see Fig. 2.6.

The resulting electrical field is measured using two other electrodes (M, N). A point electrode introducing an electrical current I will generate a potential V_r at a distance r from the source. In the case of the four-electrode array shown in Fig. 2.6, the two current electrodes (A, B) introduce a current I. When assuming a homogeneous half-space, the potential difference ΔV between the electrodes M and N can be calculated as:

$$\Delta V = \rho I \left[\frac{1}{2\pi} \left(\frac{1}{\overline{\mathrm{AM}}} - \frac{1}{\overline{\mathrm{AN}}} - \frac{1}{\overline{\mathrm{BM}}} + \frac{1}{\overline{\mathrm{BN}}} \right) \right] \tag{2.1}$$

Here, $\overline{\mathrm{P_1P_2}}$ denotes the distance between two points $\mathrm{P_1}$ and $\mathrm{P_2}$. Replacing the factor in square brackets with $1/K$, we obtain the resistivity of the homogeneous half space as follows:

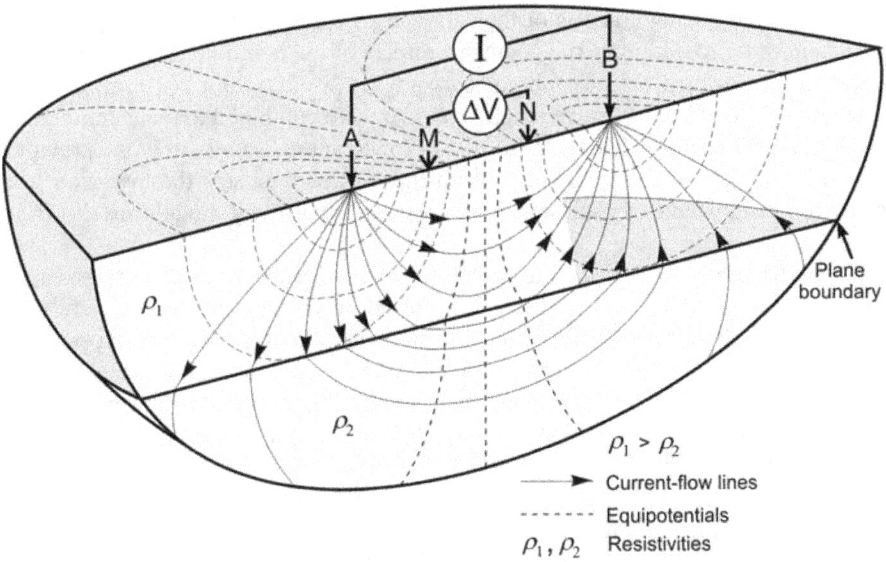

Fig. 2.6 Principle of resistivity measurement with a four-electrode array (modified after Knödel et al. [19])

$$\rho = K \frac{\Delta V}{I}. \tag{2.2}$$

The parameter K is called configuration factor or geometric factor. For inhomogeneous conditions, it yields the resistivity of an equivalent homogeneous half-space. For this value the term apparent resistivity ρ_a is introduced, which is normally assigned to the center of the electrode array. Multi-electrode resistivity meters enable the measurement of 2D resistivity surveys (2D imaging). The advantages of this kind of measurements are their high vertical and horizontal resolution along the profile. An inversion process of the measured data is necessary for the final interpretation. This process transforms the apparent resistivities into a reliable model discretised into a distinct number of elements of homogeneous resistivity. All inversions are performed using the non-commercial software BERT (Boundless Electrical Resistivity Tomography[1]) developed by Th. Günther (Leibniz Institute of Applied Geophysics, Hannover) and C. Rücker (Technical University of Berlin).

Measurements and Results

The measurements were performed on May 21st and 22nd 2019. Since the borehole is perpendicular with respect to the bedding, measurements were only taken in one orientation (0°, i.e. the electrodes are oriented upwards). Along the borehole, 35 individual profiles of 1.5 m length with half-sided overlapping were measured.

[1]https://gitlab.com/resistivity-net/bert.

The data processing consists of the following steps: First thing is the scaling of the data in order to eliminate the geometry effects of the borehole. Then data points with high statistic error (>3%) or high phase angle (absolute value > 100 mrad) were eliminated. 12 or 13 consecutive single data sets were combined to three cumulative ones (00.10–09.85 m, 08.36–18.10 m, 16.61–26.98 m; the last record of the previous section is the first of the following). With these three data sets the inversion was performed using standard parameters. The resulting resistivity models are shown in Fig. 2.7.

Up to 50 cm borehole depth, the shotcrete can be recognised as a high-resistance structure. Beyond that, the resistivity is well below 10 Ωm and rises slowly with increasing borehole depth. Embedded in this basic structure are thin layers with high resistivity (>30 Ωm). Between 18.5 and 19.5 m, there is a very extensive high-resistance range, after which the resistance level drops significantly, again with embedded high-resistance layers. Between 24.0 and 24.7 m a second distinct high-resistance range is reached, beyond which a sharp drop to values below 10 Ωm can be observed. This level is maintained down to the bottom of the borehole without further intermediate structures.

From the calculated 2D models of the resistivity, curves can be extracted for specific distances from the borehole wall. Figure 2.8 shows the corresponding resistivity curve for a distance of 5 cm. The different facies types are indicated as colour background. It can be seen that both the upper and lower shaly facies are characterised by resistivities below 10 Ωm with little variation. The mean level in the sandy facies (4.9–19.7 m) is significantly higher (about 15 Ωm), and the variability is also considerably greater. Both, the upper and the lower transition towards the carbonate-rich sandy facies are clearly indicated as a sharp drop in the resistance curve. Here, the average level of resistivity lies between the values for shaly and sandy facies, the amplitudes of the variations are highest.

Implications of the geological-geophysical investigations

Petrographic-structural studies form the basis for rock characterization and provide first indications for the compositional-structural variability, which independently is confirmed by geophysical borehole measurements.

ERT is able to characterise structures close to the borehole and to resolve rock variability with high accuracy. The individual facies types of Opalinus Clay can be distinguished based on mean resistivities and the type of heterogeneity (amplitudes and sequence). Especially the important transitions towards the carbonate-rich sandy facies can be precisely located.

The results confirm that the upper shaly facies is closely related to the most homogenous facies type of Opalinus Clay (the lower shaly facies). In contrast, the lower sandy facies and especially the carbonate-rich sandy facies represent the more heterogeneous endmembers of the investigated claystone formation. The new results are consistent with published data and support the classification of the Opalinus Clay at the Mont Terri site into several major facies types. Further investigations will focus on the characterization of intra-facies variability using the sub-facies concept [38].

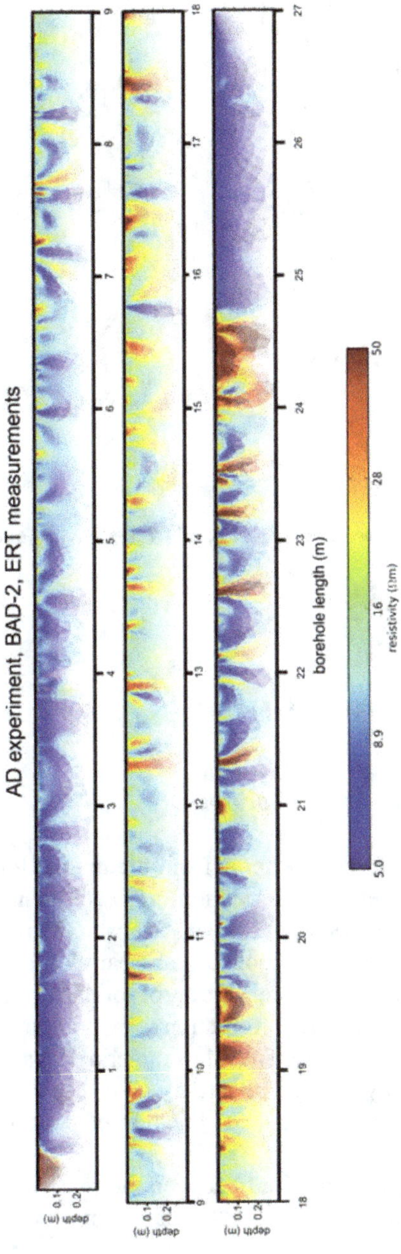

Fig. 2.7 Borehole BAD-2: Two-dimensional distribution of the resistivity as a result of the inversion calculation

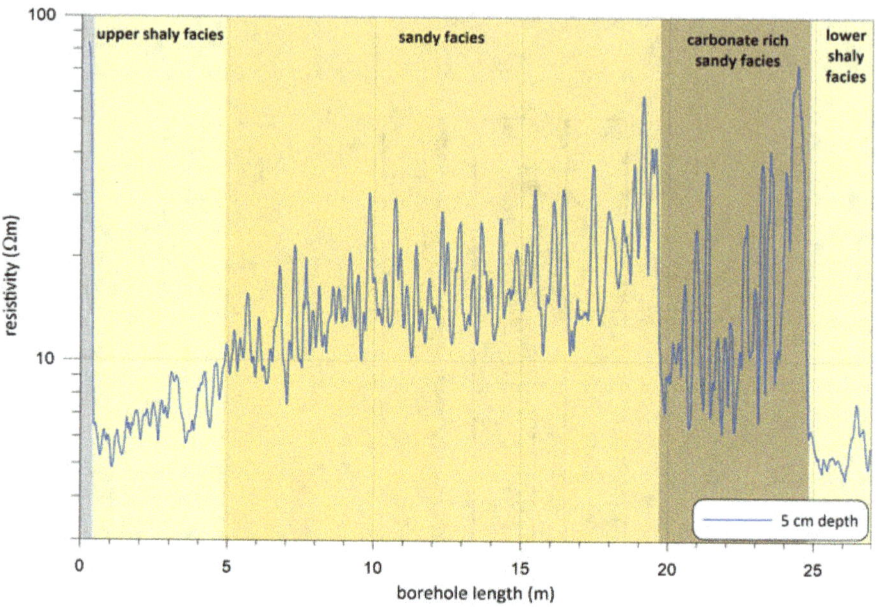

Fig. 2.8 Borehole BAD-2: Curve of the resistivity at a distance of 5 cm from the borehole wall

2.1.2 Rock Salt Samples

Mathias Nest, Dirk Naumann (IfG)

2.1.2.1 The Springen In-Situ Laboratory

The large-scale test site Merkers benefits from the unique mining situation in the bedded salt mass of the Werra salt formation (z1, Zechstein sequence) where two potash seams were mined in a room-and-pillar system at 275 m (1st floor, potash seam "Hessen", z1KH) and 360 m (2nd floor, potash seam "Thringen", z1KTh) depth, respectively. Figure 2.9 shows the preparation of an experiment on the 2nd seam. The evaporite rocks of the Zechstein formation were laid down during the Permian period around 250 million years ago. The intact mineral deposit was locally disturbed between 14 and 25 million years ago by tertiary volcanism, leading to the mutation of some potash salts to sylvinite, and the creation of pockets of CO_2 under high pressure.

The pressurized tests are conducted between the two potash seams in the very homogeneous Middle Werra rocksalt (z1Na). It consists mostly of very pure halite layers intersected by thin anhydrite lines or bands of rocksalt with finely distributed anhydrite accessories indicating the sedimentary bedding.

Fig. 2.9 Preparation of experiments on the second seam

Because the test results depend mainly on the acting stress field, i.e. the minimal stress distribution in the rock mass around the test area, it has been measured and characterized by hydro-frac measurements, and is thus well-known. The minimal stresses in the contour increase with progressive distances from the underground openings until reaching a constant value in a depth of around 15 m. The measured value of an undisturbed stress state of around 8 MPa corresponds fairly well to calculated lithostatic stresses of 7.8–8.8 MPa.

The main facility is a large borehole of nearly vertical 60 m height and 1.3 m diameter. It was drilled upwards from the second floor, ending about 20 m beneath the first floor. For access to the later sealed volume an 85 mm pilot hole has been drilled from the upper floor. The borehole was closed by a 20 m high MgO-concrete plug.

For monitoring of micro-seismic events, e.g. due to creation of an excavated damage zone around underground openings or fluid flow driven damage, a highly sensitive AE-network was installed in observation boreholes, which were drilled parallel to the main borehole. This network has constantly been updated and extended over the past years. Signals of magnitude $M < -4$ can be detected in a frequency range from 1 to 200 kHz [39]. This corresponds to intergranular microcracking on grain boundaries on a millimeter- to centimeter-scale.

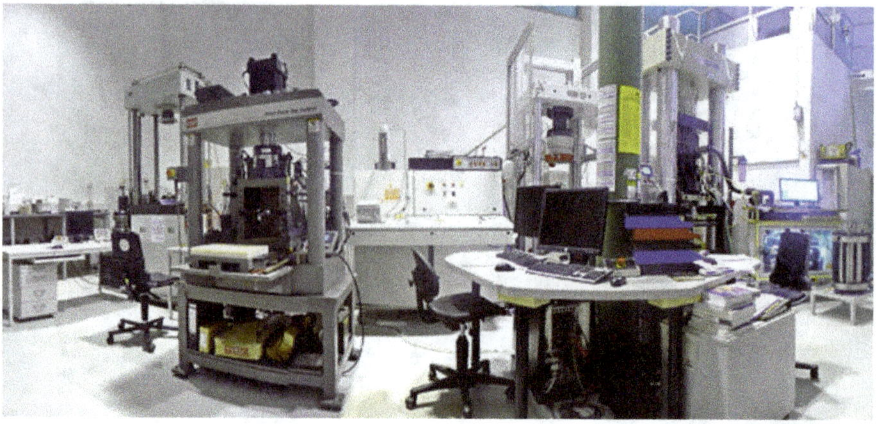

Fig. 2.10 View inside the rock mechanical lab with various servo-controlled hydraulic testing machines

2.1.2.2 Rock Salt Laboratory

The following conditions and equipment are available in the IfG labs for rock mechanical laboratory investigations:

- Climate-controlled rooms for storage of specimens at conditions which correspond to those present in situ
- Laboratory for mineral-petrographic examinations, density and moisture determination, ultra-sound measurements and photographic documentation
- Workshop for the preparation of specimens in high precision according to testing requirements (rock saws, lathes etc.)
- Test laboratory containing various servo-controlled hydraulic testing machines for conducting investigations on rock mechanics in accordance with the up-to-date state of research and development (see Fig. 2.10).

It has to be mentioned that some of the applied IfG test procedures have been developed for the requirements of the specific IfG-material laws. But generally they are in accordance to ASTM[2] and ISRM[3] standards resp. equivalent descriptions, e.g.:

- ASTM D 4543-85 Standard Practice for Preparing Rock Core Specimens and Determining Dimensional and Shape Tolerances
- ASTM D 2845-05 Standard Test Method for Laboratory Determination of Pulse Velocities and Ultrasonic Elastic Constants of Rock
- ASTM D 2664-86 Standard Test for Triaxial Compressive Strength of Undrained Rock Core Specimens without Pore Pressure Measurements

[2] www.astm.org.

[3] International Society for Rock Mechanics and Rock Engineering www.isrm.net.

- DGEG (1979): Empfehlung Nr. 2 des AK 19 der DGEG (Dreiaxiale Druckversuche).
- ASTM D4406-04 Standard Test Method for Creep of Cylindrical Rock Core Specimens in Triaxial Compression
- ASTM D7070-04 Standard Test Method for Creep of Rock Core Under Constant Stress and Temperature
- ISRM: Suggested Methods for Determining o the Uniaxial Compressive Strength and Deformability of Rock Materials
- ISRM: Suggested Methods: Part 2: 2007—Unconfined Compressive Strength with Young's Modulus and Poisson's Ratio
- ISRM: Suggested Methods: Part 2: Received 1983—Strength of Rock Material in Triaxial Compression.

2.1.2.3 Sample Characterization and Pre-investigation

Several preliminary investigations are usually done before lab testing. After preparing the cylindric samples (cutting with a rock saw and smoothing the samples with a lathe) in the IfG labs, their density is determined by measuring the geometrical dimensions and the mass. Concerning the quality of these parameters an accuracy in length determination is <0.01 mm, those in mass determination is <0.01 g.

Additionally, ultrasonic investigations are carried out to check integrity, homogeneity and isotropy of the samples. The ultrasonic pulse measurement system used for transmission of the rock specimens consists of two transducer sets for P-waves and S-waves and a receiver system for generating and evaluating the ultrasonic signals. The specimen is placed in physical contact between two piezoelectric transducers, one acts as a driver and the other one acts as a receiver. The transit time of the mechanical pulse to pass through the specimen is used to determine elastic wave velocity.

For samples where both P- and S-waves were measured in axial sample direction the elastic constants are obtained from density (ρ) and the ultrasonic velocities (v_p, v_s) using the following expressions derived from the theory of elasticity for homogeneous, isotropic solids:

$$E_{\text{dyn}} = \frac{\rho v_s^2 (3v_p^2 - 4v_s^2)}{v_p^2 - v_s^2} \tag{2.3}$$

$$\nu_{\text{dyn}} = \frac{v_p^2 - 2v_s^2}{2(v_p^2 - v_s^2)} \tag{2.4}$$

Dynamic elastic parameters of the various rock portions determined on the base of sonic wave velocity at room temperature are in a wide range representing typical values for the various materials as known from other locations.

Fig. 2.11 The Reiche Zeche in Freiberg [20]

2.1.3 Crystalline Rock Samples

Thomas Frühwirt, Daniel Pötschke (TUBAF)

2.1.3.1 URL Reiche Zeche Freiberg

The Reiche Zeche mine is located north-east of the city center of Freiberg, Saxony. It operated as a silver mine for several centuries. It became a part of the TU Freiberg 1919 when silver mining definitely was no longer profitable. Nowadays the mine is used as an underground research laboratory (URL) and for teaching purposes. The roughly 4 km^2 sized area is well documented in terms of geology, mineralogy, geophysics and geometry. Draining of the mine is done using the "Rothschönberger Stollen (tunnel)". A detailed overview about the history of the Reiche Zeche can be found in [21]. Due to its long history, the development of new technologies and the importance for the development of the whole region the mining sites and the associated infrastructure are listed as UNESCO world heritage site since 2019 (Fig. 2.11).

Current projects are for example dealing with bio-leaching or complex experiments which study the influence of hydro-fracking experiments on the stress state (STIMTEC project).

The Reiche Zeche is equipped with an underground railway system, installed electricity, water and air pressure. The experienced staff and the nearby located mining agency help to successfully conduct experiments in about 150 m depth.

2.1.3.2 Rock Material Used in the Direct Shear Tests

Two different crystalline rock types are used. Granite is a coarse-grained intrusive igneous rock. The grains are on the millimetre to centimetre scale, see Fig. 2.12a. The typical main minerals of granites are quartz, feldspar and plagioclase. The used granite origins from Kirchberg, Saxony. Basalt is an fine-grained extrusive igneous

(a) Granite sample (b) Basalt sample

Fig. 2.12 Crystalline rock samples under investigation

Table 2.1 Rock parameters of granite and basalt used in the direct shear tests

Parameter	Symbol	Granite	Basalt	Unit
Density	ρ	2.59	3.06	g/cm^3
Compressive strength	σ_c	120.54	272.92	MPa
Tensile strength	σ_t	7.02	16.61	MPa
Elastic modulus	E	49.75	105.46	GPa
Poisson's ratio	ν	0.26	0.26	–
Fracture toughness	K_I	0.95	2.61	$MPa \cdot m^{0.5}$
Friction angle (Mohr)	ϕ	52.5	44	°
Cohesion	c	22.5	25.00	MPa
Basic friction angle	ϕ_b	30	31.2	°

rock. It is rich of plagioclase. See Fig. 2.12b. Its origin is Völkershausen, Thuringia. Lab tests to evaluate basic rock parameters of the intact rock material have been conducted. The values of the granite and basalt used in the experiments can be found in Table 2.1.

The elastic modulus and the compressive strength were determined using uniaxial compression tests. For the tensile strength a tension test was conducted and the fracture toughness is evaluated by a bending test. The cohesion was determined through a shear test using a saw-cut joint. Ultrasonic measurements were used to determine the Poisson's ratio. The basic friction angles were determined based on [22]. The inner friction angle of basalt is from [23] and of granite from [24, 25].

2.1.3.3 Rock Surface Scanning

Important measurements to characterise a rock surface are surface scans. In the laboratory of the TU Freiberg a white light scanner is used, see Fig. 2.13. It is a non-contact method which uses monochromatic light. A fringe projection is done and surface scans of rock samples with a resolution of about 30–50 μm can be calculated. Further details about the scanning device can be found in [26].

2.2 Thermo-Hydro-Mechanical Laboratory Tests

In this section basic methods for rock characterization are briefly introduced such as characterization of rock structure (Sect. 2.2.1), fracture toughness as materials resistance to fracture propagation (Sect. 2.2.2), Brazilian tests to determine split tensile strength (Sect. 2.2.3), and triaxial tests to investigate the three-dimensional strain-strength behavior (Sect. 2.2.4).

2.2.1 X-ray Micro Computed Tomography

Matthias Ruf (UoS)
X-ray micro computed tomography (μXRCT) is a non-destructive imaging technique and provides the possibility to examine the inner structure of an object by creating a digital 3D image of the same. It is based on the mathematical combination, called reconstruction, of several radiograms which were acquired from different directions. Thereby, a radiogram represents the respective measured intensity values $I(L)$ of the X-rays at position $x = L$ after travelling through the object and can be related to the unattenuated X-rays intensity I_0 before the object at position $x = 0$ by the Beer-Lambert law. Under the assumption of a monochromatic X-ray beam this can be written as

$$I(L) = I_0 \exp\left(-\int_0^L \mu(x)\mathrm{d}x\right)$$

for an inhomogeneous material with the unknown material depending attenuation coefficients $\mu(x)$ which are determined during the reconstruction process, cf. [27].

In Fig. 2.14 the self-made open and modular μXRCT system at the Institute of Applied Mechanics—Chair of Continuum Mechanics of the University of Stuttgart, see [28] is depicted.

It consists of the three main components: The X-ray source (1), the X-ray detector (2) and the sample positioning system including the rotation stage in between (3). Like most industrial CT-systems the specimen is rotated during the scan and the remaining

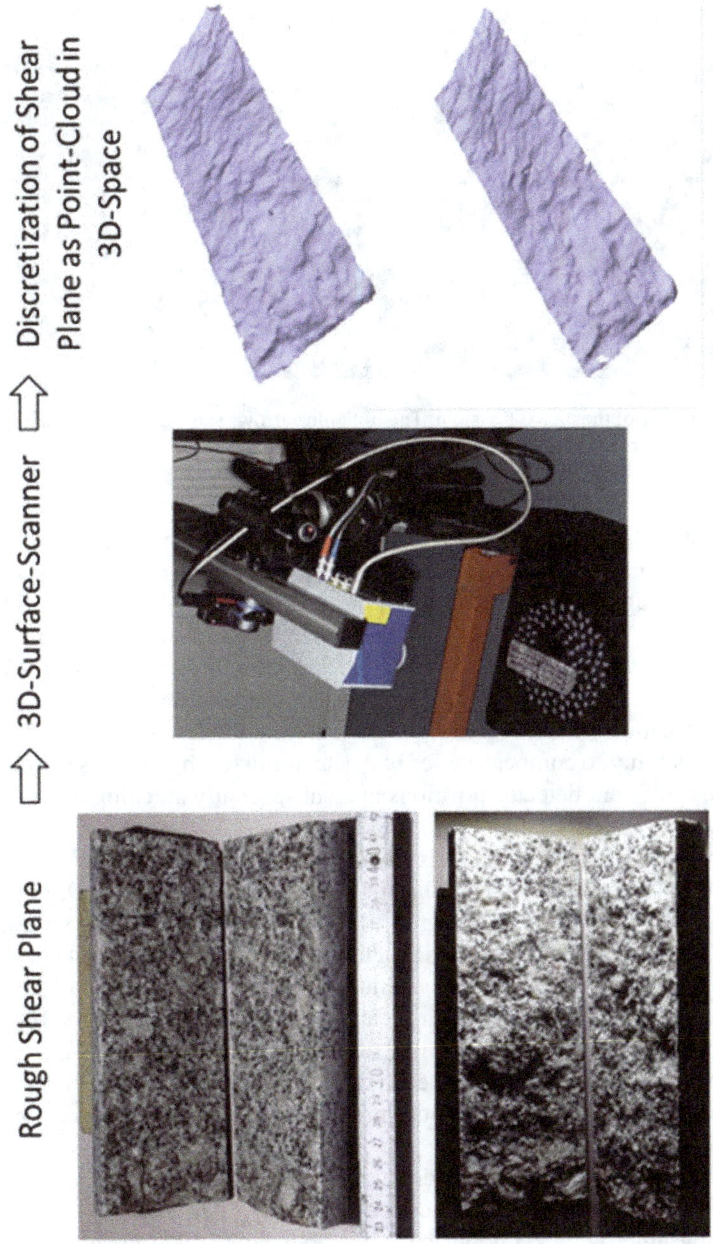

Fig. 2.13 Surface scanning. A rock surface is digitized using a surface scanning device. The result is a point cloud

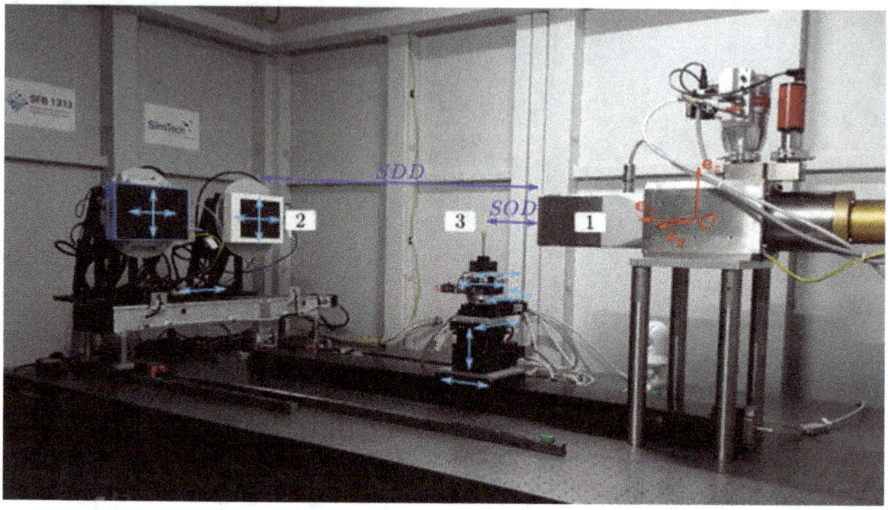

Fig. 2.14 Overview of the μXRCT-system. The light blue arrows show the possible moving directions of the motorized stages

parts are fixed. The employed X-ray tube provides a maximum power up to 80 W and at the same time a focal spot size down to 3μm for moderate power levels. The acceleration voltage of the tube can be adjusted in the range of 30 kV to 180 kV and the flux from 10 to 1000 μA. It can be chosen between two indirect conversion flat panel detectors with different characteristics and resolutions of 1944 × 1536 and 2940 × 2304 pixels. Both produce gray value images with a pixel depth of 14bit and each is separately mounted on high accurate, motorized XY stages. The latter offers the possibility to compensate for bad detector pixels by taking several images from slightly different detector positions and subsequently stitching of the same to improve the final image quality.

The geometric magnification M is given by the relation of the source detector distance SDD to the source object distance SOD, $M = SOD/SDD$, and can be adjusted in a wide range. Depending on the sample material and the smallest feature size of interest, the specimen's diameter can be up to 100 mm. The maximum achievable spatial resolution of the system is about 50lp/mm at 10% of the modulation transfer function (MTF) which means a smallest feature size of 10 μm that can be resolved. The corresponding field of view for this case is 5.88 mm in width and 4.60 mm in height. Therefore, the smallest expedient sample diameter is about 5 mm. The open and modular concept of the CT-system provides a broad range also for unconventional investigations.

In Fig. 2.15 a CT scan of a cylindrical Carrara marble core with a diameter of 5 mm for a high resolution scan is shown exemplarily.

The visible micro-cracks along the grain boundaries were created by thermal treatment and are not present in the virgin state. The geometric magnification was

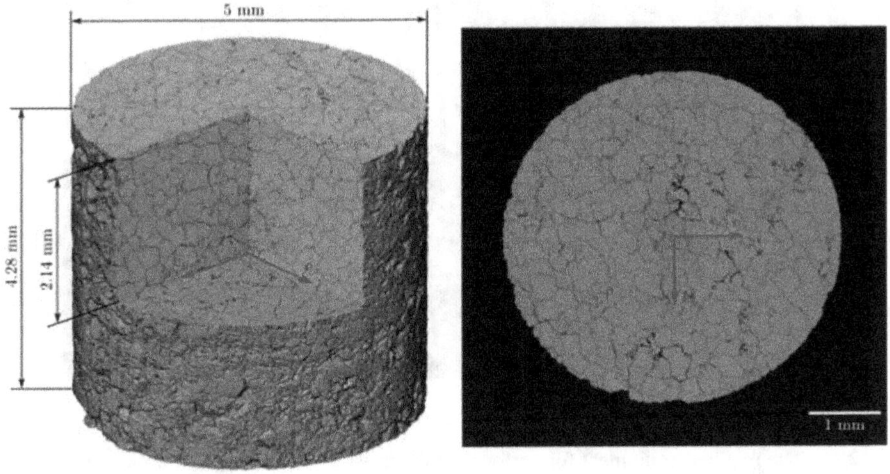

Fig. 2.15 CT scan of a Carrara marble core after thermal treatment

Fig. 2.16 The fracture toughness assessment using three-point bending test

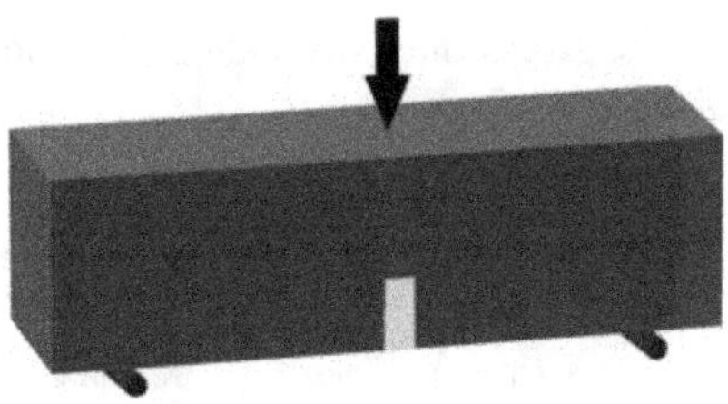

set to $M = 24.76$ which corresponds to the highest achievable spatial resolution and leads to a voxel size of $2\,\mu m$. For additional details see [28]. Besides a qualitative assessment, the data sets offer the possibility of deriving several additional information by image processing. For instance, in geosciences the 3D pore characterization, the 3D grain analysis and the fracture analysis, cf. [29].

2.2.2 Fracture Toughness of the Opalinus Clay

Amir Shoarian Sattari (CAU)

In linear elastic fracture mechanics, a materials resistance to fracture propagation is known as fracture toughness. The unit of the fracture toughness is $Pa.\sqrt{m}$ and the fracture toughness is measured under coupled or individual three different fracture modes I, II and III. In a three-point bending test, the flexural strength (σ_{flex}), flexural Young's modulus (E_{flex}) and flexural strain (ϵ_{flex}) parameters are measured. The fracture toughness test using the three-point bending test provides the Mode I fracture toughness (K_{Ic}) of the material (Fig. 2.16).

(a) (b)

Fig. 2.17 The required equipment for performing the three-point bending test **a** the loading cell and supports, **b** the climate chamber for controlling temperature and humidity

The stress intensity factor (K_I) on the crack tip of predefined notch is obtained with Eq. (2.5) [30],

$$K_I = \frac{4 f_{\text{flex}}}{B_{\text{flex}}}\sqrt{\frac{\pi}{W_{\text{flex}}}}\left(1.6\left(\frac{a_{\text{flex}}}{W_{\text{flex}}}\right)^{\frac{1}{2}} - 2.6\left(\frac{a_{\text{flex}}}{W_{\text{flex}}}\right)^{\frac{3}{2}} + 12.3\left(\frac{a_{\text{flex}}}{W_{\text{flex}}}\right)^{\frac{5}{2}} - 21.2\left(\frac{a_{\text{flex}}}{W_{\text{flex}}}\right)^{\frac{7}{2}} + 21.8\left(\frac{a_{\text{flex}}}{W_{\text{flex}}}\right)^{\frac{9}{2}}\right) \quad (2.5)$$

where, f_{flex} is the flexural load, a_{flex} is the length of the pre-defined notch, B_{flex} and W_{flex} are the thickness and height of the sample under the flexural test, respectively. During the test procedure, the load and crack mouth opening displacement (CMOD) values are measured and plotted. At the load in which the crack starts to propagate into the medium the fracture toughness K_{Ic} is calculated. In order to perform the fracture toughness test, a loading cell with three rolling supports is required (Fig. 2.17a). However, the in-situ condition can only be reached when a material is under controlled temperature and humidity conditions. Therefore, a climate chamber in a laboratory of CAU Kiel has been utilized to reach the desired temperature and humidity (Fig. 2.17b). The temperature can be controlled between 20 and 80 °C and relative humidity varies from 0 up to 100.

The claystone samples are prepared with the dimension of 140 × 30 × 30 (mm) and the notch dimension of 2 × 10 × 30 (mm)v ($L \times W \times B$) (Fig. 2.18). The span length (S_{flex}) is 120 mm, which is 4 times the size of its width and thickness. The embedded layering is perpendicular to the loading direction. The image processing technique is used to track the reference points, which are predefined prior to the

Fig. 2.18 The prepared Opalinus Clay sample with the dimension of 140 × 30 × 30 mm

Fig. 2.19 The application of image processing technique **a** the predefined reference points for measuring the CMOD, **b** the measured rough distance using optical microscope

test procedure (Fig. 2.19a). The distance between the points is measured using the optical microscopic image and is considered as an initial reading value (Fig. 2.19b). The method is able to detect the minimum displacement of 2 μm, which is possible by taking 4k video with 30fps. The load versus CMOD response of the Opalinus Clay as well as the numerical simulation outcomes are given in Sect. 4.1. The effects of anisotropy and embedded layering orientation of Opalinus Clay on fracture toughness values are not fully studied.

2.2.3 Brazilian Disk Test on Barrier Rocks

Amir Shoarian Sattari (CAU)

The tensile strength of a brittle or quasi-brittle material, such as rock, is one of the most important material properties, which in comparison to the compression strength, is much weaker. The measurement of direct tensile strength of brittle materials is difficult and time consuming. Therefore, finding the splitting tensile strength (σ_{sp}) is

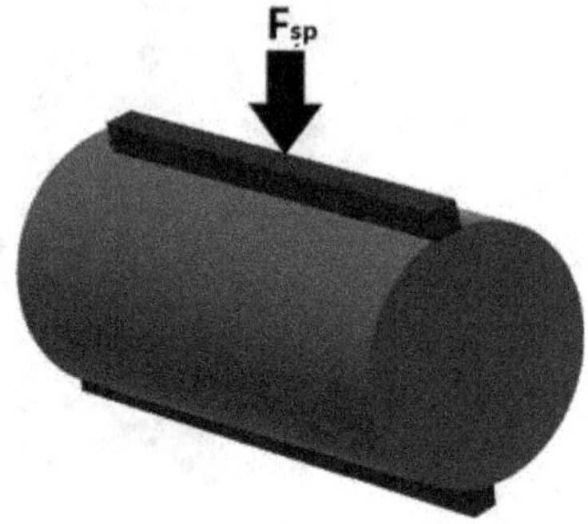

Fig. 2.20 The Brazilian disk test on a cylindrical sample with diameter and length of D_{cyl} and L_{cyl}, respectively

a fast alternative for computing the direct tensile strength. The Brazilian disk test is conducted to determine the splitting tensile strength (Fig. 2.20). The splitting strength depends on loading rate, diameter and length of the specimen (2.6). The works of [31, 32] provides a full review and a correlation between different rock strength properties.

$$\sigma_{sp} = \frac{2f_{sp}}{\pi L_{cyl} D_{cyl}} \tag{2.6}$$

In the CAU Kiel laboratory, the splitting strength under THM processes is measured. To do so, a controlled temperature and humidity climate chamber (Fig. 2.17b) and a loading frame with a displacement transducer are required. Initially, the splitting test under room temperature condition with initial humidity of the sample is conducted. Afterwards, the temperature is raised to 50 and 80 °C. The relative humidity can be controlled from 0 up to 100. Finally, the splitting strength, splitting Young modulus (E_{sp}) and load-displacement behavior are measured. The cylindrical claystone and saltstone samples are prepared in dimension of 100 × 200 mm ($D \times L$) (Fig. 2.21a) and a strain rate of 0.1 % is considered to insure a relatively fast loading rate. Figure 2.21b shows the placement of the sample under loading frame before performing the test.

Initially, the splitting strength of a saltstone samples under room temperature (20), 50 and 80 °C are measured. Due to a relatively homogeneous material property of the saltstone, the orientation of the sample will not affect the final outcomes. Figure 2.22a and b depict the failure in 20 °C and load versus displacement for different loading temperatures, respectively. The observed failure pattern for all of the setups is identical. The measured mean splitting strength using Eq. (2.6) for 20, 50 and 80 °C are 1.65, 1.59 and 1.43 MPa, respectively. As a result, the temperature has a slight influence on the splitting strength of saltstone, especially when the temperature is raised up to 50 °C.

The anisotropy of claystone and embedded layering orientation has a significant influence on material strength. To investigate the strength dependency on layering orientation, a series of tests, where the angle between the loading direction and lay-

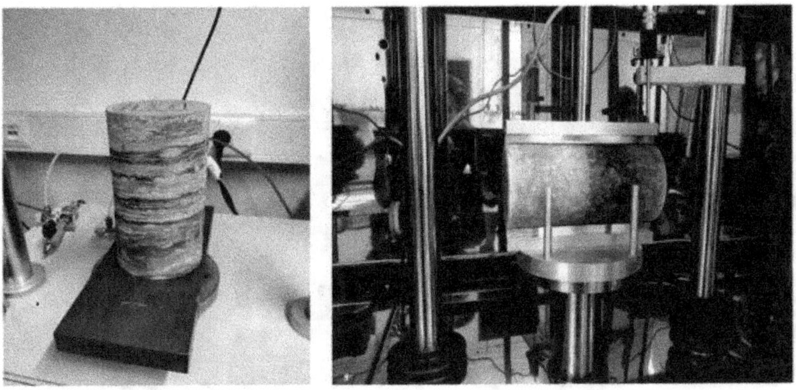

(a) (b)

Fig. 2.21 The sample preparation and test procedure **a** the claystone sample with a dimension of 100 × 200 mm, **b** the placement of saltstone inside of the climate chamber

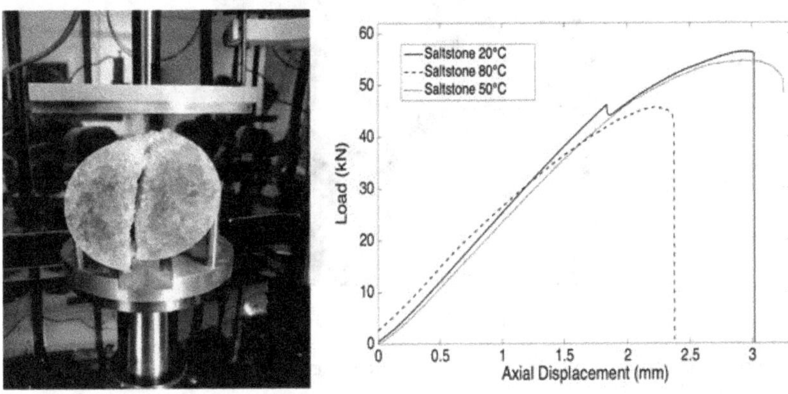

(a) (b)

Fig. 2.22 The splitting strength of a saltstone under different temperature conditions **a** the fracture path in saltstone in room temperature, **b** the load versus displacement under different temperature conditions

ering orientation is 0 (parallel), 30, 45, and 90 (perpendicular) degree, is conducted. For 20 °C, the failure pattern for 0, 45 and 90° are provided in Fig. 2.23a, b and c. Figure 2.24 depicts the load versus displacement behavior of claystone with different layering degrees.

Table 2.2 presents the mean outcome of the experimental tests for different orientation angles and temperature. The results depicts that when the loading is perpendicular to the layering orientation, the splitting strength of the claystone is almost 5 times higher than when it is parallel to the layering orientation. The temperature effect on splitting strength is negligible.

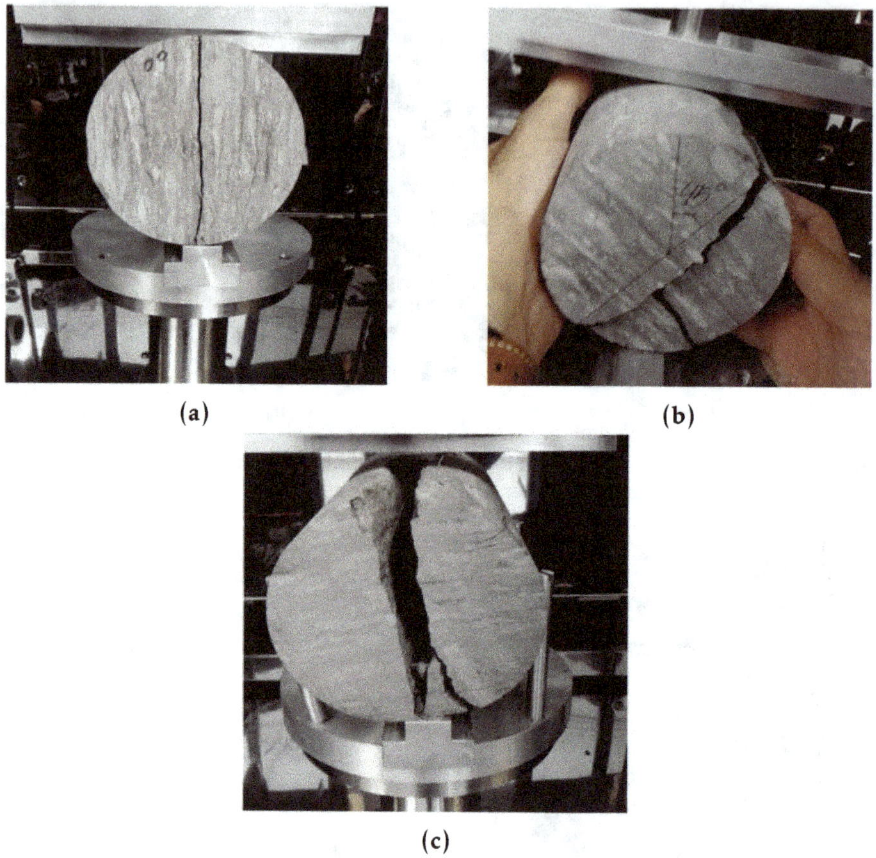

Fig. 2.23 The fracture pattern under different layering orientation **a** 0°, parallel, **b** 45°, and **c** 90°, perpendicular (20 °C)

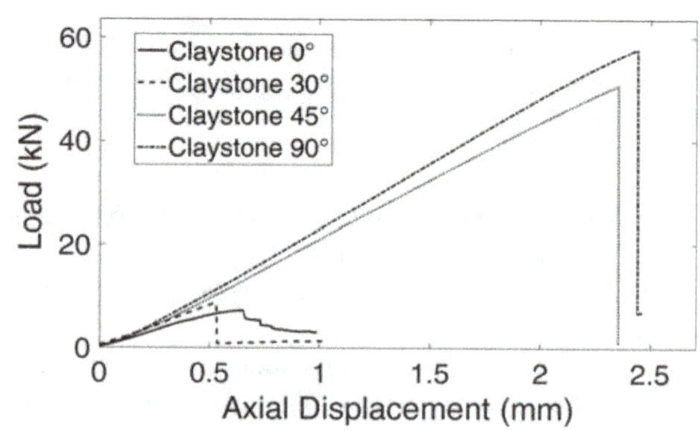

Fig. 2.24 The load versus displacement under different layering orientation, 20 °C

Table 2.2 The splitting tensile strength of the Opalinus Clay with different temperature and layering orientations

Test results	0°	30°	45°	60°	90°
σ_{sp} (MPa) 20 °C	0.47	0.68	1.13	1.45	1.92
σ_{sp} (MPa) 80 °C	0.52	0.64	1.02	1.25	1.86

(a) (b)

Fig. 2.25 The **a** true triaxial apparatus in geomechanics laboratory of CAU Kiel, and **b** prepared cubic claystone sample with curved edges

2.2.4 True Triaxial Test on the Cubic Opalinus Clay Samples

Amir Shoarian Sattari (CAU)

The true triaxial apparatus, where the stresses are controlled along three axes, is used to investigate the three-dimensional strain-strength behavior of soil or rock geomaterial (Fig. 2.25a). The true triaxial device in the CAU Kiel laboratory is able to reach a mechanical loading of 600 MPa as well as a thermal loading of up to 600 °C. The cubic samples are prepared in the side dimension of 43 mm and the edges are slightly curved in order to avoid the stress concentration and failure in the corners (Fig. 2.25b).

The coupled thermo-mechanical loading conditions are considered in order to investigate the materials anisotropic stiffness along three axis, the elastic and plastic deformation under cyclic thermal and mechanical loadings, deviatoric stress field, and material failure. A thermal loading of up to $T_{iso}^{max} = 150\,°\mathrm{C}$ is considered, which is considered to be higher than the maximum temperature that the claystone samples are subjected to in nuclear waste disposal sites. The maximum isotropic mechanical loading is considered to be $\sigma_{iso}^{max} = 100\,\mathrm{MPa}$, where max and c superscripts represent the maximum and constant values, respectively. The considered boundary conditions are:

- Sample MT-01: Mechanical Condition,5 loading cycles, $\sigma_{iso}^{max} = 100$ MPa.
- Sample MT-02: Coupled Thermo-Mechanical Condition, 4 loading cycles, $T_{iso}^{max} =$ 150 °C and $\sigma_{iso}^{c} = 12$ MPa.
- Sample MT-03: Coupled Thermo-Mechanical Condition, 4 loading cycles, $T_{iso}^{max} =$ 150 °C, and $\sigma_{iso}^{max} = 100$ MPa.
- Sample MT-04: Coupled Thermo-Mechanical Condition, 4 loading cycles, $\sigma_{dev}^{max} =$ 60 MPa, $\sigma_{con}^{c} = 20$ MPa and $T_{iso}^{c} = 150$ °C.

With the installation of the ultrasonic sensors on the pistons, the apparatus is able to measure the ultrasonic P, $S90$ and $S0$ waves along the axis. The anisotropy factor, density (ρ), the dynamic Youngs modulus (E_{dyn}), dynamic shear modulus (G_{dyn}), and dynamic Poissons ratio (ν_{dyn}) values can all be calculated according to the analytical relation which already exist in the literature [33] as given in Eqs. (2.3) and (2.4). In order to detect and analyse the ultrasonic signals in Opalinus Clay, the minimum confining mechanical stresses of 12 MPa is required. The test results of MT-01 (Fig. 2.26) depicts the increment of the mean E_{dyn}, G_{dyn} and ν_{dyn} values under the isotopic confinement stresses up to 100 MPa, which is due to the layering orientation and structure of claystones and eventually improved contact qualities. It is also noted that the rate of the increment after each loading cycle (slope of the line) is decreased and due to the plastic deformations and improved contact qualities, the materials mechanical properties in initial loading condition of 12 MPa are enhanced.

Figure 2.27 shows the deformation versus time result of MT-02 in the 1st cycle. The thermal plastic deformation after 5 cycles of loading and unloading was negligible and can be neglected. The MT-02 sample is situated in a way that the loading frame in direction of Z is parallel to the layering orientations of claystone. The volumetric thermal expansion coefficient (α_V) is calculated based on the measured volumetric strain ($\epsilon_V = \frac{\Delta V}{V}$) and temperature change (ΔT). The effect of temperature on α_V is shown at Fig. 2.28a. The rate of α_V is decreased after 80 °C. The anisotropy in linear thermal expansion coefficient along Z, Y and X axis is shown in Fig. 2.28b. The anisotropy in the measured linear expansion coefficient (α_{Li}) is substantially decreased after 80 °C.

$$\alpha_V = \frac{\epsilon_V}{\Delta T} = \frac{\Delta V}{V \Delta T} \tag{2.7}$$

The test results of MT-03 investigates the anisotropy of Opalinus Clay samples. Figure 2.29 illustrates the 1st cycle loading results of E_{Dyn} and ν_{Dyn} values in X, Y and Z directions under the isotopic confinement stresses up to 100 MPa. The G_{Dyn} has a similar behavior to the E_{Dyn} values. The results indicate a weaker stiffness in Z direction, which is parallel to the layering orientation. The thermal loading upto 150 °C results in weaker E_{Dyn} values during the heating process, which is the reason for the fluctuation of the results at the confinement stress of 100 MPa. The test results of MT-04 is similar to the MT-02 and the same material response is observed.

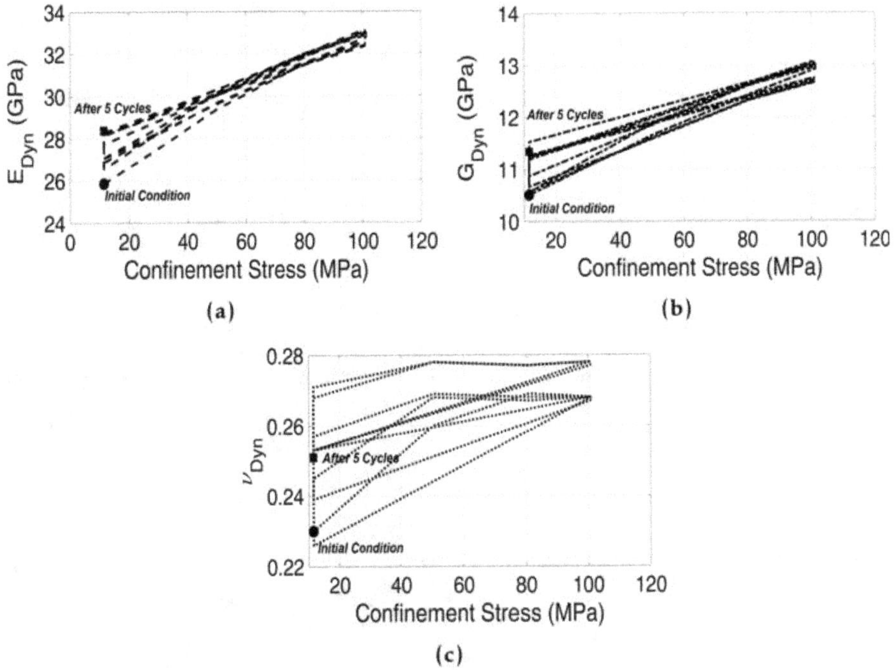

Fig. 2.26 The true triaxial results for the MT-01 sample under 5 loading cycles **a** E_{dyn} versus confinement stress, **b** G_{dyn} versus confinement stress, and **c** ν_{dyn} versus confinement stress

Fig. 2.27 The deformation versus time for sample MT-02

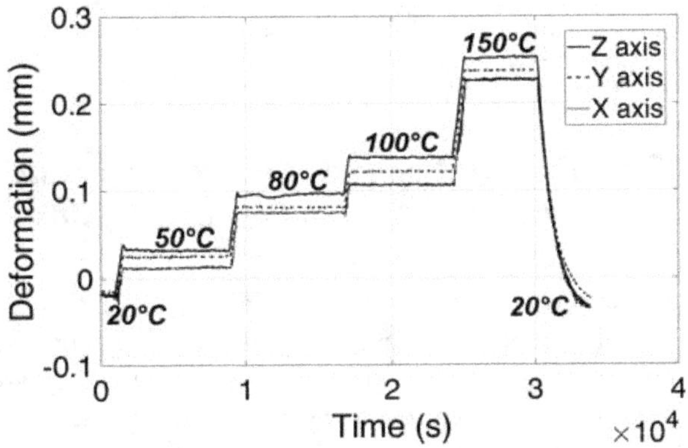

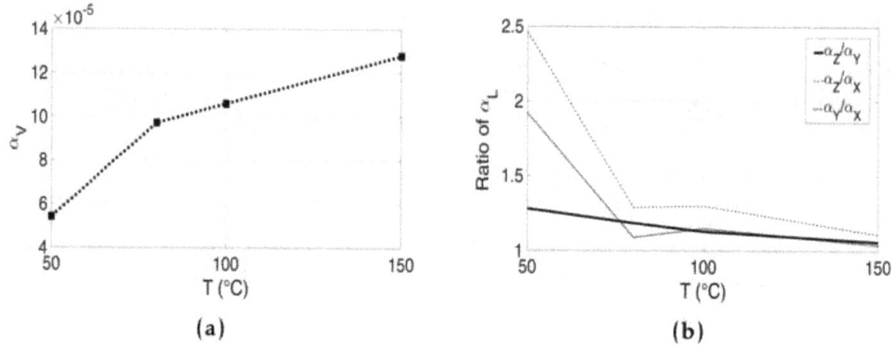

Fig. 2.28 The effect of temperature on **a** α_V, and **b** the anisotropy in the linear thermal expansion coefficient (α_{Li})

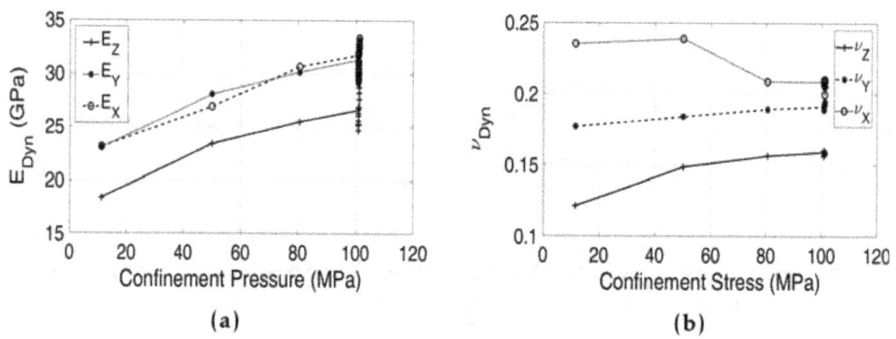

Fig. 2.29 The true triaxial results for the MT-03 sample in the 1st cycle **a** E_{Dyn} versus Confinement Stress, and **b** ν_{Dyn} versus Confinement Stress

2.2.5 Triaxial Compression Strength Tests for Salt—Methodology and Equipment

Mathias Nest, Dirk Naumann (IfG)

Triaxial compression tests (TC) are executed on cylindrical core samples with dimensions of e.g. 200*100 mm (generally a 2:1 ratio of height to diameter) in one of two available servo-hydraulic testing machines of the IfG labs (RBA 2500, Schenk/Trebel, Germany and D2000 GGL Testsystems using the MTS-TestStar software; see Fig. 2.30 generating the necessary stress σ_1 in axial direction and σ_3 in lateral direction.

The cylindrical samples are sealed with rubber tubes and oil is used as confining medium. Outside the vessel three LVDT transducers are mounted between the piston and the load frame near the sample for the measurement of the axial strain. The axial load is determined from an external load cell. Tests can also be carried out at in-situ relevant temperatures (e.g. 55 °C). During all tests the following parameters are

Fig. 2.30 The two available servo-controlled hydraulic testing machines (right: system RBA 2500 of SCHENK/TREBEL and left: D2000 of GGL TestSystems)

measured automatically: the axial deformation Δh, axial load F and the confining pressure $p = \sigma_3$. The conversion of the measured values to effective stress and strain is given as:

$$\sigma_{\text{eff}} = \left(\frac{F}{A_0} - \sigma_3\right)(1 - \epsilon_1) \tag{2.8}$$

$$\epsilon_1 = \frac{h_0 - h}{h_0} \tag{2.9}$$

with

h_0, h: length of the sample before and during deformation
A_0: cross sectional area of the undeformed sample
F: axial load
σ_3: confining pressure

The term $1 - \epsilon_1$ results from the consideration of the sample bulge during the deformation, e.g. the changes of the cross sectional area as well in compression as in extension are regarded.

In addition, as a standard technique of the IfG during the triaxial strength experiments the sample volume changes ΔV are determined by a volume balance of the mantle oil volume changes as measured via the pressure intensifier and the axial piston displacement in the cell:

$$\Delta V = \Delta h A_{pp} - \Delta S_{pi} A_{pi} \tag{2.10}$$

with

Δh: displacement of the piston of the triaxial cell
A_{pp}: cross section of the piston of the triaxial cell
ΔS_{pi}: displacement of the cylinder within the pressure intensifier
A_{pi}: cross section of the piston within the pressure intensifier

During all tests the following parameters were measured automatically: The axial deformation Δh, axial load F and the confining pressure $p = \sigma_3$.

The triaxial test procedure is divided into two steps:

- 1st step: Reconsolidation isotropic phase (applied for all samples): Because it is assumed that core material is generally dilated due the applied procedures during core recovery (stress relaxation) and subsequent weathering, the samples are firstly re-compacted. In a hydrostatic load cycle ($\sigma_1 = \sigma_3$) the samples are pressurized up to 60 MPa (at 0.01 MPa/s) and then, after 30 min. unloaded to the respective pressure state of the strength test.
- 2nd step: Triaxial test cycle deviatoric phase: At constant confining pressure the sample is axially loaded using a constant deformation rate, until the sample fails and reaches the post failure state. Common standard triaxial compression tests (TC) with defined deformation rate at constant confining pressures are between 1 and 25 MPa. A determination of the deformation dependent strength and dilatancy behavior is realized, too.

The standard triaxial strength test (TC) for salt rocks is performed with a constant deformation rate ($5 \cdot 10^{-6}\ \mathrm{s}^{-1}$) at constant confining pressure. After the failure occurs the axial load decreases with further deformation due to the decrease of the load-bearing capacity becoming nearly constant if the post-failure stage is reached. Finally the sample is unloaded.

Simultaneously with the deformation the volumetric strain is measured which allows to determine the stress-dependent onset of micro-cracking (given by the minimum) and to quantify the deformation induced damage.

2.3 Shrinkage and Swelling Laboratory Tests (WP1)

2.3.1 The Swelling and Permeability of T4 Salt Clay

Mathias Nest, Till Popp (IfG)

The transport properties of a material depend not only on the pore space, but also on the permeability, which is the decisive measure for the hydraulic integrity of a material. We measure the permeability based on the Darcy equation, which takes

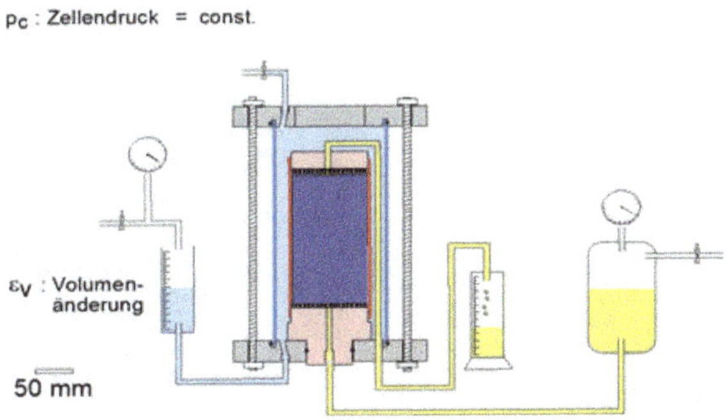

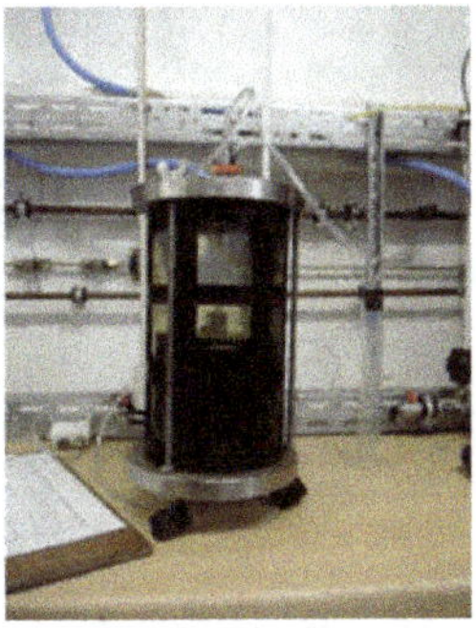

Fig. 2.31 Setup of the experiment to measure the closure of pathways in clay through swelling

into account the pressure difference, the viscosity of the fluid, and a geometry factor of the sample.

For the present project the permeability was measured under stationary stress/ pressure conditions using a saturated brine solution. The setup that we used was build at the IfG, and is shown in Fig. 2.31. The mantle of the cylindric samples were covered with rubber cuffs, and filter plates on top and below. Thus they are isolated from the medium (water) which provides the cell pressure. The amount of brine that passes through the sample is measured as a function of time, so that the permeability and its change can be calculated.

2.3.2 The Wetting and Drying Paths of the Opalinus Clay

Amir Shoarian Sattari (CAU)

The shrinkage and swelling of claystone results in micro fracking and higher permeability values, which in nuclear waste disposal sites can lead to contamination of groundwater. Micro fracking also decreases the strength of the material subjected to THM loading processes. Typically, the swelling pressure and heave of claystone is determined using Oedometer tests [34], where a constrained or unconstrained sample is subjected to the swelling process (Fig. 2.32a). During the test procedure, the swelling pressure, as well as the heave magnitude, are recorded. It is observed that the swelling pressure in sandy facies of claystone is lower than shaly facies. In contrast to the swelling tests, shrinkage tests are less common and are complicated to perform on rock materials. Minardi et al. [35] performed a shrinkage test on claystone with shaly and sandy facies using a desiccator and various salt solutions (Fig. 2.32b). During the test procedure the axial strain values obtained from the strain gauges, as well as sample weight, are measured.

Two prepared thin cylindrical sections of sandy facies of the Opalinus Clay (Mont Terri) with a dimension of 100 × 10 mm ($D \times H$) are used to determine the anisotropy in shrinkage and swelling behavior of claystone. Due to the mineral

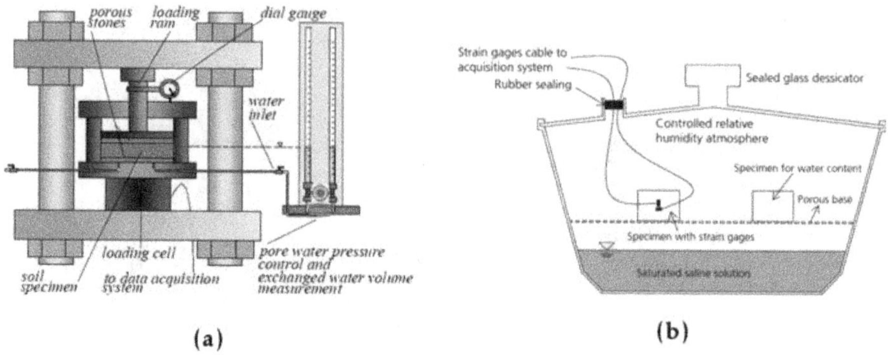

Fig. 2.32 The swelling and shrinkage tests on Opalinus Clay **a** the Oedometer test setup for constrained swelling pressure in Opalinus Clay samples [34], and **b** measuring the drying and wetting paths using desiccator [35]

structure of claystone and its layering orientation, the anisotropy factor has a significant influence on the direction of shrinkage and swelling as well as the micro fracking formation. The initial sample is used to determine the axial strains in parallel and perpendicular directions to the layering orientations. The strain gauge strips (HBM, $LY10$ (mm)/120 Ω) are glued and attached to the surface of the sample. The second sample is used to measure the weight change in the sample under different salt solutions. The saturated salt solutions are located inside the desiccator and will result in osmotic suction and eventually in the wetting and drying of the samples. The total suction(ψ_{total}) value can be measured using the Kelvin's relation, which is derived from ideal gas law:

$$\psi_{\text{total}} = \frac{RT}{V_{\text{mol}}} \ln\left(\frac{P_{\text{vap}}^{cur}}{P_{\text{vap}}^{cur=0}}\right) \tag{2.11}$$

where, $(\frac{P_{\text{vap}}^{cur}}{P_{\text{vap}}^{cur=0}})$ is the relative humidity, $P_{\text{vap}}^{cur=0}$ is the vapor pressure when the surface curvature is equal to zero (flat surface), R is the universal gas constant, T is the temperature and V_{mol} is the molecular volume of water. The equilibrium inside the desiccator is reached when the strain gauges value or water content of the samples are constant in two consecutive readings. The results obtained from the drying and wetting paths of the Opalinus Clay are given in Sect. 4.5, where a numerical simulation and comparison to the experimental data are provided. The test procedure is time consuming and after measuring the data for more than 120 d, the applied suction and water content percentages are calculated and plotted.

2.3.3 *In-situ Condition Desiccation Process*

Topology, respectively morphology investigations of fractures induced in clay stone throughout drying processes require a number of experimental sequences ranging from the drying process itself to characterization of the induced fractures by X-Ray Computed Tomography. The required samples are prepared and characterized by the University of Kiel with a dimension of radius $r = 5\,\text{mm}$ and height $h = 50\,\text{mm}$ to guarantee the realization of scans in Regions of Interests (RoIs) down to a characteristic edge length of 6 mm resulting in a resolution of 2 μm per voxel.

Two different experimental set-ups are planned; nevertheless the first proposed experimental implementation is preferred over the second.

Drying Wet Samples—From Wet to Dry

Saturated samples are prepared by the University of Kiel and submerged in a shrink tube in order to minimize the change of the desired state. Once the experiments are performed on the sample the shrink tube is removed and the probe is installed in an uni-axial testing device. The sample is installed in a heat chamber and loaded by an axial force of $f_{\text{ax}} = 500\,\text{N}/$, resulting in an axial stresses of $\sigma_{\text{ax}} = 0.25\text{MPa}$ (potentially up to $f_{ax} = 5\,\text{kN}/\sigma_{\text{ax}} = 2.5\text{MPa}$) while axial deformations are measured under controlled temperature conditions. By measuring the sample's weight at characteristic states (twice—beginning and end) the volumetric deformations can be determined and related to the change of saturation. The sample characterization is then completed by XRCT scans to determine the features of the induced fractures. Finally the stiffness degradation is determined by ultrasound experiments measuring the P-wave run times at 2, respectively 6 MHz.

Wetting dry Samples—From Dry to Wet

Since it is extremely time consuming to increase the saturation of the sample this second setting is only an alternative to the introduced set-up in case the first experiment fails. The set-up is comparable to the proposed experiment but performed in a heat-humidity chamber at rel. humidity between 10–90 % and temperatures between $5-90\,°\text{C}$. The applied axial Force would reduce to $f_{\text{ax}} = 50\,\text{N}$.

2.4 Pressure Driven Percolation Laboratory Tests (WP2)

2.4.1 Pressure Driven Percolation

Mathias Nest, Dirk Naumann (IfG)

The determination of the permeability, which quantifies the flow behavior of fluids in the pore space of a rock, is based on the Darcy equation:

$$q = \frac{kA}{L\eta}\Delta p \tag{2.12}$$

with:

q: flow rate (m^3/s)
k: permeability (m^2)
A: cross sectional area (m^2)
L: length of sample (m)
η: dynamic viscosity (Pa· s)
Δp: pressure difference (Pa)

Thereafter, the flow rate of a fluid through a sample at a given pressure differential is measured by the viscosity of the flowing medium, the geometric factor of the sample, and the permeability (with the dimension of an area). The permeability is given as SI-unit in (m^2 or, traditionally, in D (Darcy) (1 D corresponds to about 10^{-12} m^2).

The equation above is only valid for incompressible fluids, because only then is the flow rate q constant over the flow path. With sufficient accuracy, this applies to low compressible fluids. When a gas flows through the pore space of a solid, however, an expansion of the gas takes place along the flow path, so that the flow rate is not constant here. In this case, instead of the flow rate q the mean flow rate q_m is set, for which, for small pressures with sufficient accuracy according to the law of Boyle-Mariotte:

$$q_m p_m = q_0 p_0 \tag{2.13}$$

with:

q_m: mean flow rate
p_m: mean pressure
q_0: measured flow rate at p_0
p_0: pressure at flow rate measurement

If, for the mean pressure p_m, the arithmetic mean of the pressures p_1 on the high pressure side and p_2 on the low pressure side of the porous solid is used, taking into account that $\Delta p = p_1 - p_2$, the modified Darcy equation for gas flows is as follows:

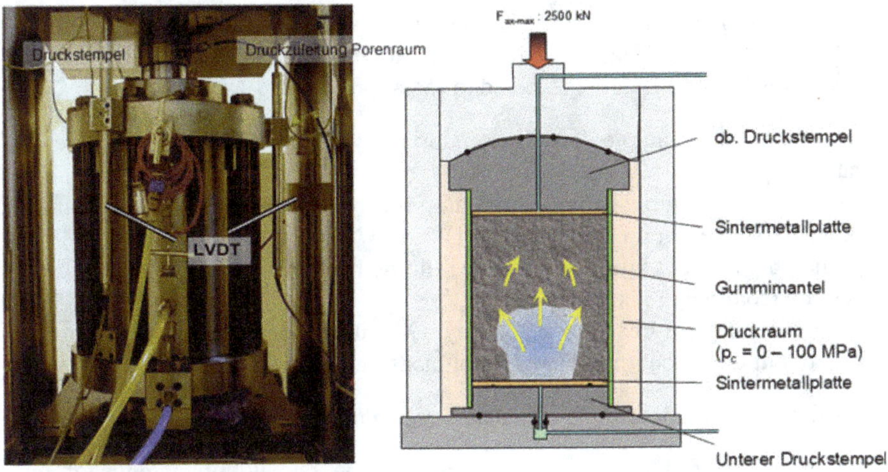

Fig. 2.33 Triaxial IfG pressure cell for flow tests with brine or gas

$$k = \frac{2p_0 q_0 \eta L}{A(p_1^2 - p_2^2)} \tag{2.14}$$

In summary, the determination of the permeability with caustic or gas with knowledge of the viscosity requires in each case an exact measurement of the flow rate after setting (quasi) stationary flow conditions and the pressure gradient.

To carry out permeability tests, the IfG Leipzig uses a servo-hydraulic testing machine with a pressure cell up to pc-max = 1000 bar, which is otherwise used for strength tests with F_{max} = 2500 kN (manufacturer: Schenk/Trebel) (Fig. 2.33). The tests routinely set hydrostatic pressure conditions ($\sigma_1 = \sigma_2 = \sigma_3$), but it is also possible to implement deviatoric stresses or defined deformations. The axial load or deformation and the jacket pressure are each controlled independently via a servo-hydraulic system.

The desired jacket pressure is generated by a pressure intensifier. From the axial deformation and the measured change in volume of the lateral pressure chamber (piston displacement of the pressure booster), the volume change of the test specimen, referred to here as dilatancy, can be determined at constant jacket pressure.

In hydrostatic loads sintered metal plates are used, which allow over the entire cross-sectional area of the sample a fluid pressurization. The pressure measurement of the measuring fluid is carried out by pressure transducers from Hottinger (accuracy class 0.2), whereby depending on the measuring range a 20 bar or 200 bar encoder is used.

The measuring principle for permeability determination under stationary conditions is based on the measurement of the flow rate q, here under atmospheric conditions (p_0), in the axial sample direction at a predetermined pressure gradient $\Delta p = p_1 - p_2$ (p_1 = inlet pressure, p_2 = outlet pressure).

The measuring arrangement used in the triaxial cell has the following advantages over the frequently used Hassler cells, in which a cylindrical sample is firmly clamped between punches and is only subjected to radial pressure.

- Determination of sample deformation during hydrostatic and deviatoric pressurization
- Use of variable sample geometries (stamp sets between 60 and 110 mm diameter are available, height up to 2 x diameter)
- The height of the gas injection pressure is limited only by the available gas cylinder pressure (routinely max: 200 bar).
- For higher gas pressures (up to 1000 bar), a Maximator pneumatic pressure booster is used, but this is only used for special measurements.

For the determination of gas permeability two methods are available:

- 1st: Injection of nitrogen at a defined injection rate within a range of min. 0.1 ml/min to max. 500 ml/min under measurement of the injection pressure when stationary conditions are reached with leakage of the fluid against the atmosphere.
- 2nd: Injecting nitrogen under a defined pre-pressure and measuring the flow rate at the exit side of the sample against the atmosphere.

The lower limit of the measuring range depends essentially on the pre-pressure or the minimum measurable flow rate, as shown schematically in Fig. 2.34. For gas permeability measurements, EL-FLOW mass flow controllers or flow meters from Bronkhorst are used with the following specifications:

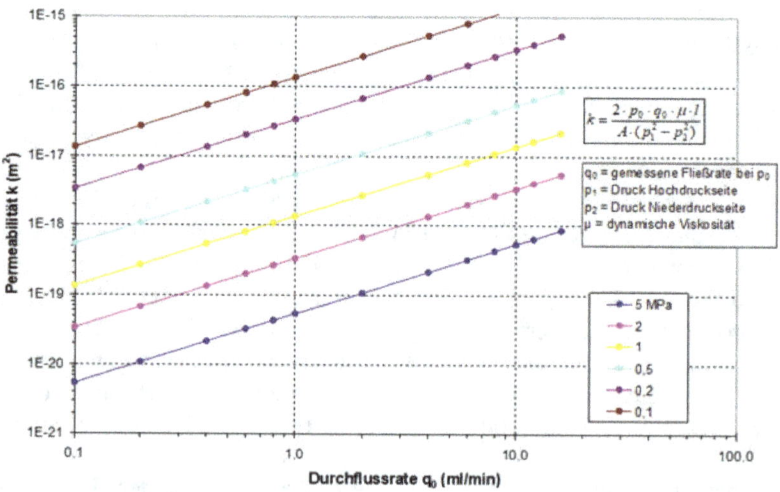

Fig. 2.34 Variation diagram of permeability versus flow rate for a sample with l = 220 mm and d = 110 mm at different injection pressures

- Mass flow controller Type: F-230M: Measuring range: (0) ... 10 ... 500 ml/min N2
- Form: 200 bar/outlet pressure: 194 bar/temperature: 20 °C
- Measuring accuracy: 1% of final value, typ. better 0.5%
- Mass flow controller Type: F-230M: Measuring range: (0) ... 0.4 ... 20 ml/min.

2.4.2 *Fluid Driven Percolation Tests on Cubic Opalinus Claystone Samples from Mont Terri*

Amir Shoarian Sattari (CAU)

The investigation of a fluid transport in claystone due to its anisotropic behavior and its role as a rock barrier in nuclear waste repositories has a great importance. Performing hydraulic fracking under pressurized fluid or storing pressurized fluids leads to the fracking of rock barrier and fluid transport through the hydraulic apertures and cavities. This can lead to pressure and volume drop in the reservoir, decrease the output and efficiency of the designed system and the contamination of ground water. In the geomechanics laboratory of CAU Kiel, the true triaxial apparatus with the maximum mechanical pressure of 600 MPa and thermal loading up to 600 °C is used to conduct the fluid driven percolation in claystone samples from Mont-Terri (Fig. 2.25a). The syringe pump, with the maximum pressure of 517 *bar*, is used to pressurize the oil fluid. The cubic samples with the side dimension of 43 mm and center hole length and diameter of 20 and 8 mm are prepared (Fig. 2.35a), respectively, and attached to the pump pipes, where the sealing is done with O-rings and epoxy glue (Fig. 2.35b).

Two different stress configurations, as well as fluid injection directions parallel or perpendicular to the layering orientation, are considered to investigate the fracking pattern as well as flow pathways. The fracking tests are carried out under a constant fluid pressure and the peak fluid pressure, where the flow rate increases and any

(a) (b)

Fig. 2.35 The fluid driven percolation test preparation **a** the prepared cubic claystone sample and the adapter and **b** the sample placement inside the true triaxial apparatus sudden drops in fluid pressure is recorded. In the initial test, a sample with a fluid injection direction perpendicular to the layering orientation of the Opalinus sample is considered. The initial stress configuration is 12, 14 and 16 MPa in three different loading axis. In order to prevent damaging the sample prior to the hydraulic fracking test, the isotropic stresses of 8 MPa is applied from all pistons and is then gradually increased to the planned stress configuration. The oil pressure is increased gradually up until the point where the borehole pressure drops and an increase in the flow rate can be seen. The test is then aborted immediately in order to avoid causing any damage to the true triaxial apparatus. In the second test setup, a sample with a fluid injection direction parallel to the layering orientation of the Opalinus sample is considered. The initial stress configuration of 16, 10 and 8 MPa in three different loading axis is applied. The results of the percolation tests on the Opalinus claystone samples are given in Sect. 4.8, where a numerical simulation and comparison to the experimental data are provided and the effect of stress distribution, anisotropy and layering orientation on frack paths and fracking pressure are discussed.

2.5 Stress Redistribution Laboratory Tests (WP3)

2.5.1 Direct Shear Test

Thomas Frühwirt, Daniel Pötschke (TUBAF)

To conduct direct shear tests special equipment is necessary. The direct shear testing device at the rock mechanical laboratory of the TU Freiberg (see Fig. 2.36) is specially developed to ensure the wanted functionality.

Some key features can be found in Table 2.3. Additionally it is possible to superimpose dynamic forces. In the tests for the GeomInt project this functionality was not used.

The sample preparation includes the cutting of the rock block in the cuboid shape. This sample is split into two parts and the rock joints are arranged in a matching position. It is arranged in the shear box, Fig. 2.37.

The sample is grouted in the shear box to avoid any unwanted movements of it, Fig. 2.38.

The finally equipped shear box is connected to the measuring units, the LVDTs (Linear Variable Differential Transformer), Fig. 2.39. The accuracy of this length measurements is in the order of a μm.

The set-up of the experiment is the same for CNL and CNS test. In the CNS test the stiffness which adds an extra load is calculated. This means if a normal displacement of the sample is measured the normal stress will be adapted according to the defined stiffness.

Fig. 2.36 The shear testing device at the rock mechanical laboratory of the TU Freiberg. (From: [36])

Table 2.3 Technical data of the shear testing device (From: [36])

Feature	Value	Unit
Max. normal force	1000	kN
Max. shear displacement	50	mm
Min. shear velocity	1e-7	mm/s
Max. shear velocity	70	mm/s
Max. sample size (rectangular)	200 × 400	mm
Max. fluid pressure	10	MPa

2.5.2 Cyclic Loading Pressure Diffusion

Matthias Ruf, Holger Steeb (UoS)

In order to study characteristic, time-dependent states of fractures under cycling loading conditions a sample is prepared with a single fracture and borehole before it is installed in a triaxial cell as shown in Fig. 2.40. The dimension of the sample are chosen to be $r = 30$ mm and height $h = 70$ mm.

The experiment is performed in three steps. After applying a confining pressure p_c the initial state is approached by deformation control while the acting normal forces are measured in a first step. Once a desired normal force is reached the deformation state is held constant and a fluid pressure of p_{fix} is applied. Finally, the fracture is stimulated by a harmonic fluid pressure with a frequency of 0.1 Hz and varying

Fig. 2.37 Sample in shear box (From: [37])

Fig. 2.38 Grouted sample before direct shear test (From: [37])

Fig. 2.39 Measuring equipment: LVDT (From: [37])

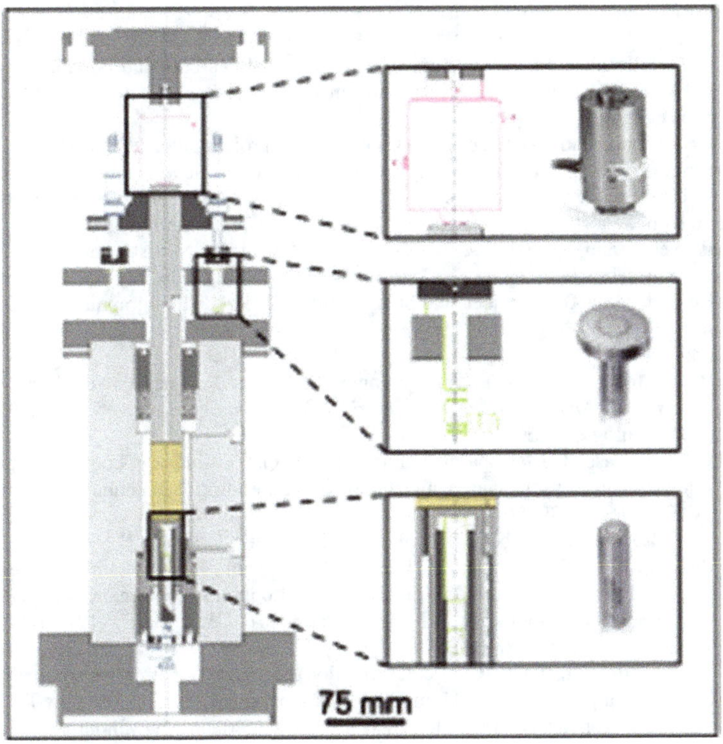

Fig. 2.40 Set-up of triaxial cell

amplitudes p_A between 0.5−3 MPa. Throughout the experiment flow and pressure are measured at the fluid induction point to study the relationship of pressure and flow under non-constant fracture permeabilities triggered by deformations.

References

1. Thomas Bolliger, Burkart Engesser, and Marc Weidmann. Première découverte de mammifères pliocènes dans le jura neuchâtelois. *Eclogae Geol. Helv*, 86(3):1031–1068, 1993.
2. Arnfried Becker. The Jura Mountains an active foreland fold-and-thrust belt? *Tectonophysics*, 321(4):381–406, 2000.
3. Christophe Nussbaum, Paul Bossart, Florian Amann, and Charles Aubourg. Analysis of tectonic structures and excavation induced fractures in the opalinus clay, mont terri underground rock laboratory (switzerland). *Swiss Journal of Geosciences*, 104(2):187, 2011.
4. M Freivogel and P Huggenberger. Modellierung bilanzierter profile im gebiet mont terri–la croix (kanton jura). *Mont Terri Project–Geology, paleohydrogeology and stress field of the Mont Terri region. Federal Office for Water and Geology Rep*, 4:7–44, 2003.
5. Annette Kaufhold, Matthias Halisch, Gerhard Zacher, and Stephan Kaufhold. X-ray computed tomography investigation of structures in opalinus clay from large-scale to small-scale after mechanical testing. *Solid Earth*, 7(4):1171–1183, 2016.
6. Bernhard Hostettler, Achim G Reisdorf, David Jaeggi, Gaudenz Deplazes, Hansruedi Bläsi, Alain Morard, Susanne Feist-Burkhardt, Anton Waltschew, Volker Dietze, and Ursula Menkveld-Gfeller. Litho-and biostratigraphy of the opalinus clay and bounding formations in the mont terri rock laboratory (switzerland). In *Mont Terri Rock Laboratory, 20 Years*, pages 23–39. Springer, 2018.
7. PETER Heitzmann and PAUL Bossart. Das mont-terri-projekt untersuchungen über den opalinuston im internationalen felslabor. *Bulletin für angewandte Geologie*, 6:183–197, 2001.
8. Gerhard Einsele. Mechanismus und tiefgang der verwitterung bei mesozoischen ton-und mergelsteinen. *Zeitschrift der Deutschen Geologischen Gesellschaft*, pages 289–315, 1983.
9. Andreas Wetzel and Vincenzo Allia. Der opalinuston in der Nordschweiz: Lithologie und Ablagerungsgeschichte. *Eclogae Geologicae Helvetiae*, 96(3):451–469, 2003.
10. Paul J Bossart, Mont Terri Project, and Suisse. Office fédéral de topographie. *Mont Terri Rock Laboratory Project: Programme 1996 to 2007 and Results*. Federal Office of Topography Swisstopo, 2008.
11. Suisse. Office fédéral des eaux et de la géologie. *Mont Terri project: Geochemistry of water in the Opalinus clay formation at the Mont Terri Rock Laboratory*, volume 5. Office fédéral des eaux et de la géologie OFEG, 2003.
12. D Traber and P Blaser. Gesteinsparameter der wirtgesteine opalinuston,'brauner dogger', effinger schichten und helvetische mergel als grundlage für die sorptionsdatenbank. *Nagra Working Report NAB*, pages 12–39, 2013.
13. G. Ziefle, P. Bossart, S. Costabel, O. Czaikowski, M. Furche, Graupner B., Hesser J., D. Jaeggi, O. Kolditz, F. Kniger, J. Maßmann, K. Rink, R. Schuhmann, K. Schuster, B. Vowinckel, and K. Wieczorek. Influence of Humidity on Cyclic and Longterm Deformations: CD-A Experiment, Mont Terri Rock Laboratory. In *Berichtsband zur 10. CMM Tagung, "Material Prozesse Systeme" in Karlsruhe, Germany*, 2019.
14. J.-M. Matray and A. Möri. CD (Cyclic deformation) experiment, petrophysical measurements on BCD-3 core samples from the Mont Terri rock laboratory. Technical Report TN 2013-49, Federal Office of Topography (swisstopo), Wabern, Switzerland, www.mont-terri.ch, 2012.
15. C. Girardin and C. Nussbaum. CD (cyclic deformation) experiment: Data report of Phase 17 (July 2011–June 2012). Technical Report TN 2012-68, Federal Office of Topography (swisstopo), Wabern, Switzerland, www.mont-terri.ch, 2012.

16. Gesa Ziefle, Jean-Michel Matray, Jobst Maßmann, and Andreas Möri. Coupled hydraulic-mechanical simulation of seasonally induced processes in the mont terri rock laboratory (switzerland). In *Mont Terri Rock Laboratory, 20 Years*, pages 197–214. Springer, 2018.
17. S. Jahn, S. Mrugalla, and L. Stark. Projekt AnSichT Methodik und Anwendungsbezug eines Sicherheits- und Nachweiskonzeptes für ein HAW-Endlager im Tonstein. Endlagerstandortmodell SD Teil II: Zusammenstellung von Gesteinseigenschaften für den Langzeitsicherheitsnachweis. Technical Report Ergebnisbericht, Bundesanstalt für Geowissenschaften und Rohstoffe (BGR), Hannover, Germany, 2016.
18. M. Galletti and D. Jaeggi. AD experiment: Drillcore documentation of boreholes BAD-1 and BAD-2. Technical Report TN 2019-06, Federal Office of Topography (swisstopo), Wabern, Switzerland, www.mont-terri.ch, 2019.
19. K. Knödel, G. Lange, and H.-J. Voigt. *Environmental Geology-Handbook of Field Methods and Case Studies*. Springer, 2007.
20. Forschungs- und Lehrbergwerk Reiche Zeche, Fuchsmühlenweg 9. https://tu-freiberg.de/universitaet/profil/campusplan/lehr-und-forschungsbergwerk-reiche-zeche-fuchsmuehlenweg-9. [Online; accessed 02-March-2020].
21. Klaus Grund and Manfred Bayer. Zur Geschichte des Freiberger Bergbaues. https://tu-freiberg.de/lfbw/geschichte. [Online; accessed 28-February-2020].
22. L.R. Alejano, J. González, and J. Muralha. Comparison of different techniques of tilt testing and basic friction angle variability assessment. *Rock Mechanics and Rock Engineering*, 45(6):1023–1035, 2012.
23. R.A. Schultz. Limits on strength and deformation properties of jointed basaltic rock masses. *Rock Mechanics and Rock Engineering*, 28(1):1–15, 1995.
24. Flavio Lanaro and Anders Fredriksson. Rock mechanics model-summary of the primary data preliminary site description forsmark area. R-05-83, 2005.
25. G.V. Ramana, P. Shashank, and H. Dev. Shear strength parameters of granite rock mass: A case study. *Lecture Notes in Civil Engineering*, 16:273–280, 2019.
26. 3D surface and body scanners. https://tu-freiberg.de/en/fakultaet3/gt/felsmechanik/rock-mechanics/3d-surface-and-body-scanners. [Online; accessed 02-March-2020].
27. Simone Carmignato, Wim Dewulf, and Richard Leach, editors. *Industrial X-Ray Computed Tomography*. Springer International Publishing, 2018.
28. Matthias Ruf and Holger Steeb. An open and flexible μXRCT system for applications in research. *Review of Scientific Instruments*, 2020. In process. Submission middle/end of February.
29. V. Cnudde and M.N. Boone. High-resolution X-ray computed tomography in geosciences: A review of the current technology and applications. *Earth-Science Reviews*, 123:1–17, aug 2013.
30. A.F. Bower. *Applied mechanics of solids*. CRC Press, 2009.
31. M.A. Perras and Diederichs M.S. A review of the tensile strength of rock: Concepts and testing. *Geotech Geol Eng*, 32:525–546, 2014.
32. D. Li and L.N.Y Wong. The brazilian disc test for rock mechanics applications: Review and new insights. *Rock mech rock eng*, 46:269–287, 2013.
33. H.B. Motra and S. Zertani. Influence of loading and heating processes on elastic and geomechanical properties of eclogites and granulites. *Journal of Rock Mechanics and Geotechnical Engineering*, 10:127–137, 2018.
34. H. Peron, S. Salager, M. Nuth, P. Marschall, and L. Laloui. Analysis of the swelling pressure development in opalinus clay - experimental and modelling aspects. *2nd International Conference on Fault and Top Seals*, 2009.
35. A. Minardi, E. Crisci, A. Ferrari, and L. Laloui. Anisotropic volumetric behaviour of opalinus clay shale upon suction variation. *Gotechnique Letters 6, 1–5*, 2016.
36. Heinz Konietzky, Thomas Frühwirt, and Hartmut Luge. A new large dynamic rockmechanical direct shear box device. *Rock Mechanics and Rock Engineering*, 45(3):427–432, May 2012.
37. Van-Manh Nguyen, Heinz Konietzky, and Thomas Frühwirt. New methodology to characterize shear behavior of joints by combination of direct shear box testing and numerical simulations. *Geotechnical and Geological Engineering*, 32(4):829–846, Aug 2014.

38. T. Kneuker and M. Furche. Capturing the structural and compositional variability of Opalinus Clay: Constraints from multidisciplinary investigations of Mont Terri drill cores (Switzerland). *Environmental Earth Sciences*, currently under review.
39. Plenkers, K., Philipp, J., Dörner, D., Minkley, W., Popp, T., Wiedemann, M. (2018): Observation of seismic and aseismic rock behavior during largescale loading experiment. Mechanical Behavior of Salt IX, 2018, Proceedings (ISBN 978-3-9814108-6-0). 961–972.

3

GeomInt Project and Numerical Methods

Keita Yoshioka, Mathias Nest, Daniel Pötschke, Amir Shoarian Sattari, Patrick Schmidt, and David Krach

An essential scientific goal of the GeomInt project is the analysis of potentials and limitations of different numerical approaches for the modelling of discontinuities in the rocks under consideration in order to improve the understanding of methods and their synergies with regard to theoretical and numerical fundamentals. As numerical methods, the "Lattice Element Method" (LEM), the non-continuous discontinuum methods "Discrete Element Method" (DEM), the "Smoothed Particle Hydrodynamics" (SPH), the "Forces on Fracture Surfaces" (FFS) as well as the continuum approaches "Phase-Field Method" (PFM), "Lower-Interface-Method" (LIE), "Non-Local Deformation" (NLD) and the "Hybrid-Dimensional Finite-Element-Method" (HDF) will be systematically investigated and appropriately extended based on experimental results (Fig. 3.1).

The numerical methods in Fig. 3.1 are displayed in accordance to the scale-ability, i.e. increasing temporal and spatial scales from right to left.

K. Yoshioka (✉)
UFZ, Helmholtz Centre for Environmental Research, Leipzig, Germany
e-mail: keita.yoshioka@ufz.de

M. Nest
IfG, Institut für Gebirgsmechanik, Leipzig, Germany

D. Pötschke
TUBAF, Technische Universität Bergakademie Freiberg, Freiberg, Germany

A. S. Sattari
CAU, Christian-Albrechts-Universität zu Kiel, Kiel, Germany

P. Schmidt · D. Krach
UoS, University of Stuttgart, Stuttgart, Germany

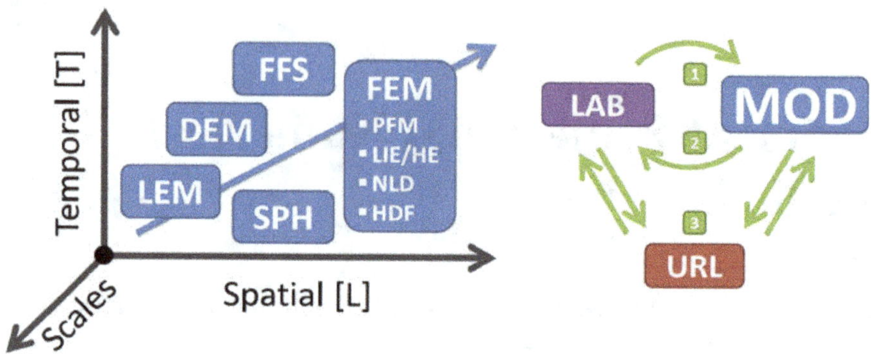

Fig. 3.1 Overview of the Numerical Platform (left) as part of the GeomInt research concept (right), see Sect. 1.2

3.1 State-of-the-Art

3.1.1 THM Simulations and Open Source Development

Process-oriented numerical simulation programmes are necessary for predicting possible environmental impacts as well as for the macroeconomic and safety design of geosystems for underground use with, if necessary, different or even multiple management. These programmes must be able to represent the running processes and their interactions. Already in the mid-eighties of the last century, specific models were developed in the USA and partly implemented in scientific simulation platforms in order to describe THM processes taking place in the geological subsoil which are connected with the thermal use of the subsoil as energy source (geothermics) or energy storage. However, these investigations primarily had a basic character. In addition, the numerical calculation tools are often oriented towards the description of special processes and only partially consider couplings of different physical processes. Geotechnical applications can also be simulated with a number of established commercial program systems. For hydraulic processes such as multiphase flow in porous media are simulators from the oil and gas industry available (e.g. ECLIPSE, STARS), for the description of mechanical processes as well (FLAC3D). All mentioned codes can only cover a part of the necessary process spectrum. Therefore, simulation programs are required which can represent thermal, hydraulic, mechanical, and chemical (THMC) processes coupled, such as TOUGH [78], HYTEC [113], DuMuX [33] or OpenGeoSys [52]. In order to be able to represent the foreseeable impact area of underground use in realistic simulation areas, efforts have been made to parallelise these codes (e.g. OpenGeoSys [125], TOUGH [135]). In particular, the simulation of systems subject to discontinuity requires high performance computing. A major limitation of commercially available numerical simulation programs is that their source codes are not accessible and therefore not transparent and that a further development of such programs is therefore only possible by the commercial developer. In the research project applied for here, the platforms OpenGeoSys (UFZ (coordinating), BGR, CAU, IfG, TUBAF), mD-LEM (CAU) and pythonSPH (Uni Stuttgart) developed by some of the applicants as open source software can be used, so that the limitations mentioned do not exist. The description of discontinuities with different approaches described in the following as well as their processing in the sense of high-performance computing (HPC) requires targeted program extensions.

3.1.2 Continuum Models (XFEM and Variational Phase Field)

In recent years, extended [10] or also known as generalized [108] finite element methods (XFEM/GFEM) and phase-field methods [14] for the description of existing and developing discontinuities and singularities within continuum mechanical approaches have established themselves ahead of all others. Both methods differ fundamentally and have their own strengths and weaknesses. XFEM locally extends the approach and test function space by formulations that can map the discontinuous course of the solution and introduces corresponding additional local degrees of freedom. Usually, this approach is combined with so-called level set methods, which help to localize the discontinuity and thus ultimately determine in which elements the solution space has to be extended. This approach allows the approximation of discontinuous solutions on comparatively coarse grids, but requires programmatic infra-structures for the treatment of flexible additional degrees of freedom, level sets and other aspects, which require a considerable implementation effort, especially in branched crack systems. In contrast, the variational phase field method was originally proposed as a generalized Griffith criterion by [34] and numerically implemented using a phase-field variable by [15]. In the variational phase-field model, cracks are represented by a smoothly varying function (phase-field variable) that transitions from intact material (phase-field variable = 1.0) to fully broken state (phase-field variable = 0.0) using a regularization parameter with the dimension of a length and the energy consumed by the cracks is computed from this diffused representation. One of the strengths of this approach is to account for arbitrary numbers of pre-existing or propagating cracks in terms of energy minimization, without any a priori assumption on their geometry or restriction on the growth to specific grid directions.

XFEM [10] was originally developed for crack propagation problems and was also applied in the geotechnical context, e.g. for multiphase flows [25, 69] and heat transport [48, 98]. Current developments of generalized and extended finite element methods in the context of hydraulic stimulation are mainly concerned with the efficient coupling of solid-state and flow-mechanical problems [65, 126, 137].

The variational phase-field model of fracture has witnessed wide ranging applicability in from dynamic fracture [13, 17, 56], to ductile fracture [3, 4, 67], to thermal and drying fracture [19, 63, 66]. The first application of the variational phase-field model to hydraulically driven crack propagation has been proposed by [18] and followed by many others [26, 43, 55, 86, 128, 130] or for land slide modeling [127] with various formulation and numerical implementation. While the reported findings are promising thus far, the method still needs more establishments for practical field scale applications. The required efforts may include validation against laboratory/field experiments, approaches to recover explicit properties such as fluid leak-off from smeared crack representation, and more complex physics phenomena such as visco-elasticity.

3.1.3 Discontinuum Models

Discontinuum models directly map forces of interaction between predefined discrete elements. The latter may themselves be discretized and mapped by continuum mechanics. Decisive for the mapping of developing discontinuities, however, are the pre-defined interfaces subject to certain interface formulations. This type of modelling was applied in geotechnics, for example, to geothermal systems [140] and has also made a decisive contribution to the simulation of the pressure-driven generation of flow paths in polycrystalline salt rocks, which is bound to the discontinuum-mechanical microstructure of the salt rocks [143]. Polycrystalline salt rocks represent a discontinuum of intergrown salt crystals on the micromechanical level [144]. In contrast to porous media, there is no cross-linked pore space in salt rocks. Only by pressure-driven opening and cross-linking of pathways, i.e.g.eneration of connectivity by opening channels along the grain boundaries of the salt crystals, cross-linked flow paths are created in salt rocks. Fluid pressure-driven percolation is direction-dependent and seeks the path of least resistance along the crystal grain boundaries in the polycrystalline salt rock under the effect of the existing stress field [145]. This mechanism of directional percolation can be simulated in coupled HM models on a discontinuity mechanical basis.

The observations on numerical models of pathogenesis by source and shrinkage processes based on a microscale based analysis must be able to map significant structural changes and discontinuity developments in nonisothermal HM coupled processes, which manifest themselves in progressive fracture or self-healing processes under pressure, saturation and temperature influence. First basics of the modelling of fracture processes on the microscale were published at the end of the 1990s with reference to self-organising fracture processes based on Voronoi discretizations [12]. By combining the approaches of HM modelling in saturated media [6] and TM modelling [80], the connection for the simulation of self-organizing fracture processes in geomaterials shall be established with consideration of complex TH^{2}M processes. Based on the elasticity theory, linear fracture models following Mode I and Mode II were developed for fragile, largely homogeneous material with few inclusions. For materials with high interference, models based on continuum fracture mechanics [110] were developed which require specific information on material microstructure and fracture behaviour. Thermal conductivity in cemented geomaterials is determined by heat transfer between mineral particles, porosity, fluid and contact quality [7, 129, 133]. The Thermal Particle Dynamics method can be used to simulate the transient heat propagation in granular media and the associated thermal expansion [115]. This was considered in a thermal DEM [116, 117], but the calculation effort is enormous and the grain or contact shape is greatly simplified [141]. In contrast to the particle methods, the heat propagation in cemented materials can be determined numerically very effectively by classical FEM, but microlevel information disappears due to the underlying homogenization. This poses a problem for the initiation of discontinuities by thermal processes in THM coupling.

The hybrid lattice models have been developed to tackle the shortages in continuum based models, such as the simplicity to define the heterogeneity/anisotropy as well as the fracture simulation and stress redistribution during the frack propagation (discontinuities) without the need to re-mesh the domain [11, 114]. The lattice model is similar to the finite volume (FVM) or finite difference (FDM) methods, with the difference that the FVM or FDM explicitly discretize the continuum [81, 84]. The simplicity and accuracy of lattice models to simulate the fracking in cemented geomaterials, such as rock and concrete [47, 58, 77], are well established. The lattice models in comparison to the continuum methods are time consuming and expensive. Therefore, their applicability and development in real engineering applications or commercial softwares are not well developed. However, with the increase of computational power during past years as well as the implementation of parallel computing or GPU computing methods, the application of lattice models in commercial softwares is imminent.

The mentioned DEM approaches, as classical discontinuum models, have the disadvantage that additional connections between the particles have to be implemented by beam elements, which contain the fracture-mechanical criteria. Lattice based models—LEM [25, 27] have been developed for modeling of fracture mechanical processes considering discontinuity and crack initiation as well as crack propagation. These include a networking of the existing heterogeneity ranges (Voronoi and Delaunay triangulation) and use simple linear fracture criteria on the microscale. The cross-linked two- and three-dimensional continuum regions are microscopically coupled by 1D elements in the center of gravity of the Voronoi cells. In the simplest case these elements are Hookesche springs with a normal stiffness [29]. In three-dimensional space these simple springs already give a good approximation of the Mode I failure model [132]. With the use of Born spring models and an additional tangential degree of freedom, shear behaviour can already be modelled [46]. By extending the spring, for example as a beam element [90], displacements, rotations and moments can be transferred to the node in addition to the forces, whereby an additional bending contact can be taken into account [85]. For the spatial lattice network thus generated, the displacements at each point are determined by generating an equilibrium or by minimizing the energy [64] or dynamic relaxation [28]. The LEM combines the advantages of simple implementation with the ability to control particle interaction in the model while simultaneously self-organizing initiation and progression organization of a discontinuity, [88, 136].

Additionally, in contrast to discrete models, the lattice models can be implemented to represent a continuum, as the lattice elements do not necessarily define the particle to particle contact mechanics [79]. The hybrid lattice model can represent a continuum or particle-to-particle contacts, depending on the objective of the simulation. In both cases, the domain is discretized into series of spring or beam elements, representing the bonds. The regularization of a lattice model grants the independency of the results from the mesh size and meshing technique [75]. The lattice models were initially emerged in order to simulate the fracture initiation and propagation in cemented geomaterials. With the time, lattice models have been extended to simulate the wide variety of the thermal [80, 82, 99], thermo-mechanical [88, 89] and hydro-mechanical [39] problems in the engineering applications. In the recent years, the hybrid lattice models have been extended to determine the granular, cemented or swelling geomaterials response under the coupled thermo-hydro-mechanical (THM) processes.

3.1.4 Smoothed Particle Hydrodynamics

Smoothed Particle Hyrodynamics (SPH) methods are reticule numerical collocation methods for solving partial differential equations. SPH methods were formulated almost 40 years ago to solve astrophysical problems and have been further developed in recent years to solve a variety of problems and models in fluid and solid state mechanics [70]. The SPH method is particularly suitable for problems with free surfaces or material interfaces such as discontinuities and cracks: SPH methods are Updated or Total Lagrange methods, i.e. boundary conditions at discontinuities can be described numerically well. In recent years, great progress has been made in the efficiency of SPH formulations, especially for questions with internal interfaces, such as for non-Darcy flows in porous media or in the multiphase fluidics of immiscible fluids in porous media [73, 112]. The so-called "Whole Domain Formulation", i.e. a numerical procedure in which the surface conservation equations (mass and momentum), such as the Young-Laplace equation in multiphase fluidics, are "smeared" by means of the kernel function and integrated into the bulk conservation equations (Continuum Surface Force—CSF), can be interpreted here as a "phase field method" which "continuously smears" the physical properties of the discontinuities. Besides the consideration of the SPH-inherent kernel function in the CSF methods and the absence of the need to artificially adduce discontinuities, the net-free SPH methods above all show great efficiency advantages when complex and small-scale (pore) geometries are to be precisely mapped [101]. In addition to small-scale direct numerical simulations on the pore space scale, SPH methods for coupled HM problems in geomechanics have already been developed [20, 21]. The two HM-coupled biotubes poroelastic equation sets for the porous solid phase and the viscous pore fluid were formulated in these works with two disjunctive particle sets which can lead to difficulties in impulse interaction modelling. A further development of the SPH method for HM processes, also taking into account propagating discontinuities such as cracks and crack networks, is therefore imperative to establish the SPH method as an efficient and reliable tool for geoscientific problems.

All the approaches described above have proved to be suitable in principle for the physical analysis of the growth of discontinuities. However, in connection with the extension of the methods to coupled THM processes, there is still a fundamental need for development in many areas. This is to be supported by an improved process understanding to be worked out, by building on it some of the numerical methods used here are to be further developed purposefully beyond the state of the art. Applications that go beyond the simulation of laboratory experiments and use the methods for solving practically motivated problems of large-scale geosystems have so far hardly been found in the literature or have not even been developed for certain essential process couplings. There is an urgent need for systematic investigations into the questions of how these methods can be translated into practical applications, what computing resources are required, and in which cases certain methods appear more suitable than others. The aim of this project is to develop such an overall view and a systematic comparison of the methods at defined benchmarks as well as their embedding in proven software, partly with the inclusion of methods of high performance computing.

3.2 Numerical Methods

3.2.1 FFS—Forces on Fracture Surfaces

This numerical method explicitly uses the geometry of a rock surface and calculations on single surface elements are executed. The main advantage of this method is the possibility to closely look inside the mechanisms which control the shear behaviour of the joint (Fig. 3.2). The drawback is the high computation time needed due to the more complex calculation scheme.

Starting point were the works by [23, 31]. The last mentioned work uses a FFS approach. The geometry of surface is represented as a triangular surface. The apparent dip angles θ^* for the elements are calculated. An iterative scheme decides whether the surface can slide over its counterpart or whether the surface elements in contact

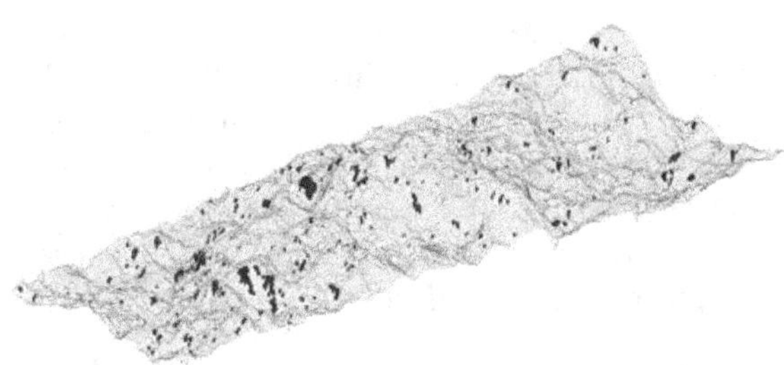

Fig. 3.2 The elements used in a shear test simulation are marked in red. The FFS approach is able to look directly inside a model and helps to deepen the understanding of the active processes

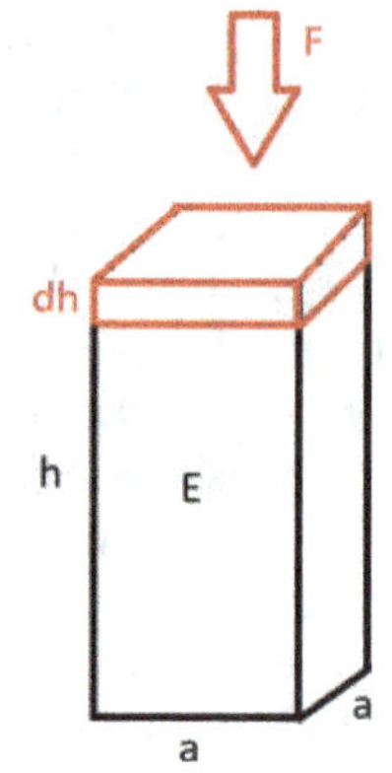

Fig. 3.3 A surface element is represented as a rock column which withstands normal forces by elastic deformation

are destroyed. In the case of destruction the geometry is corrected and the next check for sliding versus destruction starts. The two important formulas are the one for the sliding forces:

$$F_{\text{slide}} = F_{\text{loc}} \tan(\varphi_b + \theta^*) \tag{3.1}$$

where F_{loc} is the local force acting on one element, φ_b is the basic friction and θ^* the apparent dip angle of this element. The other formula is the one for shear forces, which is the force needed to destroy the surface element:

$$F_{\text{shear}} = A\ (c + \sigma_{\text{loc}} \tan(\Phi)) \tag{3.2}$$

where A is the ground area of the element, c the cohesion of the rock material, σ_{loc} the local normal stress and Φ the angle of inner friction of the rock material.

The idea of the newly developed approach is to have a physically consistent calculation scheme. Therefore the normal forces of the surface elements in contact have to be estimated. The simplest approach was chosen to keep things manageable. An elastic stress-displacement behaviour was basically used (Fig. 3.3). The resulting formula is:

$$F_n = \sum_i \left(E\, a^2 \frac{\Delta h_i}{h} \right) \tag{3.3}$$

For all i surface elements in contact the relative height change $\frac{\Delta h_i}{h}$, the ground area a^2 and the Young's modulus E were used. For a specific rock joint the two surfaces are moved towards each other until the force created by the elastic deformation equals the force which is applied to the fracture. Another simplification compared to [23] is the usage of quadratic grid elements. This allows to store the height values in a matrix form which is easy to handle.

3.2.2 *LEM—Lattice-Element-Method*

The application of the lattice element in modeling the fracture initiation and propagation in geomaterials has been well established [58, 77, 114]. The main advantage of the LEM over other numerical methods is the ability to model the stress redistribution and concentration upon the fracking process. The application of the LEM is extended to model the heat transfer in cemented geomaterials [89] as well as non-cohesive granular particles [83]. The thermo-mechanical lattice model based on the integration of the interface element is able to model the expansion and shrinkage processes during the heating and cooling cycles [88]. The LEM is also implemented to model the foam concrete behavior under dynamic loading [81]. In the past decade, using the dual lattice model to simulate the coupled hydro-mechanical loadings in geomaterials has developed [39]. In these models, the dual mesh grid for the fluid transport is generated. The short description of the implemented coupled thermo-hydro-mechanical lattice method is given below.

Discretization of the Domain

The domain is discretized into a series of Voronoi cells to represent the individual particles or a continuum depending on the purpose of the investigation. With the application of the vectorized random lattice (VRL), the irregularity factor known as the randomness factor (α_R), which varies between 0 and 1, is introduced [74]. When the randomness factor is 0, the generated mesh is regular and when it is equal to 1, it reaches the maximum irregularity for VRL model. Afterward, the Voronoi tessellation is implemented and polygonal cells are generated (see Fig. 3.4a, b). The Delaunay triangulation process results in the Voronoi cell connectivity, which are defined as the bond elements between two adjacent nodes.

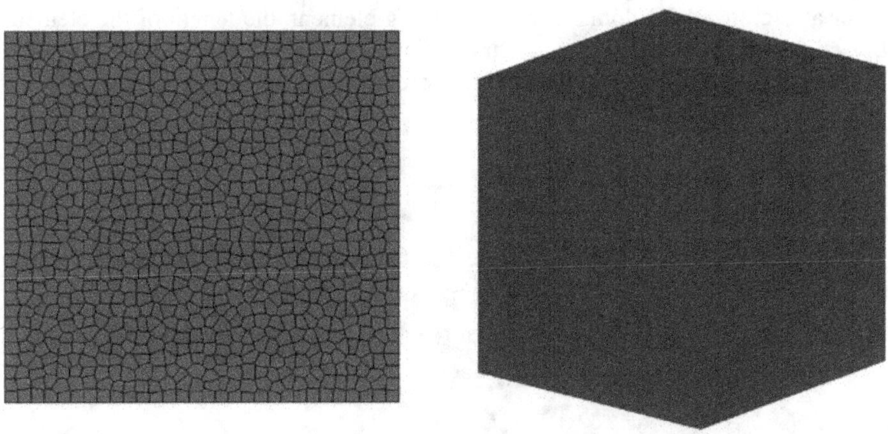

(a) (b)

Fig. 3.4 The generated domain with $\alpha_R = 0.5$. **a** 2D discretization, and **b** 3D discretization

Mechanical Lattice Model

The mechanical lattice model is based on the assumption of the Mode I and II linear elastic fracture mechanics. The simulation of the fracture in LEM is based on the removal of the bond elements between the neighboring Voronoi cells [79]. The elements strength threshold is defined based on the critical strain energy or the fracture toughness for Mode I and II. In a different approach, the strength threshold is defined based on the Mohr-Coulomb's tension cutoff model [11]. The lattice elements are represented by a series of spring (1DOF), Euler-Bernoulli beam (3DOF) (Fig. 3.5) or Timoshenko beam elements (4DOF). The regularization of the regular lattice model, such as a triangular or square discretization technique, is carried out and a relationship between the continuum and element properties is presented [47, 75]. This regularization assumes that the stored strain energy of a continuum, $U_{\mathbb{R}}$, is equal to the stored strain energies in each individual Voronoi cells, U_{Cell}. The strain energy stored in a unit cell depends on the total number of each cells bond elements (N_b), the elements response forces (F_b) and the response displacements (u_b). For a continuum, the stored energy depends on the continuum stresses ($\sigma_{\mathbb{R}}$) and continuum strains ($\varepsilon_{\mathbb{R}}$) throughout the continuum volume ($V_{\mathbb{R}}$).

$$U_{\text{cell}} = U_{\mathbb{R}} \tag{3.4}$$

$$U_{\text{cell}} = \frac{1}{2}\sum_{b=1}^{b=N_b} F_b\, u_b \tag{3.5}$$

$$U_{\mathbb{R}} = \frac{1}{2}\int_{V_{\mathbb{R}}} \sigma_{\mathbb{R}}\, \varepsilon_{\mathbb{R}}\, dV \tag{3.6}$$

For a discretized 2D domain with the spring element, the length of the element (L_b), alignment orientation ($n_{i,j,k,m}$), first stiffness coefficient ($(R)'$), continuum stiffness matrix $C_{\mathbb{R}}$, and strains of $\varepsilon_{i,j,k,m}$ are correlated as,

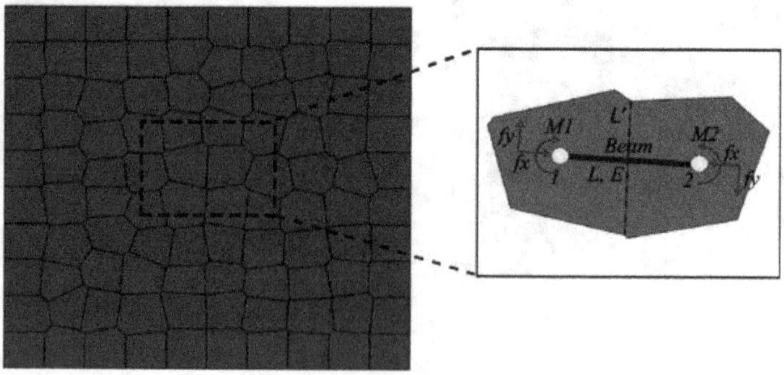

Fig. 3.5 The Euler-Bernoulli beam element representing the bond between two cells

$$U_{\text{cell}} = \frac{1}{2}\sum_{b=1}^{b=N_b} L_b^2 \left((R)' \, n_i \, n_j \, n_k \, n_m \, \varepsilon_{ij} \, \varepsilon_{km}\right)_b \tag{3.7}$$

$$U_{\mathbb{R}} = \frac{1}{2}\varepsilon_{\mathbb{R}} \, C_{\mathbb{R}} \, \varepsilon_{\mathbb{R}} \tag{3.8}$$

For a Euler-Bernoulli beam element in 2D, the curvature strain ($\kappa_{i,j}$), curvature stiffness ($D_{i,j}$), stiffens matrix ($C_{i,j,k,m}$) and second stiffness coefficient ($(R)''$) are related as,

$$U_{\mathbb{R}} = \frac{V}{2}\varepsilon_{ij}C_{ijkm}\varepsilon_{km} + \frac{V}{2}\kappa_i D_{ij}\kappa_j \tag{3.9}$$

$$C_{ijkm} = \sum_{b=1}^{b=N_b} \left(n_i \, n_k \left(n_j \, n_m \, (R)'\right) + n_j \, n_m \, (R)''\right)\Big)_b \tag{3.10}$$

After the regularization of the lattice model and with the minimization of the potential energy of the system, the load versus displacement relation in each time step is determined. For a single element, the stored total strain energy (U_t^b) is equal to the sum of axial (U_a^b), shear (U_s^b) and moment (U_m^b) strain energies. Eventually, the total strain energy depends on the axial force (f_x), shear force (f_y) and moment (M_b) along the element's length of $z = 0 : L_b$, the area of elements (A_b), element's shear modulus (G_b), element's Young's modulus (E_b), and moment of inertia (I_b). The bi-linear softening scheme is implemented to model the quasi-brittle material behavior existing in rock and concrete geomaterials [45]. The measured E_b values depends on the peak strain (ε_p), failure strain (ε_f), current element strain (ε_b) and peak load (f_p) where the stiffness degradation starts.

$$U_t^b(z) = U_a^b(z) + U_s^b(z) + U_m^b(z) = \frac{1}{2}\int_0^{L_b} \left(\frac{f_x(z)^2}{E_b A_b} + \frac{f_y(z)^2}{G_b A_b} + \frac{M_b(z)^2}{E_b I_b}\right) \cdot dz \tag{3.11}$$

$$E_b = \frac{f_p}{\varepsilon_f - \varepsilon_p}\left(\frac{\varepsilon_f}{\varepsilon_b} - 1\right) \tag{3.12}$$

Thermo-mechanical Lattice Model

The thermo-mechanical lattice model is based on the weak coupling scheme between the thermal and mechanical models, which decreases the computational costs. The thermal lattice model is based on the discrete thermal lattice model (TDEM) [32, 142], where the Hertzian contact model is implemented to account for the heat conductance (h_b) between the particles. The axial compression force increment (f_x)

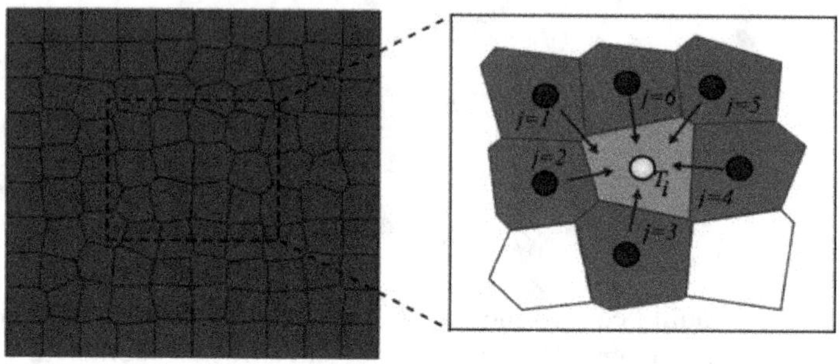

Fig. 3.6 The heat flow into i_{th} cell from surrounding boundaries

results in higher thermal conductance between the particles, which eventually leads to a higher effective thermal conductivity (K_{eff}). The regularization of the thermal lattice model is based on the relationship between the heat conductivity of elements and the continuum [83]. The h_b depends on the contact length (L'_b) (or area in 3D domain), contact forces and assigned elements thermal conductivities (k_b).

$$h_b = k_b \left((L'_b) + \left(\frac{3 f_x L_b}{4 E_b} \right)^{\frac{1}{3}} \right) \tag{3.13}$$

In a steady state, the amount of the heat in- and outflow (q_b) from a Voronoi cell (Fig. 3.6) is equal to zero,

$$\rho_i c_i v_i \frac{dT_i}{dt} - \nabla \cdot (k_i \nabla T_i) - \rho_i \dot{q}_i = 0 \tag{3.14}$$

$$\nabla \cdot (k \nabla T_i) = \sum_{b=1}^{b=N_b} q_b = \sum_{b=1}^{b=N_b} h_b (T_i - T_j)_b = 0 \tag{3.15}$$

where, $\dot{q}$ is heat density (assumption: $\dot{q} = 0$), t is time, ρ_i is density, c_i is heat capacity and v_i is the volume of each Voronoi cell (i). In a transient case,

$$\sum_{b=1}^{b=N_b} q_b = \rho_i c_i v_i \frac{dT_i}{dt} \tag{3.16}$$

The effective thermal conductivity is calculated based on the average volume technique, where q_{ave} is the average heat flow, q^b_{Cell} is the heat flow through the assigned cells located in the boundary (N_C), $\dot{T}$ is the temperature gradient and $\hat{x}_{cell}$ is the relative coordinates of each cell.

$$q_{ave} = \frac{\sum_{b=1}^{b=N_C} q^b_{Cell} \cdot \hat{x}_{Cell}}{V} \tag{3.17}$$

$$q_{ave} = K_{eff} \cdot \dot{T} \tag{3.18}$$

The thermal strain is calculated based on the linear expansion of the lattice elements and the given heat expansion coefficient. The implementation of the thermal expansion into the mechanical model results in a fully coupled thermo-mechanical model [88].

Hydro-mechanical Lattice Model

The existing hydro-mechanical lattice models are based on the assumption of the dual lattice network, where the mechanical lattice elements transfer the mechanical loads between the two nodes and the conduct elements perpendicular to the alignment of the mechanical elements transfer the fluid or gas flow between the conduct nodes [39, 40]. The implemented hydro-mechanical lattice model is based on the mass conservation (m_f) of the fluids in the continuum. The hydraulic aperture (a_h), fluid density (ρ_f), fluid viscosity (ν_f), flow length (L_b'), hydraulic resistance (R_h), saturation degree (Sr) and bulk modulus (K_f) are the main parameters used to determine the hydraulic pressures (P_f) and transferred fluid masses (Δm_f) between the conduct nodes.

$$m_f^{t+1} = m_f^t + \Delta m_f \tag{3.19}$$

$$m_f^{t=0} = \mathrm{Sr}^{t=0} V_{cav} \rho_f \left(1 + \frac{P_f^{t=0}}{K_f}\right) \tag{3.20}$$

$$\Delta m_{f,ij} = f(Sr) \frac{P_{f,j} - P_{f,i} - \rho_f g \left(Z_j - Z_i\right)}{R_h} \Delta t \tag{3.21}$$

where, Z is the relative coordinate of the i, j conduct nodes, V_{cav} is the volume of the cavity, g is the gravity and $f(Sr)$ is the saturation function which is equal to 0 and 1 in a dry and saturated conditions, respectively. According to the finite-discrete element method (FDEM) [57], the fluid mass is stored within defined physical and artificial cavities. Each conduct node represents an artificial cavity connected through conductive elements (Fig. 3.7), where the hydraulic conductivity is governed based on the parallel plate cubic flow rule.

$$R_h = \frac{12\nu_f}{a_h^3} L_b' = 12\nu_f \int_{z_i}^{z_j} \frac{1}{a_h(z^3)} dz = \frac{6\nu_f(a_{h,j} + a_{h,i})}{(a_{h,i} a_{h,j})^2} L_b' \tag{3.22}$$

When an artificial cavity is saturated, the amount of excessive fluid mass flowing inside the cavity builds the pore pressure, which then is transmitted into the mechanical nodes. If the cavity is not saturated, then the pore pressure is assumed to be zero.

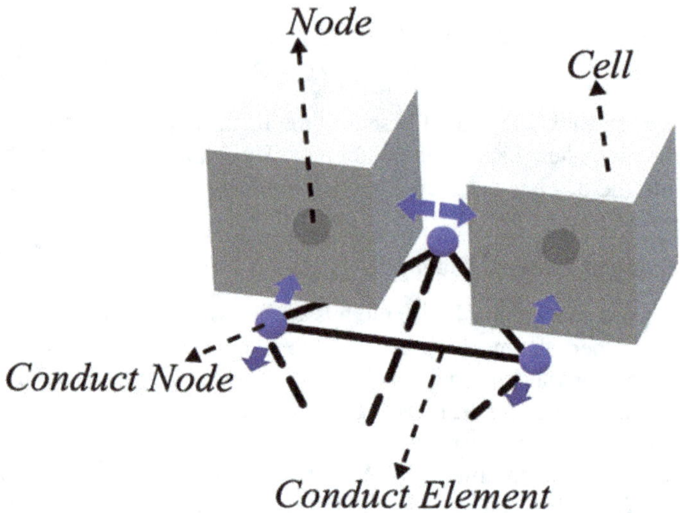

Fig. 3.7 The schematic representation of the implemented hydro-mechanical model

$$P_f^t = P_f^{t-1} + K_f \frac{\Delta m_f}{\rho_f V_{\text{cav}}^t} \quad \text{if} \quad \text{Sr}^t = 1 \tag{3.23}$$

With the implementation of the pore pressures into the mechanical lattice nodes, the pore pressure diffusion and the change of the hydraulic conductivity with the crack opening and closure are measured. The flow simulation is implemented under both the pressure- and flowrate-controlled boundary conditions.

Shrinkage and Swelling Lattice Model

The simulation of the shrinkage and swelling using the lattice element method is based on the particle shrinkage model [100], which is mainly considered in the discrete models. In contrast to the DEM, the shrinkage in LEM is implemented into the lattice elements (Fig. 3.8). To do so, the interface elements are generated to represent the bond between the particles [88]. The shrinkage and swelling coefficients (α_s) are temperature dependent. According to the initial water content ($\omega^{t=0}$) and the change of the water content during the shrinkage and swelling process, the linear strain in the lattice elements is determined and implemented into the mechanical model.

$$L_b^t = L_b^{t=0} e^{-\alpha_s \cdot \frac{t}{t=\infty}} \tag{3.24}$$

$$\alpha_s = -\frac{1}{3} \ln\left(1 - \frac{\omega^{t=0} - \omega^t}{1 + e_0} \cdot G_s\right) \tag{3.25}$$

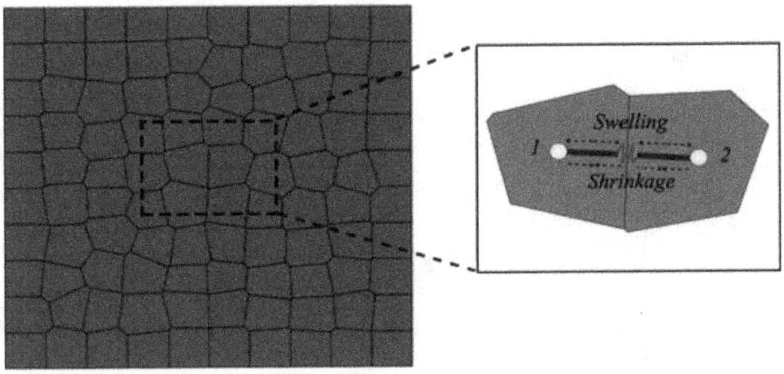

Fig. 3.8 The implementation of the interface element to simulate the shrinkage and swelling processes

The shrinkage and swelling coefficient are time, temperature and depth dependent. Therefore, graphs representing the evaporation rate as well as the soil water characteristic curves to assess the applied suction and the water content during the wetting and drying paths are required [124]. $\bar{\bar{\sigma}}$ and $\bar{\bar{\varepsilon}}$ are the stress and strain tensors, respectively.

$$\bar{\bar{\sigma}} = C : \bar{\bar{\varepsilon}} - \text{Sr}\, P_f \bar{\bar{\delta}} \tag{3.26}$$

The elements shrinkage and expansion results in the axial compression and tensile stresses in the lattice elements, which when they exceed their predefined strength threshold are removed, is simulated as well as the micro fracking process under shrinkage and swelling conditions.

3.2.3 DEM—Distinct-Element-Method

The distinct element method (DEM) extends the capabilities of continuum-mechanical approaches by introducing a new level of discretization, which allows it to describe independent deformable bodies that can interact via their contact points and surfaces, see Fig. 3.9. The behaviour of these contacts can be modelled using joint constitutive models, which are typically formulated in terms increments. This approach is especially suitable for materials with a pronounced grain structure, such as salt rocks.

The models of rock salt, from laboratory scale samples to entire potash mines, therefore need to be built from randomised assemblies of polyhedral grains, in order to simulate the discontinuous and granular nature. The generation of the randomised structures is based on Voronoi-discretization, which allows to divide arbitrary volumes (areas) into polyhedral parts. The program generates pseudo-random point clouds using a Monte-Carlo-method, which can then be refined to avoid clustering.

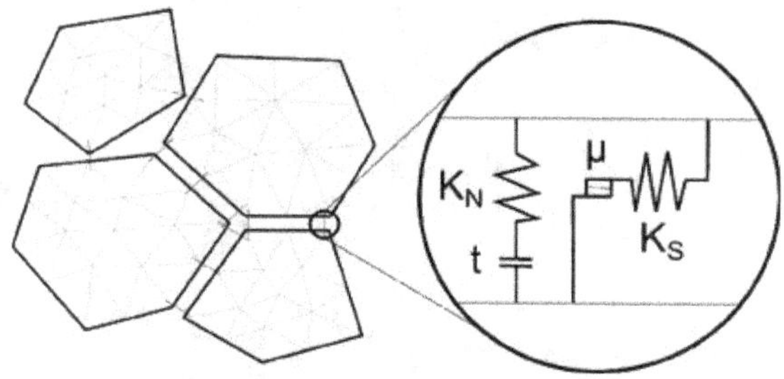

Fig. 3.9 Interaction between blocks and contacts in the DEM

It is also possible to introduce a local variation of the grains size this way. The thermo-hydro-mechanical behaviour of a model set up in this way is then a combination of the behaviour of the of bulk of the grains, and of the contact properties. For both, constitutive laws were derived at IfG [146].

Since it would be computationally unfeasible to represent a large geological structure with realistic grain sizes of millimeters or centimeters, a coarse-grained approach is taken. This approach was validated by carrying out discretization studies with a variation of Voronoi sizes. The blocks/grains dominate the hardening and the creep behaviour, while the softening occurs predominantly by shear and tensile failure on grain boundaries [147].

Another advantage of the discontinuum mechanical approach lies in its capabilities to model the pressure driven percolation of gases and fluids on discrete flow paths on opened grain boundaries. The undamaged grain boundaries start out as impermeable but can be opened due to plastic failure.

The constitutive laws for both bulk and contacts were implemented as DLLs for the programs UDEC and 3DEC of Itasca CG, Inc.

3.2.4 SPH—Smoothed-Particle-Hydrodynamics

Direct Numerical Simulations (DNS) of effective physical properties of single-phase flow through porous or fractured solid materials can be performed directly on the pore scale of the porous soil or rock. Morphological information, the basis for the subsequent DNS, is obtained as segmented (binarized) voxel-data from μ X-Ray Computed Tomography (XRCT). In general, numerical simulations of flow processes on XRCT-data at small to moderate Reynolds (Re) numbers could be performed by mesh-based Finite Element, Finite Differences or Finite Volume methods.

Here, we haven chosen the mesh-less Smoothed Particle Hydrodynamics methods as an alternative simulation technique. SPH is a Lagrangian simulation tool used to solve Partial Differential Equations (PDE) and was originally developed for astrophysical problems [36, 59]. In recent years, due to it's flexibility and scalability and applicability on HPC architectures, especially in an explicit formulation, it became attractive for various problems fluid dynamics like single and multi-phase fluid mechanics with internal interfaces, suspension flow, and single and multi-phase flow in porous media [61, 102–105].

Within the framework of this method, the discretisation of the PDEs spans a set of interacting collocation points $\mathcal{P}_i$ with position vectors $\mathbf{x}_i$, referred to as particles. The positions of the particles represent integration points at which field functions $\Phi(\mathbf{x}, t)$ are interpolated by convolution with the Dirac-Delta function δ:

$$\Phi(\mathbf{x}, t) = \int_\Omega \Phi(\mathbf{x}', t)\, \delta(\mathbf{x} - \mathbf{x}')\, \mathrm{d}v. \tag{3.27}$$

Replacing $\delta(\mathbf{x} - \mathbf{x}')$ with the kernel function $W(\mathbf{x} - \mathbf{x}', h)$ results in the approximation

$$\Phi(\mathbf{x}, t) \approx \int_\Omega \Phi(\mathbf{x}', t)\, W(\mathbf{x} - \mathbf{x}', h)\, \mathrm{d}v, \tag{3.28}$$

where the support h of the kernel determines a sphere of influence and likewise declares neighbouring particles $\mathcal{P}_j$ with position vector $\mathbf{x}_j$. Subsequently, the discretisation (numerical integration) of the integral formulation converts continuous field functions into discrete particle properties $\Phi(\mathbf{x}_i) = \Phi_i$, kernel representations into spatial discretisation and Eq. (3.28) yields

$$\Phi_i = \sum_j^N \Phi_j\, W(\mathbf{x}_i - \mathbf{x}_j, h)\, V_j. \tag{3.29}$$

Herein, V_j is introduced as the discrete representation of $\mathrm{d}v$ and $j = 1, 2, \ldots, N$ indicates the neighbour particles within the kernel support of particle $\mathcal{P}_i$. Analogously, differential operators turn into short-range interaction forces. For more technical details, also accounting to the necessary time integration we refer to [72, 103, 138].

3.2.4.1 Single-Phase Flow of a Newtonian Fluid

A SPH implementation of single-phase flow of a Newtonian fluid is based on the solution of the balance of momentum in the present local form

$$\rho^\mathfrak{f}\, \dot{\mathbf{v}}_\mathfrak{f} = \mu^\mathfrak{f} \operatorname{div}(\operatorname{grad} \mathbf{v}_\mathfrak{f}) - \operatorname{grad} p + \rho^\mathfrak{f}\, \mathbf{b} \tag{3.30}$$

and the balance of mass

$$\dot{\rho}^\mathfrak{f} = -\rho^\mathfrak{f} \operatorname{div} \mathbf{v}_\mathfrak{f}, \tag{3.31}$$

which are known as the Navier-Stokes equations. We adopted the notation used in mixture theory [106, 107] where a subscript is used for kinematic quantities and a superscript elsewhere. $\rho^\mathfrak{f}(\mathbf{x}, t)$ is the mass density field, $\mathbf{v}_\mathfrak{f}$ is the velocity vector, $\mu^\mathfrak{f}$

is the dynamic viscosity, $p(\mathbf{x}, t)$ is the pressure field and $\mathbf{b}$ are body force densities. The "dot" operator, cf. Eqs. (3.30) and (3.31), is denoting the material or substantial time derivative $\cdot(\bullet) = \partial_t \bullet + \operatorname{grad}(\bullet) \cdot \mathbf{v}_\mathfrak{f}$. In order to solve the quasi-incompressible (weakly compressible) character of the Navier-Stokes equations an equation of state for the pressure in the form $p(\rho^\mathfrak{f})$ has to be formulated. Therefore, either a linear model or the Tait equation [42]

$$p(\rho^\mathfrak{f}) = \frac{\rho_0^\mathfrak{f} c_\mathfrak{f}^2}{\gamma}\left[\left(\frac{\rho^\mathfrak{f}}{\rho_0^\mathfrak{f}}\right)^\gamma - 1\right] \tag{3.32}$$

is commonly employed, wherein $c_\mathfrak{f} = \sqrt{K^\mathfrak{f}/\rho^\mathfrak{f}}$ is the speed of sound of the fluid with the bulk modulus $K^\mathfrak{f}$ and γ is a constant, specific to the modeled problem (usual $\gamma = 7$ for quasi-incompressible fluids).

3.2.4.2 Discrete Equations

By applying the above introduced transformation from continuous field equations to discrete algebraic SPH equations, the total force on each fluid particle $\mathcal{P}_i$ is obtained as the sum of the discrete body forces $\mathbf{F}_i^B = m_i\,\mathbf{b}$, viscous interaction forces $\mathbf{F}_{ij}^V$ and pressure interaction forces $\mathbf{F}_{ij}^P$

$$m_i \dot{\mathbf{v}}_i = \sum_j^N \mathbf{F}_{ij}^P + \sum_j^N \mathbf{F}_{ij}^V + \sum_j^N \mathbf{F}^B, \tag{3.33}$$

which leads to a relation for the particle velocity update

$$\dot{\mathbf{v}}_i = -\sum_j^N m_j \left(\frac{p_i}{{\rho_i}^2} + \frac{p_j}{{\rho_j}^2}\right) \frac{\mathbf{x}_{ij}}{r_{ij}} \frac{\partial W_{ij}}{\partial r_{ij}} \tag{3.34}$$

$$+ \sum_j^N \frac{m_j(\mu_i + \mu_j)(\mathbf{v}_i - \mathbf{v}_j)}{\rho_i \rho_j} \left(\frac{1}{r_{ij}} \frac{\partial W_{ij}}{\partial r_{ij}}\right) + \mathbf{b}. \tag{3.35}$$

Reconfiguration of the balance of mass, cf. Eq. (3.31) yields

$$\dot{\rho}_i = \sum_j^N m_j (\mathbf{v}_i - \mathbf{v}_j) \cdot \frac{\mathbf{x}_{ij}}{r_{ij}} \frac{\partial W_{ij}}{\partial r_{ij}}. \tag{3.36}$$

However, the density field can also be calculated by an accumulative kernel interpolation $\rho_i = \sum m_j W_{ij}$.

3.2.4.3 Boundary Conditions, Time Integration and Artificial Viscosity

For the application of single-phase flow though porous media, the solid skeleton is considered to be rigid and fluid-solid interfaces Γ^{FS} generally satisfy no-slip and no-penetration boundary conditions. More general solid-fluid interaction phenomena could be considered in SPH formulations, e.g. to mimic rough rock surfaces with asperities, but are not further discussed in the following. To that end, the solid domain is populated by so-called "dummy" particles [103, 138]. The velocity and pressure of these dummy particles is extrapolated by the fluid phase and computed by the balance of momentum, respectively, following the method proposed by Adami et al. [1]. Resulting velocity and pressure fields of dummy particles are counteracting those of the fluid phase and thus create no-slip and no-penetration conditions on Γ^{FS}. The discrete particle properties are updated by the Velocity Verlet time integration method [109, 119] which is a common time-integration scheme in particle methods. Further, a dissipative artificial viscosity term is used to reduce non-physical oscillations, e.g. [71, 72, 138].

FEM—Finite-Element-Method

Figure 3.10 shows a conceptual illustration of three different approaches for modeling displacement discontinuities: (a) cohesive zone model using lower-dimensional (i.e. co-dimension 1) interface elements with local enrichment to represent a strong displacement discontinuity (see Sect. 3.2.4.3); (b) phase-field models of brittle fracture in which a crack surface density per unit volume is introduced for regularisation (see Sect. 3.2.5); and (c) non-local elasto-plastic damage models, in which a kernel function with a specified support region is used to characterize a fracture process zone. In the latter two approaches the discontinuities are smeared over a zone characterized by a length-scale parameter. All models (a–c) are implemented in OpenGeoSys.

In the following we briefly introduce two of the Finite-Element-Method based approaches. The first approach is the variational phase-field implemented in Open-

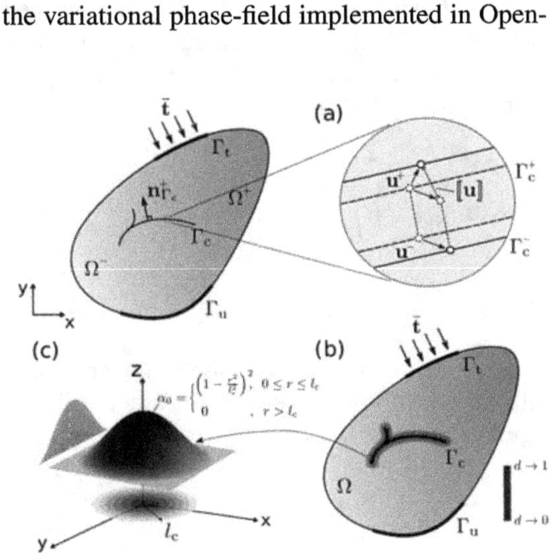

Fig. 3.10 Smeared and explicit numerical representations of fracture. Figure reproduced from [139]

GeoSys and is particularly suited for simulaiton of fracturing process. The second is the hybrid dimensional formulation and is specifically designed for numerical stable fractured-porous media analysis.

3.2.5 PFM—Variational Phase-Field Method

Phase-field models have become one of the most established numerical techniques for fracturing simulation in the last two decades. The approach was first introduced by Bourdin et al. [15] as a numerical implementation method to the variational approach to fractures proposed by Francfort and Marigo [34]. Since the initial inception of the method, models for brittle and cohesive fracture have been further studied by many others [4, 5, 16, 41, 53, 60, 76, 87, 111, 118, 120, 134] including advanced numerical solution schemes [30, 35]. Lately, its application ranges from ductile fracturing [3, 4, 54, 66] to fatigue [2, 96], desiccation fracture [22, 63], and dynamic fracturing [13, 17, 44, 56, 91].

Regularisation of the Total Energy Functional

In the variational approach to fractures in [34], the total energy in the system is defined as the sum of the elastic strain energy, the potential of the external forces and the surface energy as:

$$E(\mathbf{u}, \Gamma_c) := \int_{\Omega\backslash\Gamma_c} \psi(\mathbf{u})\,\mathrm{d}\Omega - \int_{\partial_N\Omega} \bar{\mathbf{t}}\cdot\mathbf{u}\,\mathrm{d}\Gamma - \int_{\Omega\backslash\Gamma_c} \varrho\mathbf{b}\cdot\mathbf{u}\,\mathrm{d}\Omega + G_c\int_{\Gamma_c}\mathrm{d}\Gamma, \tag{3.37}$$

where ψ is the strain energy density, $\bar{\mathbf{t}}$ is the boundary traction force, $\mathbf{u}$ is the displacement, ϱ is the mass density of the porous medium, $\mathbf{b}$ is the applied specific body force, and G_c is the critical energy release rate. For hydraulic fracturing, the work done by the fluid pressure within the fracture p needs to be added to the energy, and the total energy becomes:

$$E(\mathbf{u}, \Gamma_c) := \int_{\Omega\backslash\Gamma_c} \psi(\mathbf{u})\,\mathrm{d}\Omega - \int_{\partial_N\Omega} \bar{\mathbf{t}}\cdot\mathbf{u}\,\mathrm{d}\Gamma - \int_{\Omega\backslash\Gamma_c} \varrho\mathbf{b}\cdot\mathbf{u}\,\mathrm{d}\Omega + G_c\int_{\Gamma_c}\mathrm{d}\Gamma + \int_{\Gamma_c} p[\![\mathbf{u}]\!]\cdot\mathbf{n}_\Gamma\,\mathrm{d}\Gamma \tag{3.38}$$

where $\mathbf{n}_\Gamma$ is the normal direction to the crack set Γ. Following Bourdin et al. [18], with introduction of a phase-field order variable (or damage variable) d (3.38) is regularised as:

$$\begin{aligned} E(\mathbf{u}, d, p) = & \int_\Omega (1-d)^2 \psi(\mathbf{u}) \mathrm{d}\Omega - \int_{\partial_N \Omega} \bar{\mathbf{t}} \cdot \mathbf{u} \, \mathbf{ds} - \int_\Omega \varrho \mathbf{b} \cdot \mathbf{u} \, \mathrm{d}\Omega \\ & + \frac{G_c}{4c_w} \int_\Omega \left(\frac{w(d)}{\ell} + \ell |\nabla d|^2 \right) \mathrm{d}\Omega + \int_\Omega p\mathbf{u} \cdot \nabla d \, \mathrm{d}\Omega. \end{aligned} \tag{3.39}$$

where d is the damage variable that equals 0 when undamaged and 1 for a fully damaged state, c_w is a normalization parameter defined as $c_w := \int_0^1 \sqrt{w(s)} \mathrm{d}s$, ℓ is a regularisation length, and $w(d)$ is the dissipative energy function. Various possible dissipative energy functions $w(d)$ are discussed in [60]. The most widely used function is a quadratic form [15, 51, 53, 68] $w(d) = d^2$. The variational phase-field model with this choice of the dissipative function is called AT_2 model in [19], and we follow this terminology in this report. Another choice for $w(d)$ is a linear form $w(d) = d$ and is known as AT_1. A thorough study comparing these two models in terms of fracture nucleation and propagation can be found in [111]. Note that in arriving at (3.39), the work done by the fracture pressure in the crack is approximated as $\int_\Omega p\mathbf{u} \cdot \nabla d \, \mathrm{d}\Omega$.

Numerical Implementation

The solution of (3.39) follows the alternate minimisation scheme introduced in [15] with respect to $\mathbf{u}$ and d given a time-evolving p. Thus, the minimisation problem can be stated as

$$(\mathbf{u}, d)^* = \underset{\begin{cases} \mathbf{u} \in H^1 \\ d \in H^1, d^t \subset d^{t+\Delta t} \end{cases}}{\arg\min} \tilde{E}(\mathbf{u}, d, p) \ , \tag{3.40}$$

The first variation of the energy functional with respect to $\mathbf{u}$ is

$$\delta E(\mathbf{u}, d, p; \delta\mathbf{u}) = \frac{1}{2} \int_\Omega (1-d)^2 \mathbf{C}(\mathbf{u}) : \boldsymbol{\epsilon}(\delta\mathbf{u}) \, \mathrm{d}\Omega - \tag{3.41}$$

$$- \int_{\partial_N \Omega} \mathbf{t} \cdot \delta\mathbf{u} \, \mathrm{d}\Gamma - \int_\Omega \varrho \mathbf{b} \cdot \delta\mathbf{u} \, \mathrm{d}\Omega + \int_\Omega p \delta\mathbf{u} \cdot \nabla d \, \mathrm{d}\Omega, \tag{3.42}$$

where $\mathbf{C}$ is the constitutive relation that satisfies $\psi = \mathbf{C}(\boldsymbol{\epsilon}) : \boldsymbol{\epsilon}/2$. The first variation of the energy functional with respect to d of AT_1 is:

$$\begin{aligned} \delta E(\mathbf{u}, d, p; \delta d) = & - \int_\Omega d \delta d \mathbf{C}(\boldsymbol{\epsilon}(\mathbf{u})) : \boldsymbol{\epsilon}(\mathbf{u}) \mathrm{d}\Omega \\ & + \frac{3G_c}{8} \int_\Omega \left(\frac{\delta d}{\ell} + 2\ell \nabla d \cdot \nabla \delta d \right) \mathrm{d}\Omega + \int_\Omega p\mathbf{u} \cdot \nabla \delta d \, \mathrm{d}\Omega, \end{aligned} \tag{3.43}$$

and AT_2 is

$$\delta E(\mathbf{u}, d, p; \delta d) = -\int_\Omega d\delta d\mathbf{C}(\boldsymbol{\epsilon}(\mathbf{u})) : \boldsymbol{\epsilon}(\mathbf{u})\mathrm{d}\Omega + G_\mathrm{c}\int_\Omega \left(\frac{d}{\ell}\delta d + \ell\nabla d \cdot \nabla\delta d\right)\mathrm{d}\Omega + \int_\Omega p\mathbf{u} \cdot \nabla\delta d\, \mathrm{d}\Omega. \tag{3.44}$$

It should be noted that the solution of (3.43) or (3.44) for d is not bounded in $[1, 0]$. Therefore, the solution scheme would require enforcement of the variational inequality $(0 \leq d \leq 1)$. For this study, a variational inequality non-linear solver of the PETSc library [8, 9] has been applied to fulfill the bound of d.

3.2.6 HDF—Hybrid-Dimensional-Formulation

Hydro-mechanical modeling of fluid flow in deformable high aspect-ratio fractures (aperture $\ll$ fracture length) requires special numerical treatment of the governing equations. Discretization of high aspect-ratio fractures is challenging, and small absolute fracture deformations might lead to non-linear changes in the flow conditions within the fracture domain. Hybrid-dimensional elements were designed to overcome these difficulties and implicitly couple flow in deformable fractures with the deformation state of the surrounding matrix [24, 37, 38, 49, 50, 92–95, 97, 121–123].

Governing Equations of the Hybrid-Dimensional Formulation

Flow in hydraulic transmissive high aspect-ratio fractures is based on the physics of viscous fluids $Re \ll 1$ and creeping flow conditions resulting in a Poiseuille-type flow description [131]. The introduced assumptions simplify the balance of momentum to a pressure driven flow formulation where the relative fluid velocity

$$\mathbf{w}_\mathfrak{f} = -\frac{\delta(\mathbf{u})^2}{12\,\eta^{\mathfrak{f}R}}\,\mathrm{grad}\,p =: -\frac{k^\mathfrak{s}_{Fr}}{\eta^{\mathfrak{f}R}}\,\mathrm{grad}\,p, \tag{3.45}$$

is obtained from the parabolic velocity profile. In (3.45) δ denotes the fracture aperture, $\eta^{\mathfrak{f}R}$ the effective dynamic viscosity, $\mathbf{u}$ the fracture deformation, p the fluid pressure and $k^\mathfrak{s}_{Fr}$ the introduced local fracture permeability. To determine relations between injected fluid and fracture volume, fluid density changes and fluid velocity the balance of mass is evaluated for a compressible fluid in deformable fractures. Hence,

$$\dot{\overline{(\rho^{\mathfrak{f}R}\,\delta)}} + \mathrm{div}\left(\mathbf{w}_\mathfrak{f}\,\rho^{\mathfrak{f}R}\,\delta\right) = 0 \tag{3.46}$$

provides the condition to fulfill the conservation of mass in a deformable fracture setting, where $\rho^{\mathfrak{f}R}$ is the fluid density. Once the fracture is surrounded by a porous medium leak-off can occur and is simply taken into account by an additional source term on the right hand side of (3.46). Evaluation of the balance of mass and momentum and assuming a linear constitutive relationship between fluid pressure and effective fluid density for barotropic fluids provides the scalar, lower-dimensional governing equation for fluid flow in deformable fractures

$$\underbrace{\dot{p}}_{(I)} \underbrace{- \frac{\delta^2}{12\,\eta^{\mathfrak{f}R}} \operatorname{grad} p \cdot \operatorname{grad} p}_{(II)} \underbrace{- \frac{\delta}{12\,\eta^{\mathfrak{f}R}\,\beta^{\mathfrak{f}}} \operatorname{grad} \delta \cdot \operatorname{grad} p}_{(III)} \underbrace{- \frac{1}{12\,\eta^{\mathfrak{f}R}\beta^{\mathfrak{f}}} \operatorname{div}\left(\delta^2 \operatorname{grad} p\right)}_{(IV)} + \underbrace{\frac{1}{\delta\,\beta^{\mathfrak{f}}} \frac{\partial \delta}{\partial t}}_{(V)} = \underbrace{q_{lk}}_{(VI)} \tag{3.47}$$

Equation (3.47) consists of a transient (*I*), a quadratic (*II*), a convection (*III*), a diffusion (*IV*), a coupling (*V*) and a leak-off (*VI*) term. Due to their minor contribution to the overall solution terms (*II*) and (*III*) will be neglected in the following.

Numerical Implementation

Different strategies to solve (3.47) have been proposed in the literature. Due to the stiff nature of the global system, implicit coupling of the rock matrix and fracture domain is mandatory. Solution strategies can be distinguished between staggered and monolithic approaches. Staggered algorithms allow calculations on independent domains and non-conformal meshing of fluid and solid domains [38], but require a higher number of iterations compared to the monolithic approaches [92]. Monolithic approaches such as zero-thickness interface elements are numerically stable and allow modeling of constitutive contact behavior of fracture surfaces since connectivity between both surfaces is guaranteed [95]. Here we focus on the element formulation of zero-thickness interface elements. The interested reader is referred to [37, 38, 92] for a detailed description of weak coupling schemes.

Interface elements allow longitudinal and transversal fracture flow as well as pressure jumps along both fracture surfaces. Numerically it is realized by auxiliary lower dimensional elements and averaging of the balance of mass and momentum. For the averaging process the already introduced balances of mass and momentum are extended to capture longitudinal and transversal fracture flow by decomposition of the seepage velocity

$$\mathbf{w}_{\mathfrak{f}} = \mathbf{w}_{\mathfrak{f}}^{l} + \mathbf{w}_{\mathfrak{f}}^{t}. \tag{3.48}$$

In order to allow integration of aligning node values on auxiliary element level the balance of mass is averaged over the fracture height and reads

$$\frac{\delta}{K^{\mathfrak{f}}}\dot{P}_{\mathfrak{f}} + \dot{\delta} + \mathrm{div}_l\left(\mathbf{W}_{\mathfrak{f}}\delta\right) = \mathbf{w}_{\mathfrak{f}}^{+}\cdot\mathbf{n}^{+} + \mathbf{w}_{\mathfrak{f}}^{-}\cdot\mathbf{n}^{-} \tag{3.49}$$

where $P_{\mathfrak{f}} = \int_{-\delta/2}^{\delta/2} \hat{p}\,\mathbf{dn}$ is the integral pressure, $\mathbf{W}_{\mathfrak{f}} = \int_{-\delta/2}^{\delta/2} \mathbf{w}_{\mathfrak{f}}^{t}\,\mathbf{dn}$ the integral seepage velocity, $K^{\mathfrak{f}}$ the fluid's bulk modulus, div_l the divergence evaluated in longitudinal direction and the superscripts $^+$ and $^-$ denote the fracture surfaces. Averaging of the balance of momentum is conducted for the longitudinal and transversal fracture flow separately. In the longitudinal direction, the averaging process results in

$$\mathbf{W}_{\mathfrak{f}} = -\frac{k^{\mathfrak{s}}_{Fr,l}(\mathbf{x},t)}{\eta^{\mathfrak{f}R}}\,\mathrm{grad}_l\, P_{\mathfrak{f}}. \tag{3.50}$$

Description of the balance of momentum for transversal flow is realized by introducing and averaging a Darcy-like pressure-flow relationship which is approximated by the trapezoidal rule and finally given by

$$\frac{1}{2}\left(-\mathbf{w}_{\mathfrak{f}}^{+}\cdot\mathbf{n}^{+} + \mathbf{w}_{\mathfrak{f}}^{-}\cdot\mathbf{n}^{-}\right) = -\frac{k^{\mathfrak{s}}_{Fr,t}}{\eta^{\mathfrak{f}R}}\,\frac{-p^{+}+p^{-}}{\delta}. \tag{3.51}$$

A boundary condition for pressure $P_{\mathfrak{f}}$ closes the set of averaged balance equations. The formulation of the averaged quantity $P_{\mathfrak{f}}$ depends on the domain composition of the fracture height and reads

$$\begin{aligned}
-\xi\,\mathbf{w}_{\mathfrak{f}}^{+}\cdot\mathbf{n}^{+} + \frac{2\,k^{\mathfrak{s}}_{Fr,t}}{\delta\,\eta^{\mathfrak{f}R}}\,p^{+} &= -(1-\xi)\,\mathbf{w}_{\mathfrak{f}}^{-}\cdot\mathbf{n}^{-} + \frac{2\,k^{\mathfrak{s}}_{Fr,t}}{\delta\,\eta^{\mathfrak{f}R}}\,P\\
-\xi\,\mathbf{w}_{\mathfrak{f}}^{-}\cdot\mathbf{n}^{-} + \frac{2\,k^{\mathfrak{s}}_{Fr,t}}{\delta\,\eta^{\mathfrak{f}R}}\,p^{-} &= -(1-\xi)\,\mathbf{w}_{\mathfrak{f}}^{+}\cdot\mathbf{n}^{+} + \frac{2\,k^{\mathfrak{s}}_{Fr,t}}{\delta\,\eta^{\mathfrak{f}R}}\,P
\end{aligned} \tag{3.52}$$

in its general form. Here the stability limit $\xi = 1/2$ [93] is chosen [62] so that the pressure jump along the fracture height and averaged pressure P are given by

$$\begin{aligned}
P_{\mathfrak{f}} &= \frac{p^{+}+p^{-}}{2}\\
\mathbf{w}_{\mathfrak{f}}^{-}\cdot\mathbf{n}^{-} - \mathbf{w}_{\mathfrak{f}}^{+}\cdot\mathbf{n}^{+} &= 2\,k^{\mathfrak{s}}_{Fr,t}\,\frac{p^{-}-p^{+}}{\delta}.
\end{aligned} \tag{3.53}$$

The averaged governing equation for fluid flow in a deformable fracture at auxiliary element level is finally obtained by addition/subtraction of (3.49) and (3.53) and leads to

$$
\begin{aligned}
&\frac{1}{2}\left[\frac{1}{K^{\mathfrak{f}}}\dot{P}_{\mathfrak{f}} + \dot{\delta} + \operatorname{div}_l\left(\mathbf{W}_{\mathfrak{f}}\delta\right)\right] - k^{\mathfrak{s}}_{Fr,t}P^t_{\mathfrak{f}} = \mathbf{w}^+_{\mathfrak{f}} \cdot \mathbf{n}^+ \quad \text{on } \Gamma^{Fr}_+,\\
&\frac{1}{2}\left[\frac{1}{K^{\mathfrak{f}}}\dot{P}_{\mathfrak{f}} + \dot{\delta} + \operatorname{div}_l\left(\mathbf{W}_{\mathfrak{f}}\delta\right)\right] + k^{\mathfrak{s}}_{Fr,t}P^t_{\mathfrak{f}} = \mathbf{w}^-_{\mathfrak{f}} \cdot \mathbf{n}^- \quad \text{on } \Gamma^{Fr}_-.
\end{aligned} \tag{3.54}
$$

To solve the governing equation (3.54) in a finite element (FE) framework the weak form of the zero-thickness element formulation results in

$$
\begin{aligned}
&\int_{\Gamma^{Fr}_+}\left[\frac{1}{2}\left(\dot{P}_{\mathfrak{f}}\, w_a + \frac{\delta^2}{12\,\eta^{\mathfrak{f}R}\beta^{\mathfrak{f}}}\operatorname{grad}_l P_{\mathfrak{f}} \cdot \operatorname{grad}_l w_a + \frac{1}{\delta\,\beta^{\mathfrak{f}}}\frac{\partial\delta}{\partial t}\,w_a\right) - \frac{1}{\delta\,\beta^{\mathfrak{f}}}k^{\mathfrak{s}}_{Fr,t}\;P^t_{\mathfrak{f}}\;w_a\right]\mathrm{d}v\\
&= \frac{1}{\delta\,\beta^{\mathfrak{f}}}\,\mathbf{w}^+_{\mathfrak{f}} \cdot \mathbf{n}^+,\\
&\int_{\Gamma^{Fr}_-}\left[\frac{1}{2}\left(\dot{P}_{\mathfrak{f}}\, w_a + \frac{\delta^2}{12\,\eta^{\mathfrak{f}R}\beta^{\mathfrak{f}}}\operatorname{grad}_l P_{\mathfrak{f}} \cdot \operatorname{grad}_l w_a + \frac{1}{\delta\,\beta^{\mathfrak{f}}}\frac{\partial\delta}{\partial t}\,w_a\right) + \frac{1}{\delta\,\beta^{\mathfrak{f}}}k^{\mathfrak{s}}_{Fr,t}\;P^t_{\mathfrak{f}}\;w_a\right]\mathrm{d}v\\
&= \frac{1}{\delta\,\beta^{\mathfrak{f}}}\,\mathbf{w}^-_{\mathfrak{f}} \cdot \mathbf{n}^-
\end{aligned} \tag{3.55}
$$

where w_a is the lower dimensional test function used for the integration of the auxiliary elements. The form provided by Eq. (3.55) is implemented in a FE framework using a Newton-Raphson scheme to solve non-linear, transient pressure diffusion problems in deformable fractures surrounded by a porous matrix which is captured by Biot's theory.

References

1. S. Adami, X.Y. Hu, and Nikolaus A. Adams. A generalized wall boundary condition for smoothed particle hydrodynamics. *Journal of Computational Physics*, 231(21):7057–7075, 2012.
2. R. Alessi, M. Ambati, T. Gerasimov, S. Vidoli, and L. De Lorenzis. *Comparison of Phase-Field Models of Fracture Coupled with Plasticity*, pages 1–21. Springer International Publishing, Cham, 2018.
3. R. Alessi, J. J. Marigo, C. Maurini, and S. Vidoli. Coupling damage and plasticity for a phase-field regularisation of brittle, cohesive and ductile fracture: One-dimensional examples. *International Journal of Mechanical Sciences*, pages 1–18, 2017.
4. M. Ambati, T. Gerasimov, and L. De Lorenzis. Phase-field modeling of ductile fracture. *Computational Mechanics*, 55(5):1017–1040, 2015.
5. Hanen Amor, Jean-jacques Marigo, and Corrado Maurini. Regularized formulation of the variational brittle fracture with unilateral contact: numerical experiments. *Journal of Mechanics and Physics of Solids*, 57(8):1209–1229, 2009.
6. D. Asahina, J.E. Houseworth, J.T. Birkholzer, J. Rutqvist, and J.E. Bolander. Hydromechanical model for wetting/drying and fracture development in geomaterials. *Computers and Geosciences*, 65:13–23, 2014. cited By 30.
7. M. Bahrami, M.M. Yovanovich, and J.R. Culham. Effective thermal conductivity of rough spherical packed beds. *International Journal of Heat and Mass Transfer*, 49(19–20):3691–3701, 2006. cited By 110.

8. S. Balay, S. Abhyankar, M. F. Adams, J. Brown, P. Brune, K. Buschelman, L Dalcin, A. Dener, V. Eijkhout, W. D. Gropp, D. Karpeyev, D. Kaushik, M. G. Knepley, D. A. May, L. C. McInnes, R. T. Mills, T. Munson, K. Rupp, P. Sanan, B. F. Smith, S. Zampini, H. Zhang, and H. Zhang. PETSc Web page, 2019.
9. S. Balay, S. Abhyankar, M. F. Adams, J. Brown, P. Brune, K. Buschelman, L Dalcin, A. Dener, V. Eijkhout, W. D. Gropp, D. Karpeyev, D. Kaushik, M. G. Knepley, D. A. May, L. C. McInnes, R. T. Mills, T. Munson, K. Rupp, P. Sanan, B. F. Smith, S. Zampini, H. Zhang, and H. Zhang. PETSc users manual. Technical Report ANL-95/11—Revision 3.11, Argonne National Laboratory, 2019.
10. T. Belytschko and T. Black. Elastic crack growth in finite elements with minimal remeshing. *International Journal for Numerical Methods in Engineering*, 45(5):601–620, 1999. cited By 2659.
11. J.E. Bolander and S. Saito. Fracture analyses using spring networks with random geometry. *Engineering Fracture Mechanics*, 6:1569–1591, 1998.
12. J.E. Bolander Jr. and S. Saito. Fracture analyses using spring networks with random geometry. *Engineering Fracture Mechanics*, 61(5–6):569–591, 1998. cited By 322.
13. M.J. Borden, C.V. Verhoosel, M.A. Scott, T.J.R. Hughes, and C.M. Landis. A phase-field description of dynamic brittle fracture. *Computer Methods in Applied Mechanics and Engineering*, 217–220:77–95, 2012.
14. B. Bourdin, G.A. Francfort, and J.-J. Marigo. Numerical experiments in revisited brittle fracture. *Journal of the Mechanics and Physics of Solids*, 48(4):797–826, 2000. cited By 456.
15. B. Bourdin, G.A. Francfort, and J.-J. Marigo. Numerical experiments in revisited brittle fracture. *J. Mech. and Phys. of Solids*, 48(4):797–826, 2000.
16. B. Bourdin, G.A. Francfort, and J.-J. Marigo. The variational approach to fracture. *Journal of Elasticity*, 91(1-3):5–148, 2008.
17. B. Bourdin, C.J. Larsen, and C.L. Richardson. A time-discrete model for dynamic fracture based on crack regularization. *International Journal of Fracture*, 168(2):133–143, 2011.
18. Blaise Bourdin, Chukwudi P. Chukwudozie, and Keita Yoshioka. A Variational Approach to the Numerical Simulation of Hydraulic Fracturing. In *the 2012 SPE Annual Technical Conference and Exhibition*, 2012.
19. Blaise Bourdin, Jean Jacques Marigo, Corrado Maurini, and Paul Sicsic. Morphogenesis and propagation of complex cracks induced by thermal shocks. *Physical Review Letters*, 112(1):1–5, 2014.
20. H.H. Bui, J.K. Kodikara, A. Bouazza, A. Haque, and P.G. Ranjith. A novel computational approach for large deformation and post-failure analyses of segmental retaining wall systems. *International Journal for Numerical and Analytical Methods in Geomechanics*, 38(13):1321–1340, 2014. cited By 15.
21. H.H. Bui, K. Sako, and R. Fukagawa. Numerical simulation of soil-water interaction using smoothed particle hydrodynamics (sph) method. *Journal of Terramechanics*, 44(5):339–346, 2007. cited By 94.
22. T. Cajuhi, L. Sanavia, and Laura De Lorenzis. Phase-field modeling of fracture in variably saturated porous media. *Computational Mechanics*, pages 1–20, 2017.
23. Davide Casagrande, O Buzzi, Anna Giacomini, Cedric Lambert, and G Fenton. A new stochastic approach to predict peak and residual shear strength of natural rock discontinuities. *Rock Mechanics and Rock Engineering*, pages 69–99, 08 2017.
24. N. Castelletto, J. A. White, and H. A. Tchelepi. Accuracy and convergence properties of the fixed-stress iterative solution of two-way coupled poromechanics. *International Journal for Numerical and Analytical Methods in Geomechanics*, 39(14):1593–1618, 2015.
25. J. Chessa and T. Belytschko. An extended finite element method for two-phase fluids. *Journal of Applied Mechanics, Transactions ASME*, 70(1):10–17, 2003. cited By 195.
26. C. Chukwudozie, B. Bourdin, and K. Yoshioka. A variational phase-field model for hydraulic fracturing in porous media. *Computer Methods in Applied Mechanics and Engineering*, 347:957–982, 2019. cited By 6.

27. J. Chung, A. Roos, J.M. De Hosson, and E. van der Giessen. Fracture of disordered three-dimensional spring networks: A computer simulation methodology. *Physical Review B—Condensed Matter and Materials Physics*, 54(21):15094–15100, 1996. cited By 23.
28. P.A. Cundall and O.D.L. Strack. A discrete numerical model for granular assemblies. *Geotechnique*, 29(1):47–65, 1979. cited By 9142.
29. W.A. Curtin and H. Scher. Brittle fracture in disordered materials: A spring network model. *Journal of Materials Research*, 5(3):535–553, 1990. cited By 117.
30. P. Farrell and C. Maurini. Linear and nonlinear solvers for variational phase-field models of brittle fracture. *International Journal for Numerical Methods in Engineering*, 109(5):648–667, 2017.
31. Ali Fathi, Zabihallah Moradian, Patrice Rivard, Gérard Ballivy, and Andrew J. Boyd. Geometric effect of asperities on shear mechanism of rock joints. *Rock Mechanics and Rock Engineering*, 49(3):801–820, Mar 2016.
32. Y.T. Feng, K. Han, C.F. Li, and D.R.J. Owen. Discrete thermal element modeling of heat conduction in particle systems: Basic formulations. *Journal of Computational Physics*, 227:5072–5089, 2008.
33. B. Flemisch, M. Darcis, K. Erbertseder, B. Faigle, A. Lauser, K. Mosthaf, S. Müthing, P. Nuske, A. Tatomir, M. Wolff, and R. Helmig. Dumux: Dune for multi-phase,component,scale,physics,... flow and transport in porous media. *Advances in Water Resources*, 34(9):1102–1112, 2011. cited By 138.
34. G.A. Francfort and J.-J Marigo. Revisiting brittle fracture as an energy minimization problem. *J. Mech. and Phys. of Solids*, 46(8):1319–1342, 1998.
35. T. Gerasimov and L. De Lorenzis. A line search assisted monolithic approach for phase-field computing of brittle fracture. *Computer Methods in Applied Mechanics and Engineering*, 312:276 – 303, 2016. Phase Field Approaches to Fracture.
36. Robert A Gingold and Joseph J Monaghan. Smoothed particle hydrodynamics: theory and application to non-spherical stars. *Monthly notices of the royal astronomical society*, 181(3):375–389, 1977.
37. V. Girault, M. F. Wheeler, B. Ganis, and M. E. Mear. A lubrication fracture model in a poro-elastic medium. *Mathematical Models and Methods in Applied Sciences*, 25(4):587–645, 2015.
38. Vivette Girault, Kundan Kumar, and Mary F. Wheeler. Convergence of iterative coupling of geomechanics with flow in a fractured poroelastic medium. *Computational Geosciences*, 20(5):997–1011, Oct 2016.
39. P. Grassl. A lattice approach to model flow in cracked concrete. *Cement & Concrete Composites*, 31:454–460, 2009.
40. P. Grassl, C. Fahy, D. Gallipoli, and J. Bolander. A lattice model for liquid transport in cracked unsaturated heterogeneous porous materials. *VIII International Conference on Fracture Mechanics of Concrete and Concrete Structures*, 2013.
41. V. Hakim and A. Karma. Laws of crack motion and phase-field models of fracture. *Journal of the Mechanics and Physics of Solids*, 57(2):342–368, 2009.
42. Alan Thomas Joseph Hayward. Compressibility equations for liquids: a comparative study. *British Journal of Applied Physics*, 18(7):965, 1967.
43. Y. Heider and B. Markert. A phase-field modeling approach of hydraulic fracture in saturated porous media. *Mechanics Research Communications*, 80:38–46, 2017. cited By 34.
44. M. Hofacker and C. Miehe. Continuum phase field modeling of dynamic fracture: Variational principles and staggered FE implementation. *International Journal of Fracture*, 178(1–2):113–129, 2012.
45. R. Ince, A. Arslan, and B.L. Karihaloo. Lattice modelling of size effect in concrete strength. *Engineering Fracture Mechanics*, 70:2307–2320, 2003.
46. A. Jagota and G.W. Scherer. Viscosities and sintering rates of a two-dimensional granular composite. *Journal of the American Ceramic Society*, 76(12):3123–3135, 1993. cited By 40.
47. B.L. Karihaloo, P.F. Shao, and Q.Z. Xiao. Lattice modelling of the failure of particle composites. *Engineering Fracture Mechanics*, 70:2385–2406, 2003.

48. A.R. Khoei, S. Moallemi, and E. Haghighat. Thermo-hydro-mechanical modeling of impermeable discontinuity in saturated porous media with x-fem technique. *Engineering Fracture Mechanics*, 96:701–723, 2012. cited By 28.
49. J. Kim, H.A. Tchelepi, and R. Juanes. Stability and convergence of sequential methods for coupled flow and geomechanics: Drained and undrained splits. *Computer Methods in Applied Mechanics and Engineering*, 200(23):2094 – 2116, 2011.
50. J. Kim, H.A. Tchelepi, and R. Juanes. Stability and convergence of sequential methods for coupled flow and geomechanics: Fixed-stress and fixed-strain splits. *Computer Methods in Applied Mechanics and Engineering*, 200(13):1591 – 1606, 2011.
51. Markus Klinsmann, Daniele Rosato, Marc Kamlah, and Robert M. McMeeking. An assessment of the phase field formulation for crack growth. *Computer Methods in Applied Mechanics and Engineering*, 294(Supplement C):313 – 330, 2015.
52. O. Kolditz, S. Bauer, C. Beyer, N. Böttcher, P. Dietrich, U.-J. Görke, T. Kalbacher, C.-H. Park, U. Sauer, C. Schütze, H. Shao, A. Singh, J. Taron, W. Wang, and N. Watanabe. A systematic benchmarking approach for geologic CO2 injection and storage. *Environmental Earth Sciences*, 67(2):613–632, 2012.
53. C. Kuhn and R. Müller. A continuum phase field model for fracture. *Engineering Fracture Mechanics*, 77(18):3625–3634, 2010.
54. C. Kuhn, T. Noll, and R. Müller. On phase field modeling of ductile fracture. *GAMM Mitteilungen*, 39(1):35–54, 2016.
55. K. Li and S. Zhou. Numerical investigation of multizone hydraulic fracture propagation in porous media: New insights from a phase field method. *Journal of Natural Gas Science and Engineering*, 66:42–59, 2019. cited By 2.
56. T. Li, J.-J. Marigo, D. Guilbaud, and S. Potapov. Gradient damage modeling of brittle fracture in an explicit dynamic context. *International Journal for Numerical Methods in Engineering*, 00(March):1–25, 2016.
57. A. Lisjak, P. Kaifosh, L. Hea, B.S.A. Tatone, O.K. Mahabadi, and G. Grasselli. A 2d, fully-coupled, hydro-mechanical, fdem formulation for modelling fracturing processes in discontinuous, porous rock masses. *Computers and Geotechnics*, 81:1–18, 2017.
58. J.X. Liu, S.C. Deng, J. Zhang, and N.G Liang. Lattice type of fracture model for concrete. *Theoretical and Applied Fracture Mechanics*, 48:269–284, 2007.
59. Leon B Lucy. A numerical approach to the testing of the fission hypothesis. *The astronomical journal*, 82:1013–1024, 1977.
60. J.-J. Marigo, C. Maurini, and K. Pham. An overview of the modelling of fracture by gradient damage models. *Meccanica*, 51(12):3107–3128, 2016.
61. D Markauskas, H Kruggel-Emden, R Sivanesapillai, and H Steeb. Comparative study on mesh-based and mesh-less coupled cfd-dem methods to model particle-laden flow. *Powder Technology*, 305:78–88, 2017.
62. V. Martin, J. Jaffré, and J. Roberts. Modeling fractures and barriers as interfaces for flow in porous media. *SIAM Journal on Scientific Computing*, 26(5):1667–1691, 2005.
63. C. Maurini, B. Bourdin, G. Gauthier, and V. Lazarus. Crack patterns obtained by unidirectional drying of a colloidal suspension in a capillary tube: Experiments and numerical simulations using a two-dimensional variational approach. *International Journal of Fracture*, 184(1–2):75–91, 2013.
64. P. Meakin. Models for material failure and deformation. *Science*, 252(5003):226–234, 1991. cited By 104.
65. G. Meschke and D. Leonhart. A generalized finite element method for hydro-mechanically coupled analysis of hydraulic fracturing problems using space-time variant enrichment functions. *Computer Methods in Applied Mechanics and Engineering*, 290:438–465, 2015. cited By 19.
66. C. Miehe, M. Hofacker, L. M. Schänzel, and F. Aldakheel. Phase field modeling of fracture in multi-physics problems. Part II. Coupled brittle-to-ductile failure criteria and crack propagation in thermo-elastic-plastic solids. *Computer Methods in Applied Mechanics and Engineering*, 294:486–522, 2015.

67. C. Miehe, M. Hofacker, L.-M. Schänzel, and F. Aldakheel. Phase field modeling of fracture in multi-physics problems. part ii. coupled brittle-to-ductile failure criteria and crack propagation in thermo-elastic-plastic solids. *Computer Methods in Applied Mechanics and Engineering*, 294:486–522, 2015. cited By 103.
68. C. Miehe, F. Welschinger, and M. Hofacker. Thermodynamically consistent phase-field models of fracture: Variational principles and multi-field fe implementations. *International Journal for Numerical Methods in Engineering*, 83(10):1273–1311, 2010.
69. T. Mohammadnejad and A.R. Khoei. An extended finite element method for fluid flow in partially saturated porous media with weak discontinuities; the convergence analysis of local enrichment strategies. *Computational Mechanics*, 51(3):327–345, 2013. cited By 26.
70. J.J. Monaghan. Smoothed particle hydrodynamics and its diverse applications. *Annual Review of Fluid Mechanics*, 44:323–346, 2011. cited By 306.
71. Joe J Monaghan. Smoothed particle hydrodynamics. *Annual review of astronomy and astrophysics*, 30(1):543–574, 1992.
72. Joseph J Monaghan. Smoothed particle hydrodynamics and its diverse applications. *Annual Review of Fluid Mechanics*, 44:323–346, 2012.
73. J.P. Morris. Simulating surface tension with smoothed particle hydrodynamics. *International Journal for Numerical Methods in Fluids*, 33(3):333–353, 2000. cited By 272.
74. C. Moukarzel and H. J. Herrmann. A vectorizable random lattice. *J. Stat. Phys.*, 68:911–923, 1992.
75. M. Ostoja-Starzewski. Lattice models in micromechanics. *Applied mechanics*, 55(1):35–60, 2002.
76. K. Pham, H. Amor, J.-J. Marigo, and C. Maurini. Gradient damage models and their use to approximate brittle fracture. *Int. J. Damage Mech.*, 20(4, SI):618–652, 2011.
77. E.P. Prado and J.G.M. van Mier. Effect of particle structure on mode i fracture process in concrete. *Engineering Fracture Mechanics*, 70:1793–1807, 2003.
78. K. Pruess. The tough codes-a family of simulation tools for multiphase flow and transport processes in permeable media. *Vadose Zone Journal*, 3(3):738–746, 2004. cited By 201.
79. Z. H. Rizvi, M. Nikolic, and F. Wuttke. Lattice element method for simulations of failure in bio-cemented sands. *Granular Matter*, 21(18), 2019.
80. Z.H. Rizvi, A.S. Sattari, and F. Wuttke. Numerical analysis of heat conduction in granular geo-material using lattice element method. *Energy Geotechnics—Proceedings of the 1st International Conference on Energy Geotechnics, ICEGT*, pages 367–372, 2016.
81. Z.H. Rizvi, A.S. Sattari, and F. Wuttke. Meso scale modelling of infill foam concrete wall for earthquake loads. *16th European Conference on Earthquake Engineering (16ECEE), Thessaloniki, Greece*, 2018.
82. Z.H. Rizvi, K. Sembdner, A. Suman, M.J. Giri Prasad, and F. Wuttke. Experimental and numerical investigation of thermo-mechanical properties for nano-geocomposite. *International Journal of Thermophysics*, 40(5), 2019. cited By 1.
83. Z.H. Rizvi, D. Shrestha, A.S. Sattari, and F. Wuttke. Numerical modelling of effective thermal conductivity for modified geomaterial using lattice element method. *Heat Mass Transf*, 54(2):483–499, 2018.
84. Z.H. Rizvi, F. Wuttke, and A.S. Sattari. Dynamic analysis by lattice element method simulation. *Springer Series in Geomechanics and Geoengineering*, (216849):405–409, 2018. cited By 2.
85. M. Sahimi and S. Arbabi. Mechanics of disordered solids. iii. fracture properties. *Physical Review B*, 47(2):713–722, 1993. cited By 80.
86. D. Santillán, R. Juanes, and L. Cueto-Felgueroso. Phase field model of fluid-driven fracture in elastic media: Immersed-fracture formulation and validation with analytical solutions. *Journal of Geophysical Research: Solid Earth*, 2017.
87. J.M. Sargado, E. Keilegavlen, I. Berre, and J.M. Nordbotten. High-accuracy phase-field models for brittle fracture based on a new family of degradation functions. *Journal of the Mechanics and Physics of Solids*, 111:458–489, 2018.

88. A. S. Sattari, H. B. Motra, Z. H. Rizvi, and F. Wuttke. A new lattice element method (lem) with integrated interface elements for determining the effective thermal conductivity of rock solids under thermo-mechanical processes. *International Symposium on Energy Geotechnics (SEG), Energy Geotechnics*, pages 266–275, 2019.
89. A.S. Sattari, Z.H. Rizvi, H.B. Motra, and F. Wuttke. Meso-scale modeling of heat transport in a heterogeneous cemented geomaterial by lattice element method. *Granular Matter*, 19(66), 2017.
90. E. Schlangen and J.G.M. Van Mier. Micromechanical analysis of fracture of concrete. *International Journal of Damage Mechanics*, 1(4):435–454, 1992. cited By 33.
91. A. Schlüter, A. Willenbücher, C. Kuhn, and R. Müller. Phase field approximation of dynamic brittle fracture. *Computational Mechanics*, 54(5):1141–1161, 2014.
92. P. Schmidt and H. Steeb. Numerical aspects of hydro-mechanical coupling of fluid-filled fractures using hybrid-dimensional element formulations and non-conformal meshes. *Int J Geomath*, 10(14), 2019.
93. J. M. Segura and I. Carol. On zero-thickness interface elements for diffusion problems. *International Journal for Numerical and Analytical Methods in Geomechanics*, 28(9):947–962, 2004.
94. J Ma Segura and I Carol. Coupled hm analysis using zero-thickness interface elements with double nodes. Part I: Theoretical model. *International journal for numerical and analytical methods in geomechanics*, 32(18):2083–2101, 2008.
95. JM Segura and I Carol. Coupled hm analysis using zero-thickness interface elements with double nodes. Part II: Verification and application. *International Journal for Numerical and Analytical Methods in Geomechanics*, 32(18):2103–2123, 2008.
96. M. Seiler, P. Hantschke, A. Brosius, and M. Kästner. A numerically efficient phase-field model for fatigue fracture - 1D analysis. *Pamm*, 18(1):e201800207, 2018.
97. Randolph R Settgast, Pengcheng Fu, Stuart DC Walsh, Joshua A White, Chandrasekhar Annavarapu, and Frederick J Ryerson. A fully coupled method for massively parallel simulation of hydraulically driven fractures in 3-dimensions. *International Journal for Numerical and Analytical Methods in Geomechanics*, 41(5):627–653, 2017.
98. Q. Shao, L. Bouhala, A. Younes, P. Núñez, A. Makradi, and S. Belouettar. An xfem model for cracked porous media: Effects of fluid flow and heat transfer. *International Journal of Fracture*, 185(1–2):155–169, 2014. cited By 9.
99. D. Shrestha, Z.H. Rizvi, and F. Wuttke. Effective thermal conductivity of unsaturated granular geocomposite using lattice element method. *Heat and Mass Transfer*, 55(6):1671–1683, 2019.
100. J. Sima, M. Jiang, and C. Zhou. Modelling desiccation cracking in thin clay layer using three-dimensional discrete element method. *AIP Conference Proceedings*, 1542(1), 2013.
101. R. Sivanesapillai, N. Falkner, A. Hartmaier, and H. Steeb. A csf-sph method for simulating drainage and imbibition at pore-scale resolution while tracking interfacial areas. *Advances in Water Resources*, 95:212–234, 2016. cited By 16.
102. R Sivanesapillai, H Steeb, and A Hartmaier. Transition of effective hydraulic properties from low to high reynolds number flow in porous media. *Geophysical Research Letters*, 41(14):4920–4928, 2014.
103. Rakulan Sivanesapillai. Pore-scale study of non-darcian fluid flow in porous media using smoothed-particle hydrodynamics. 2016.
104. Rakulan Sivanesapillai, Nadine Falkner, Alexander Hartmaier, and Holger Steeb. A csf-sph method for simulating drainage and imbibition at pore-scale resolution while tracking interfacial areas. *Advances in water resources*, 95:212–234, 2016.
105. Rakulan Sivanesapillai and Holger Steeb. Fluid interfaces during viscous-dominated primary drainage in 2d micromodels using pore-scale sph simulations. *Geofluids*, 2018, 2018.
106. H Steeb. Hydro-mechanics of porous and granular material—Poro-elasticity and beyond. In *Proceedings of Alert Workshop and School*. 2019.
107. H Steeb and J Renner. Mechanics of poro-elastic media: A review with emphasis on foundational state variables. *Transport in Porous Media*, 130:437–461, 2019.

108. T. Strouboulis, I. Babuška, and K. Copps. The design and analysis of the generalized finite element method. *Computer Methods in Applied Mechanics and Engineering*, 181(1–3):43–69, 2000. cited By 543.
109. William C Swope, Hans C Andersen, Peter H Berens, and Kent R Wilson. A computer simulation method for the calculation of equilibrium constants for the formation of physical clusters of molecules: Application to small water clusters. *The Journal of Chemical Physics*, 76(1):637–649, 1982.
110. R. Talreja. Continuum modelling of damage in ceramic matrix composites. *Mechanics of Materials*, 12(2):165–180, 1991. cited By 84.
111. E. Tanné, T. Li, B. Bourdin, J-J. Marigo, and C. Maurini. Crack nucleation in variational phase-field models of brittle fracture. *J. Mech. Phys. Solids*, 110:80–99, 2018.
112. A.M. Tartakovsky and P. Meakin. A smoothed particle hydrodynamics model for miscible flow in three-dimensional fractures and the two-dimensional rayleigh-taylor instability. *Journal of Computational Physics*, 207(2):610–624, 2005. cited By 127.
113. J. van der Lee, L. De Windt, V. Lagneau, and P. Goblet. Presentation and application of the reactive transport code hytec. *Developments in Water Science*, 47(C):599–606, 2002. cited By 17.
114. J.G.M. van Mier, M.R.A. van Vliet, and K. Wang Tai. Fracture mechanisms in particle composites: statistical aspects in lattice type analysis. *Mechanics of Materials*, 34, 2002.
115. W.L. Vargas and J.J. McCarthy. Heat conduction in granular materials. *AIChE Journal*, 47(5):1052–1059, 2001. cited By 154.
116. W.L. Vargas and J.J. McCarthy. Stress effects on the conductivity of particulate beds. *Chemical Engineering Science*, 57(15):3119–3131, 2002. cited By 61.
117. W.L. Vargas and J.J. McCarthy. Thermal expansion effects and heat conduction in granular materials. *Physical Review E—Statistical, Nonlinear, and Soft Matter Physics*, 76(4), 2007. cited By 43.
118. C.V. Verhoosel and R. de Borst. A phase-field model for cohesive fracture. *International Journal for Numerical Methods in Engineering*, 00:1–20, 2010.
119. Loup Verlet. Computer "experiments" on classical fluids. i. thermodynamical properties of lennard-jones molecules. *Physical review*, 159(1):98, 1967.
120. Julien Vignollet, Stefan May, René de Borst, and Clemens V. Verhoosel. Phase-field models for brittle and cohesive fracture. *Meccanica*, 49(11):2587–2601, 2014.
121. C Vinci, J Renner, and H Steeb. A hybrid-dimensional approach for an efficient numerical modeling of the hydro-mechanics of fractures. *Water Resources Research*, 50(2):1616–1635, 2014.
122. C. Vinci, H. Steeb, and J. Renner. The imprint of hydro-mechanics of fractures in periodic pumping tests. *Geophysical Journal International*, 202(3):1613–1626, 2015.
123. Carlo Vinci. *Hydro-mechanical coupling in fractured rocks: modeling and numerical simulations*. PhD thesis, Ruhr-University Bochum, 2014.
124. T.D. Vo, A. Pouya, S. Hemmati, and A.M. Tang. Numerical modelling of desiccation cracking of clayey soil using a cohesive fracture method. *Computers and Geotechnics*, 85:15–27, 2017.
125. W. Wang, T. Fischer, B. Zehner, N. Böttcher, U.-J. Görke, and O. Kolditz. A parallel finite element method for two-phase flow processes in porous media: Opengeosys with petsc. *Environmental Earth Sciences*, 73(5):2269–2285, 2015. cited By 12.
126. N. Watanabe, W. Wang, J. Taron, U.J. Görke, and O. Kolditz. Lower-dimensional interface elements with local enrichment: Application to coupled hydro-mechanical problems in discretely fractured porous media. *International Journal for Numerical Methods in Engineering*, 90(8):1010–1034, 2012. cited By 52.
127. H. Wei, J.-S. Chen, F. Beckwith, and J. Baek. A naturally stabilized semi-lagrangian meshfree formulation for multiphase porous media with application to landslide modeling. *Journal of Engineering Mechanics*, 146(4), 2020. cited By 1.
128. M.F. Wheeler, T. Wick, and W Wollner. An augmented-Lagrangian method for the phase-field approach for pressurized fractures. *Computer Methods in Applied Mechanics and Engineering*, 271:69–85, 2014.

129. G. Widenfeld, Y. Weiss, and H. Kalman. The effect of compression and preconsolidation on the effective thermal conductivity of particulate beds. *Powder Technology*, 133(1–3):15–22, 2003. cited By 19.
130. Z.A. Wilson and C.M. Landis. Phase-field modeling of hydraulic fracture. *Journal of the Mechanics and Physics of Solids*, 96:264–290, 2016.
131. P. A. Witherspoon, C. H. Amick, J. E. Gale, and K. Iwai. Observations of a potential size effect in experimental determination of the hydraulic properties of fractures. *Water Resources Research*, 15(5):1142–1146, 1979.
132. J.K.-W. Wong, K. Soga, X. Xu, and J.-Y. Delenne. Modelling fracturing process of geomaterial using lattice element method. *Geomechanics from Micro to Macro—Proceedings of the TC105 ISSMGE International Symposium on Geomechanics from Micro to Macro, IS-Cambridge 2014*, 1:417–422, 2015. cited By 2.
133. W. Woodside and J.H. Messmer. Thermal conductivity of porous media. i. unconsolidated sands. *Journal of Applied Physics*, 32(9):1688–1699, 1961. cited By 543.
134. J.-Y. Wu. A unified phase-field theory for the mechanics of damage and quasi-brittle failure. *Journal of the Mechanics and Physics of Solids*, 103:72–99, 2017.
135. Y.-S. Wu, K. Zhang, C. Ding, K. Pruess, E. Elmroth, and G.S. Bodvarsson. An efficient parallel-computing method for modeling nonisothermal multiphase flow and multicomponent transport in porous and fractured media. *Advances in Water Resources*, 25(3):243–261, 2002. cited By 51.
136. F. Wuttke, A.S. Sattari, Z.H. Rizvi, and H.B. Motra. Advanced meso-scale modelling to study the effective thermo-mechanical parameter in solid geomaterial. *Springer Series in Geomechanics and Geoengineering*, pages 85–95, 2017. cited By 10.
137. A. Yazid, N. Abdelkader, and H. Abdelmadjid. A state-of-the-art review of the x-fem for computational fracture mechanics. *Applied Mathematical Modelling*, 33(12):4269–4282, 2009. cited By 83.
138. T. Ye, D. Pan, C. Huang, and M. Liu. Smoothed particle hydrodynamics (SPH) for complex fluid flows: Recent developments in methodology and applications. *Physics of Fluids*, 31(011301), 2019.
139. K. Yoshioka, F. Parisio, D. Naumov, R. Lu, O. Kolditz, and T. Nagel. Comparative verification of discrete and smeared numerical approaches for the simulation of hydraulic fracturing. *GEM - International Journal on Geomathematics*, 10(1), 2019.
140. C. Zeeb and H. Konietzky. Simulating the hydraulic stimulation of multiple fractures in an anisotropic stress field applying the discrete element method. *Energy Procedia*, 76:264–272, 2015. cited By 10.
141. H.W. Zhang, Q. Zhou, H.L. Xing, and H. Muhlhaus. A dem study on the effective thermal conductivity of granular assemblies. *Powder Technology*, 205(1–3):172–183, 2011. cited By 55.
142. H.W. Zhang, Q. Zhou, H.L. Xing, and H. Muhlhaus. A dem study on the effective thermal conductivity of granular assemblies. *Powder Technology*, 205:172–183, 2011.
143. Minkley, W., Knauth, M., Wüste, U. (2012): Integrity of salinar barriers under consideration of discontinuum-mechanical aspects. Mechanical Behavior of Salt VII, Paris, France, 16–19 April 2012, 469–478. Taylor & Francis Group, London, ISBN 978-0-415-62122-9
144. Knauth,M. and Minkley,W. (2014): Modeling of themechanical and hydraulic behavior of salt rocks on discontinuum mechanical basis. Copyright 2014 ARMA, American Rock Mechanics Association. This paper was prepared for presentation at the 48thUS Rock Mechanics / Geomechanics Symposium held in Minneapolis, MN, USA, 1–4 June 2014, ARMA 14-7036
145. Minkley, W., Knauth, M., Brückner, D. (2013): Discontinuum-mechanical behaviour of salt rocks and the practical relevance for the integrity of salinar barriers. ARMA, 47th US Rock Mechanics Symposium, San Francisco, CA, USA, 23–26 June 2013. Paper ARMA 13-388
146. Minkley, W. and Mühlbauer, J. (2007): Constitutive models to describe themechanical behavior of salt rocks and the imbedded weakness planes. In: M. Wallner, K.H. Lux, W. Minkley & H. R. Hardy. The Mechanical Behavior of Salt - Understanding of THMC Processes in Salt: 6th Conference (SaltMech6), Hannover, Germany, 22–25 May 2007. Publ.: Taylor and Francis, ISBN: 9780415443982, 119–127.

4

GeomInt Project and MEX

Berhard Vowinckel, Thomas Frühwirt, Jobst Maßmann, Thomas Nagel, Mathias Nest, Daniel Pötschke, Christopher Rölke, Amir Shoarian Sattari, Patrick Schmidt, Holger Steeb, Keita Yoshioka, Gesa Ziefle, and Olaf Kolditz

---

B. Vowinckel (✉) · J. Maßmann · G. Ziefle
BGR, Federal Institute for Geosciences and Natural Resources, Hannover, Germany
e-mail: b.vowinckel@tu-braunschweig.de

J. Maßmann
e-mail: jobst.massmann@bgr.de

G. Ziefle
e-mail: gesa.ziefle@bgr.de

T. Frühwirt · T. Nagel · D. Pötschke
TUBAF, Technische Universität Bergakademie Freiberg, Freiberg, Germany
e-mail: thomas.fruehwirt@ifgt.tu-freiberg.de

T. Nagel
e-mail: thomas.nagel@ifgt.tu-freiberg.de

D. Pötschke
e-mail: daniel.poetschke@ifgt.tu-freiberg.de

T. Nagel
Geotechnical Institute, TU Bergakademie Freiberg, Freiberg, Sachsen, Germany

M. Nest · C. Rölke
IfG, Institut für Gebirgsmechanik, Leipzig, Germany
e-mail: mathias.nest@ifg-leipzig.de

A. S. Sattari
Geomechanics and Geotechnics, Christian-Albrechts-Universität zu Kiel, Kiel, Germany
e-mail: amir.shoarian-sattari@ifg.uni-kiel.de

P. Schmidt · H. Steeb
UoS, University of Stuttgart, Stuttgart, Germany
e-mail: patrick.schmidt@mechbau.uni-stuttgart.de

H. Steeb
e-mail: holger.steeb@mechbau.uni-stuttgart.de

K. Yoshioka · O. Kolditz
UFZ, Helmholtz Centre for Environmental Research, Leipzig, Germany
e-mail: keita.yoshioka@ufz.de

Fig. 4.1 MEX concept

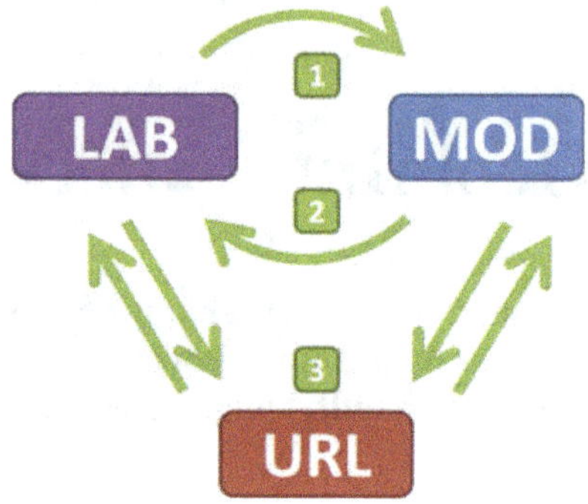

The basic idea of Model-Experiment-Exercises (MEX) is to link modelling and experimental works from the very beginning i.e. in the conceptual phase. Due to the complexity of each part in the systems analysis, this combination is sometimes lost. Moreover, both models and experiments require highly sophisticated tools and equipment as well as highly specialized professionals, which also necessitate adequate measures and incentives for collaboration. GeomInt is introducing the MEX concept exactly for this purpose. Therefore, the following MEX studies occupy the largest part of the GeomInt book and feed most of the publications with research material.

In order to illustrate the MEX concept, Fig. 4.1 depicts the dependencies between laboratory (LAB) and field experiments (URL) as well as modelling work (MOD). Lab experiments (LAB) will be analysed by models (MOD) in order to calibrate parameters and validate them [1]. This step in the analysis loop should proof that the models are able to represent the experiments (validation). Models will be also used for planning experimental work in order to improve the general process understanding [2] (experimental design). Both lab experiments and models must be scaleable to field experiments [3].

The generic approach for systems analysis described above is applied to GeomInt through the individual work packages (WP1-3) linking experimental works in the laboratories (lab and field) and numerical modelling (Fig. 4.2). Due to the limited project time this is being demonstrated for selected cases. A specific challenge for the GeomInt project is the development and implementation of generic numerical methods which are able to simulated THM coupled processes for various rock types—ductile and brittle materials (thermodynamic consistency) with an emphasis of discontinuities. This is the concept of building GeomInt frameworks—experimental (see Chap. 2) and numerical platforms (see Chap. 3).

O. Kolditz
e-mail: olaf.kolditz@ufz.de

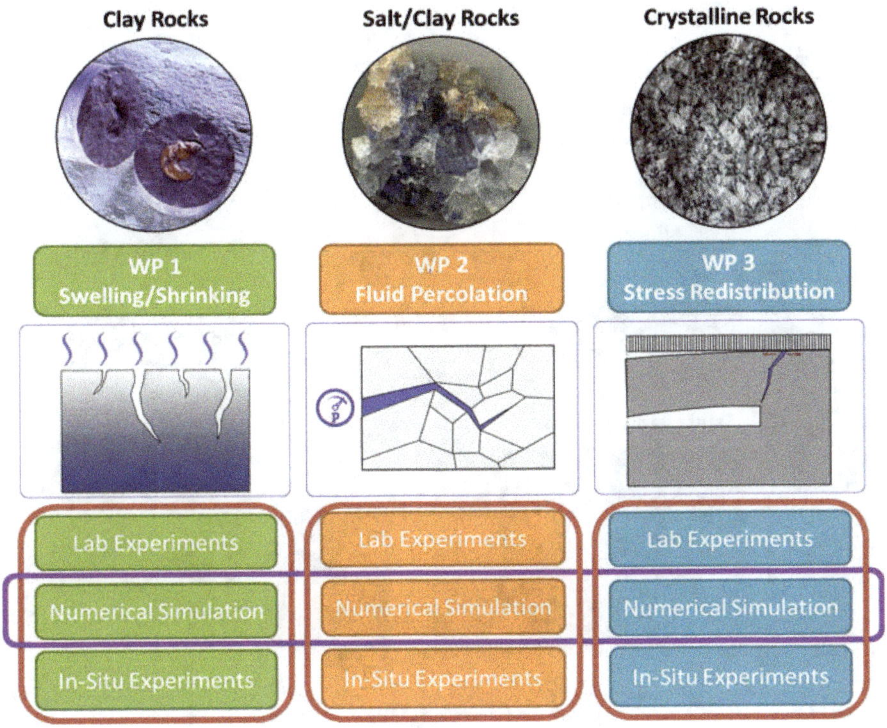

Fig. 4.2 MEX in WPs

Before WP related MEX are described, a prerequisite of two exercises dealing with bending test are introduced for granite and clay. Here the fracture process is initiated by external loads (i.e. not by fluid injection). These bending tests are providing important information on the elastic material behavior before fracturing the specimen (MEX 0-1a). Moreover, for clay samples, particularly from sandy facies, the lamination of the material is largely influencing fracturing processes. A concept for a humidity controlled bending test is presented in MEX 0-2 with is related to the planned CD-A experiment in Mt. Terri.

Figure 4.3 illustrates an overview of the forthcoming Model-Experiment-Exercises (MEX) combining both theoretical work and experiments at various scales. The MEX concept is the central synthesis element of GeomInt as it is directly linking models (MOD) with lab experiments (LAB) and paving the way towards the analysis of in-situ experiments (URL) (see also Chap. 6).

MEX are organized mainly along the projects work packages (lines) as well as experiments (LAB/URL) and various modelling approaches (columns) (Table 4.1).

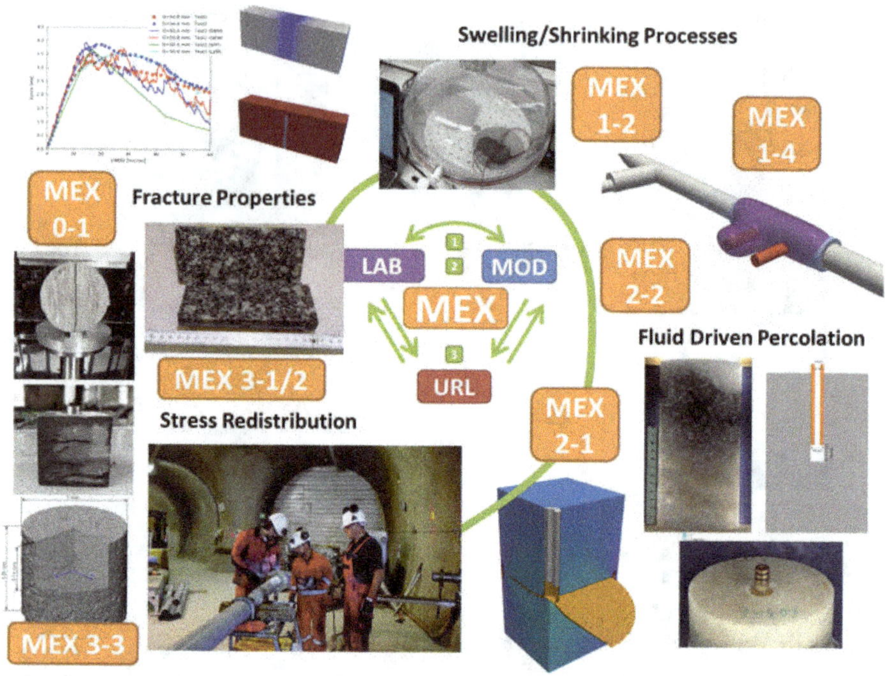

Fig. 4.3 Overview of Model-Experiment-Exercises (MEX)

4.1 Model-Experiment-Exercise MEX 0-1: Bending Fracture Test

Keita Yoshioka (UFZ), Amir Shoarian Sattari (CAU), and Mathias Nest (IfG)

4.1.1 Experimental Set-Up

Model Exercise 1 (ME 1) is concerned with fracture propagation by external loads (i.e. without internal hydraulic force). Experimental data are taken from Three Point Bending (TPB) tests conducted on Rockville Granite samples [1]. TPB experiments are performed by applying a load on the top of the samples and the load is adjusted by keeping the rate of the Crack Mouth Opening Displacement (CMOD) fixed at 0.05 μm/s. The schematic of TPB is shown in Fig. 4.4.

The width of the notch is 1.2 mm at the center of the sample and the ratio of notch length to the sample height is 0.2. Other parameters used for simulation are listed in Table 4.2.

The experimental force-displacement results from [1] for two samples are shown in Fig. 4.5.

Table 4.1 Model-Experiment-Exercises (MEX) matrix

MEX	TOP	EXP	MOD				
WP			LEM	DEM	FEM	HDF	FFS
0-1a	Bending fracture test	LIT					
0-1b	Bending fracture test (aniso)	LAB					
0-2	Humidity controlled bending	Concept					
1-1a	Swelling of clay	LAB					
1-1b	Swelling of clay	LAB					
1-2	Shrinkage of clay	LAB					
1-3	Desiccation of clay	Concept					
1-4	CD/LP experiment	URL					
2-1a	Pressure driven percolation	LIT					
2-1b	Pressure driven percolation	LAB					
2-2	Healing/closure	LAB					
2-3	Compressible fluids	LIT					
2-4	URL Springen	URL					
3-1	CNL test	LAB					
3-2	CNS test	LAB					
3-3	Cyclic loading	LAB					

Legende (EXP) experimental data from literature (LIT), LAB and field experiments (URL). Various modelling approaches (MOD): lattice (LEM), distinct (DEM), finite elements (FEM), hybrid-dimensional (HDF), and forces on fracture surfaces (FFS) are developed and compared against experimental results

4.1.2 Model Approach

Discrete-Element-Model (DEM)

Two models were set up to represent granite samples with different grain structures. As reported in Ref. [1], the average grain size of the Rockville granite in the experiment was 10 mm. As they are embedded in more finely grained material, a discrete element model with a Voronoi-block diameter of 5.4 mm has been chosen. For the Voronoi-discretization random points were inserted in the sample volume. The serve as the center coordinates of the Voronois. After 10^5 insertion attempts, this lead to 1236 grains for the first sample, and 1226 for the second sample, see Fig. 4.6. Finally a notch was cut from the models. The behaviour of the model follows from the properties of both the grains (bulk modulus K and shear modulus G), and the interface parameters (normal stiffness k_n, shear stiffness k_s, and tensile strength σ_z). These can be combined in several ways to reproduce the experimental results in Fig. 4.5.

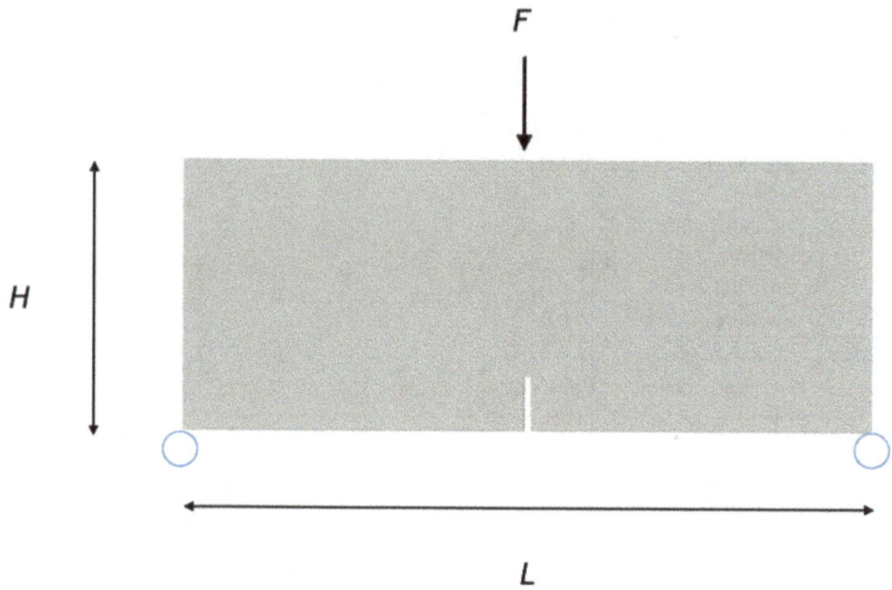

Fig. 4.4 Three Point Bending experimental set up

Table 4.2 Parameters used for the simulation of Three Point Bending tests

w	L (span)	H	E	ν	UCS	σ_T
30 mm	127 mm	50.8 mm	27.5 GPa	0.175	106 MPa	8.1 MPa

In this model we have chosen $K = 14.1$ GPa, $G = 11.7$ GPa, $k_n = 1700$ GPa/m, $k_s =$ 170 GPa/m, and $\sigma_z = 2.5$ MPa.

Lattice-Element-Model (LEM)

The lattice element method (LEM) with the refined mesh is implemented to model the three-point bending test according to the experimental data given in Table 4.2. The regularization of the parameters is carried out to assure the mesh independence results (Sect. 3.2.2). The total number of the elements is around 500,000 elements, where the mesh distance in the refined domain is 1.2 mm, granting the experimental notch width (Fig. 4.7). The considered Young Modulus is 35 GPa to match the experimental linear elastic response.

Finite-Element-Model: Variational Phase-Field (VPF)

The finite element mesh constructed for the 50.8 mm test is shown in Fig. 4.8, which consists of 1,682,308 three dimensional tetrahedral elements with 316,636 nodes. The notch with the width of 1.2 mm is explicitly meshed in the domain and the top load is applied at the row of the nodes on the top. The average tetrahedral size is 0.3 mm, which is 1/4 of the notch width, in the refined region. The Young's modulus

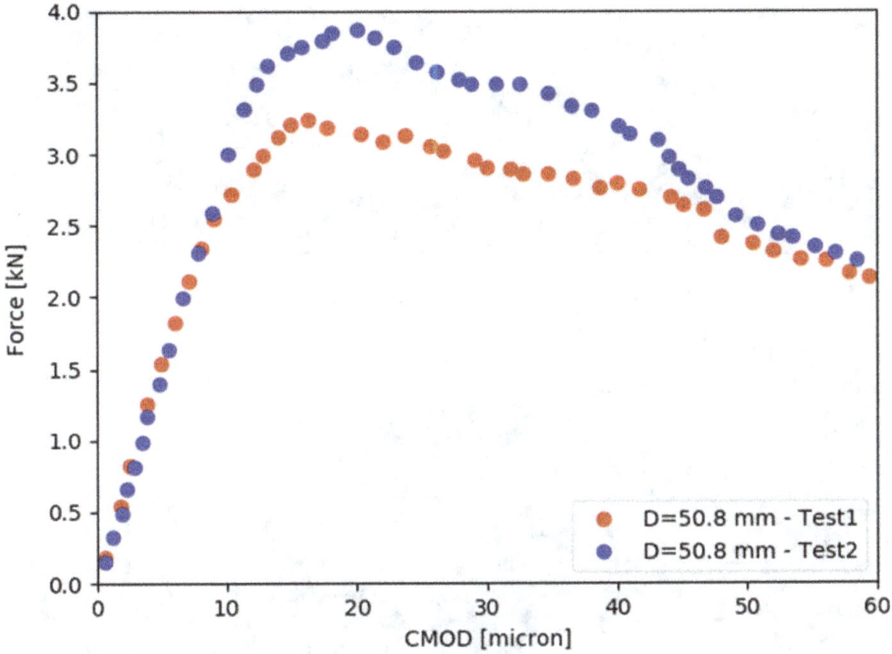

Fig. 4.5 Three Point Bending test results

Fig. 4.6 Discrete element model computation domain

Fig. 4.7 The lattice setup for the simulation of the fracture toughness

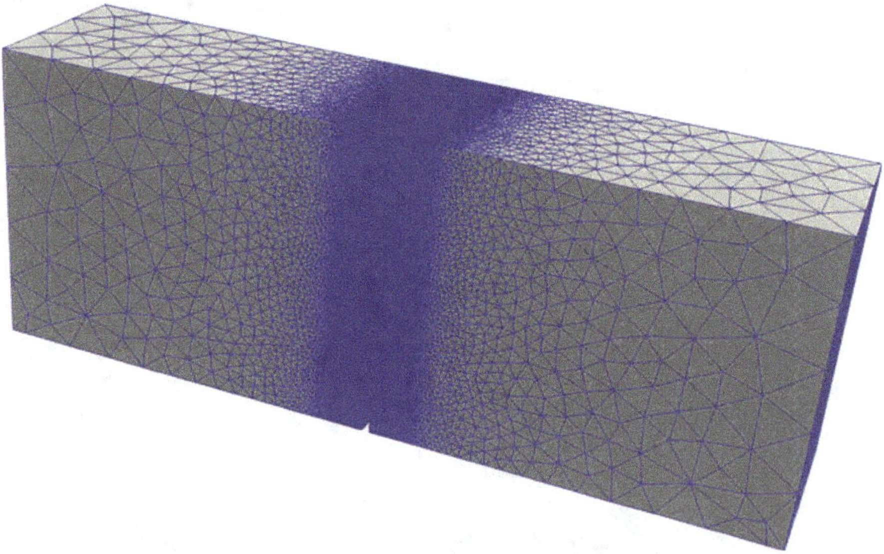

Fig. 4.8 Variational phase-field computation domain

in the simulation was increased to 40 GPa from the one listed in Table 4.2 in order to match the linear elastic response. Another property required in the variational phase-field model is the fracture toughness rather than macroscopic yield strengths as the methodology is based on fracture mechanics. The fracture toughness, G_c of 35 Pa m was chosen to match the peak force from the experiment.

4.1.3 Results and Discussion

Simulation results from all the simulations are shown in Fig. 4.10. All the models exhibit linear elastic behaviors, before failure sets in as in the experimental data. First, we can see that all the models reproduce the linear elastic response. It is straightfor-

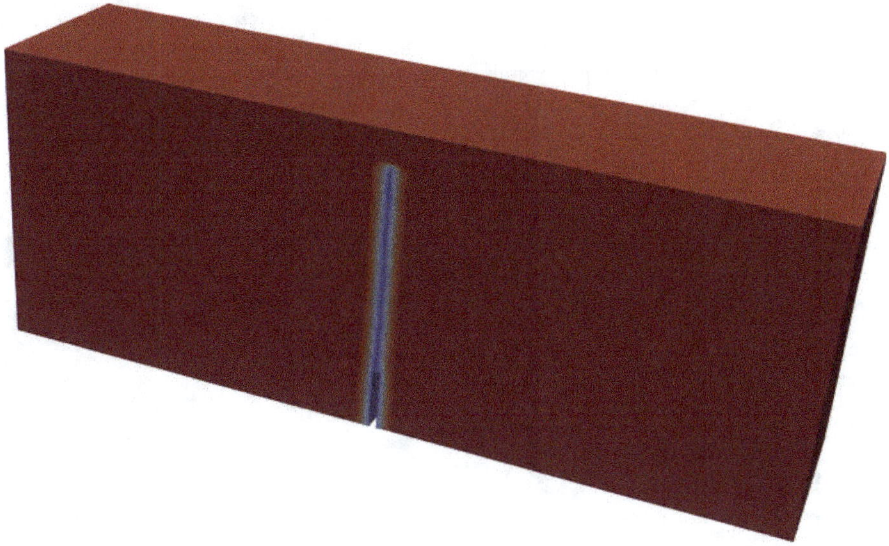

Fig. 4.9 Variational phase-field computation result

ward for the variational phase-field model as it only requires the descriptions of the linear elastic properties (i.e. the Young's modulus and the Poisson's ratio). However, for the lattice or discrete element methods, these bulk properties need to be obtained through the micro mechanical properties. In lattice method the strain energy stored in unit cell should be equal to the continuum strain energy. Therefore, for a Euler-Bernoulli beam elements and while considering the stored energies, the micro to macro (bulk) transformation of properties is carried out, see [2] (Fig. 4.9).

The main difference between the discrete/lattice element method and the variational phase-field model is the responses after the failure. While the current implementation in the variational phase-field used in this study is based on the linear elastic brittle fracture mechanics, and plasticity or softening is not considered.[1] For this reason, the result from the variational phase-field model fails elastically and the force ceases to zero quickly after the peak. On the other hand, the lattice and discrete element methods can model a softening of the bulk material. The implemented softening scheme in lattice element is based on bi-linear softening behavior found in [4]. This post-failure behavior agrees with the real rock response better than the approach based on the linear elastic fracture mechanics.

All three methods reproduce a similar pattern of horizontal displacements. The result from the DEM is shown in Fig. 4.11.

After modeling the crack mouth opening displacement (CMOD) under applied bending load using LEM, the horizontal displacement of the nodes are shown in

[1]The linear elastic fracture mechanics formulation is not necessary constrained by the variational phase-field formulation. Models that include ductility, plasticity, and softening behaviors exist. For example, see [3].

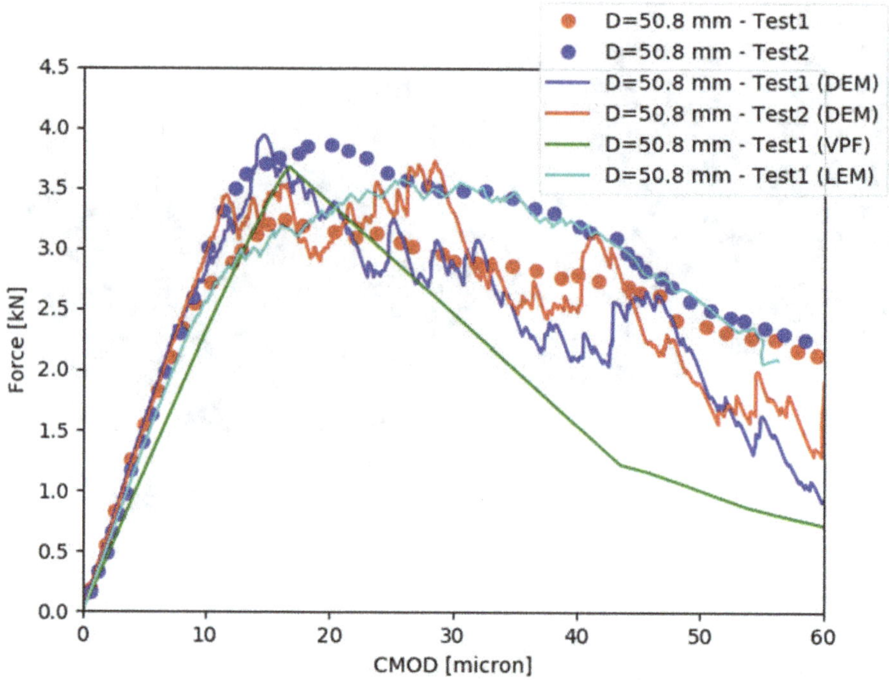

Fig. 4.10 Model comparison

Fig. 4.11 Final Horizontal displacements (DEM)

Fig. 4.12. The developed crack pattern is visible in mid-section of the simulation, where the neighboring nodes on crack tip are moving in the opposite directions.

The article by Tarokh et al. [1] also reports measurements of acoustic emissions, which serve to localize inter-granular micro-cracks. In the DEM method this could be reproduced by using the event of tensile contact failures to call a FISH function which writes the location of that contact into a file. The result, the red dots in Fig. 4.13, agrees very well with the experimental result for the process zone (dashed rectangle).

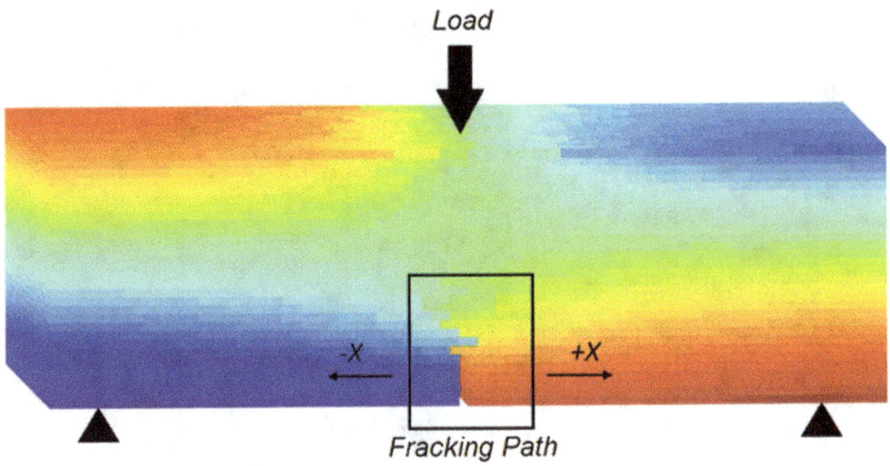

Fig. 4.12 The node displacements in X-axis during the fracking process under the applied bending load

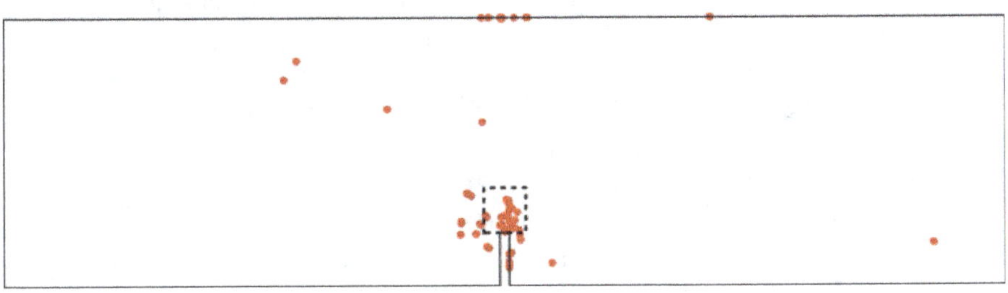

Fig. 4.13 Locations of acoustic emissions (DEM)

4.2 Model-Experiments-Exercise MEX 0-1 (01): Bending Fracture Test (OPA)

Amir Shoarian Sattari (CAU) and Keita Yoshioka (UFZ)

4.2.1 Experimental Set-Up and Results

In this model exercise, the effect of the anisotropy on the fracture toughness of the Opalinus Clay experimentally and numerically is investigated. The description of the experimental setup and the test procedure is explained in Sect. 2.2.2, where a fracture toughness test on a sample with the dimension of $140 \times 30 \times 30$ (mm) $(LxWXT)$ is carried out. The notch dimension is $2 \times 10 \times 30$ (mm) $(LxWXT)$ and the span length is 120 mm. The anisotropy of a claystone depends mainly on the orientation of the embedded layers. When the loading direction is perpendicular (⊥) to the layering orientation the materials strength is highest. In contrast, when the

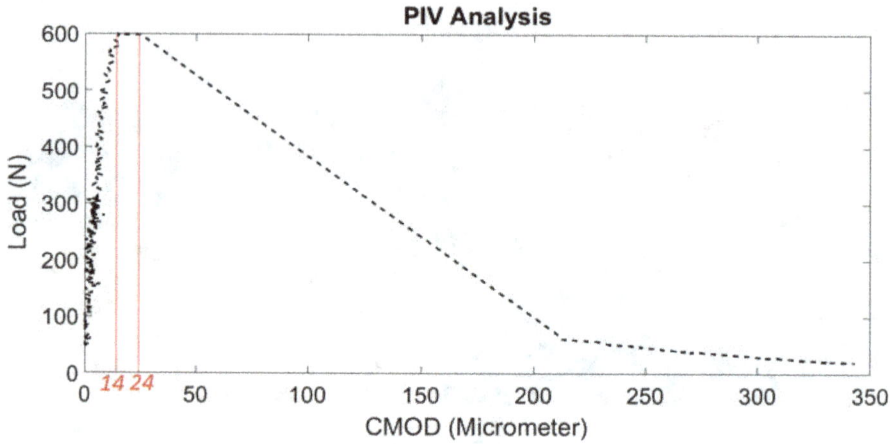

Fig. 4.14 The load versus CMOD for the Opalinus Clay sample under the room temperature

Table 4.3 The Fracture toughness under different temperature conditions

Test results	20°C	50°C	80°C
Peak force (N)	559	482	455
Fracture toughness $MPa\sqrt{m}$.	0.697	0.601	0.582

loading direction is parallel (||) to the layering orientation, the materials strength is the lowest. The peak load attained from the experimental data (⊥) is 598 N and the bending stiffness is around 3.3 GPa (Fig. 4.14). The crack mouth opening before the failure is 24 μm. The 4K video with the 30fps is used to track the reference points on the claystone. Afterward, the image analyzing technique is implemented to determine the CMOD. The 30 fps video was not able to detect the brittle failure of the sample and therefore in the experimental data, the post failure response is not well represented. The measured fracture toughness (⊥) is 0.746 MPa $\sqrt{m}$. The anisotropy effect of claystone is observed based on the fracking pattern of different samples under mechanical loading without temperature and humidity control (Fig. 4.15a). The propagation of fracks parallel to the layering orientation, where the sample is weakest, is observed (Fig. 4.15b). Similarly, inside the climate chamber, and under 50 and 80 °C temperature conditions, the fracture toughness of the claystone is measured (Table 4.3).

(a) (b)

Fig. 4.15 The anisotropy of the frack propagation in claystone **a** the frack propagation parallel to the loading direction, **b** the frack propagation through the weak interface

4.2.2 Model Approach

Lattice-Element-Model (LEM)

The LEM is implemented to investigate the anisotropy of claystone in two cases, where the loading direction is parallel or perpendicular to the layering orientation. The LEM setup (⊥) is generated in 2D with the back calculation of the outputs from the fracture toughness Sect. 2.2.2 and splitting Sect. 2.2.3 experimental results. The materials strength in perpendicular direction is considered to be 5 times higher than parallel case. The fracture paths for both (⊥) and (||) cases are shown in Fig. 4.16a, b, respectively. Figure 4.17 illustrates the comparison between the experimental and numerical data. As discussed before, the post failure behavior does not match due to the experimental shortage of capturing the true load-displacements.

Finite-Element-Approach: Variational Phase-Field (VPF)

The anisotropy claystone sample simulations were also performed with the variational phase-field model. The material strength contrast in the laminations is realized through a contrast in the fracture toughness. For the "weaker" layer, the fracture toughness of 20 Pa m is assigned while 100 Pa m is for the "stronger" layer. Consistently to the LEM simulations, the strength was alternated every 1 mm. Initial strength set-up for the lamination parallel and orthogonal can be found in Fig. 4.18 and Fig. 4.19 respectively. Fracture simulation results are in Figs. 4.20 and 4.21, and the force versus CMOD is in Fig. 4.22.

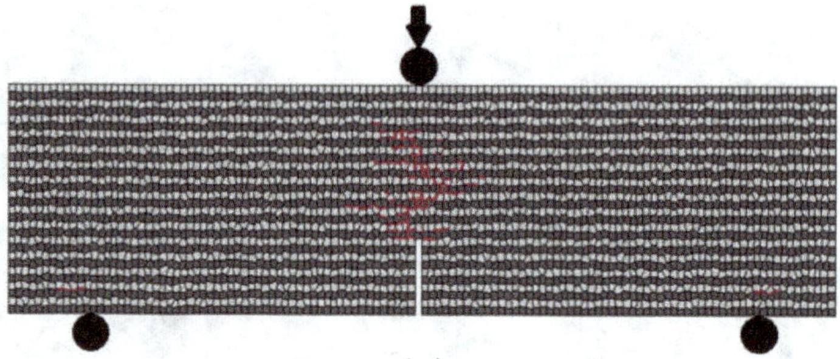

(a)

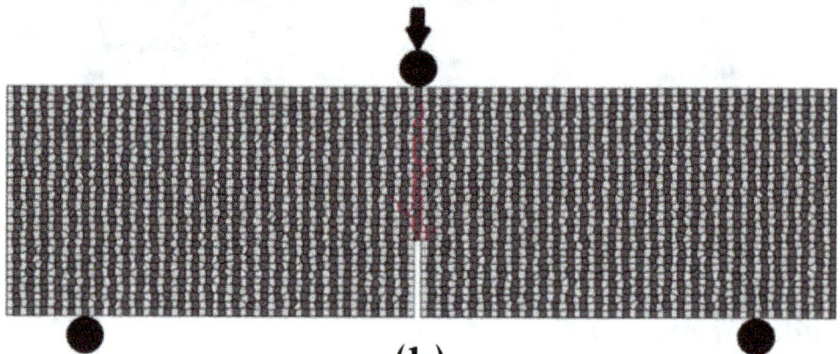

(b)

Fig. 4.16 The fracking path under loading direction **a** perpendicular, and **b** parallel to the layering orientation

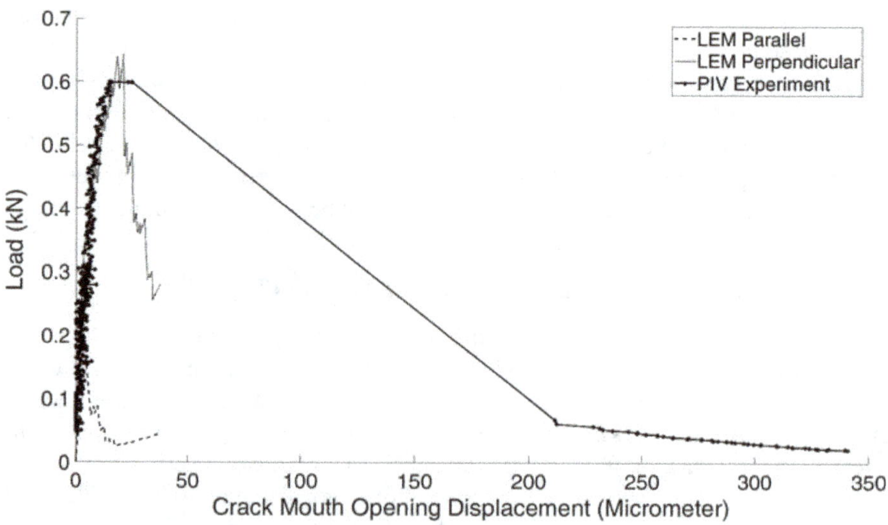

Fig. 4.17 The comparison of experimental and numerical data for effect of anisotropy in Opalinus Clay

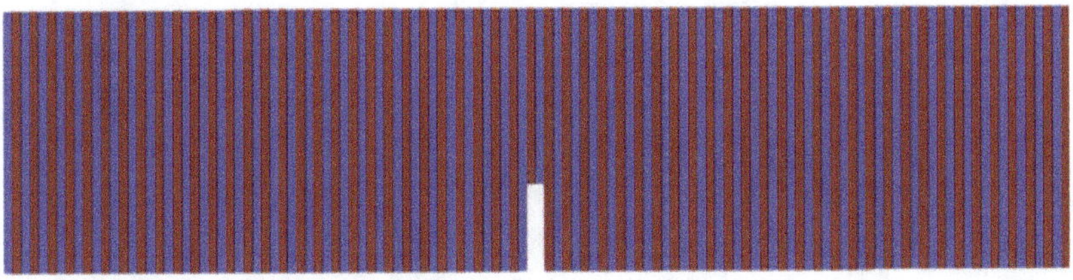

Fig. 4.18 Sample for parallel to the lamination

Fig. 4.19 Sample for orthogonal to the lamination

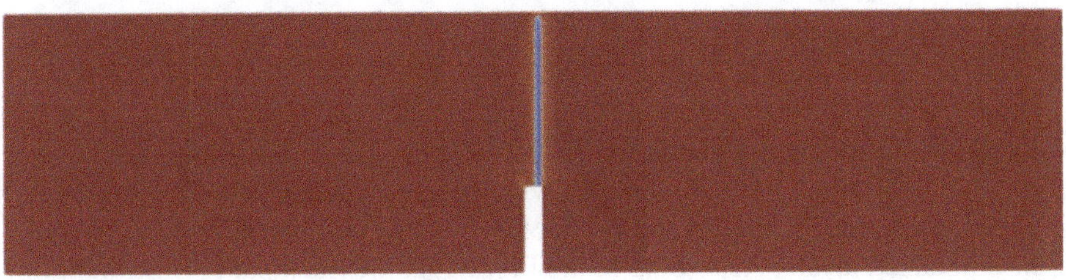

Fig. 4.20 Result of parallel to the lamination

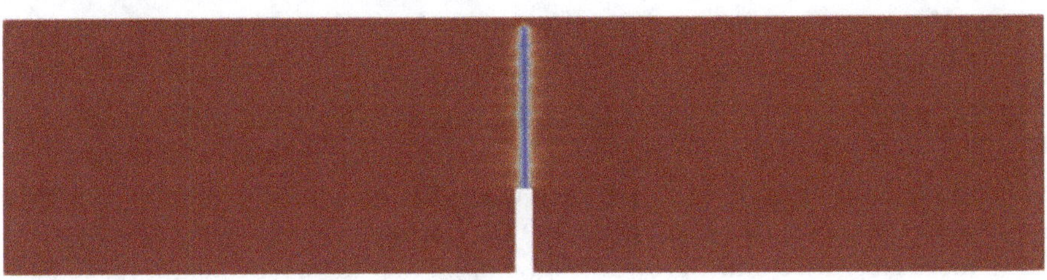

Fig. 4.21 Result of orthogonal to the lamination

4.2.2.1 Results and Discussion

The lattice model is able to capture the existing anisotropy in the Opalinus claystone material. The fracking path dependence on the orientation of the embedded layers is shown in Fig. 4.16a, b, which matches the experimental data given in Fig. 4.15b. Due to the layering orientation and stress distributions (in $\perp$ case), the fracking path has zigzag pattern following the weakest interface bonds. As mentioned before, the

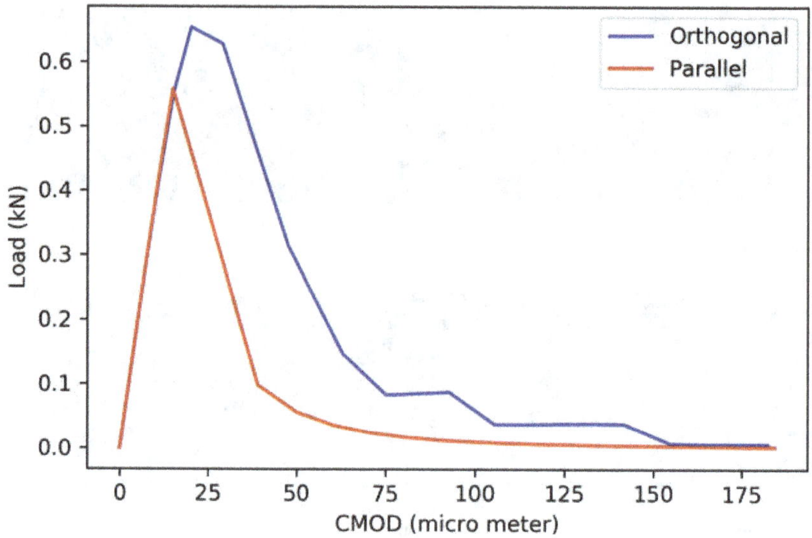

Fig. 4.22 Force versus CMOD

post failure results do not match experimental data due to the experimental tests limitation. The lattice model is validated with the experimental data exist for the (⊥) case and is extended to model the (||) case, where the loading direction is parallel to the embedded layers. According to the Fig. 4.15a, the frack path is a straight line through the weak interface bond. More investigation needs to be done to determine and investigate the Opalinus Clay fracture toughness under different loading to the embedded layers orientation and the effect of the humidity and temperature on the fracture toughness.

The simulations by VPF show similar trends observed in the lattice model. For the lamination parallel case (||), the crack is nucleated at a lamination with the lower fracture toughness and propagated straight through. The force response is also brittle and it declined very quickly after the failure. The lamination orthogonal case (⊥) shows more step–wise crack growth (Fig. 4.20). After the nucleation, it repeatedly stopped at the interfaces before the strong lamination and propagated through the weak lamination with ease (Fig. 4.21). The force versus CMOD curve (Fig. 4.22) also exhibits the non-smooth crack propagation response of the lamination orthogonal case. Additionally, the quantitative comparison of the results both for lattice and VPF models should be carried out.

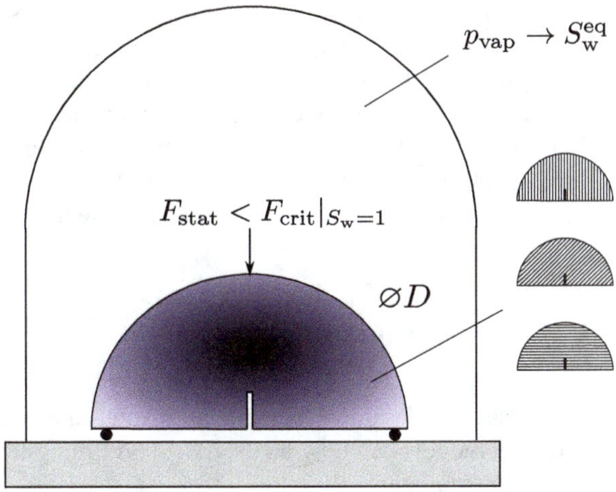

Fig. 4.23 Concept for a humidity controlled long-term bending test setup

4.3 Model-Experiment-Exercise MEX 0-2: Humidity Controlled Long-Term Bending Test

Thomas Nagel (TUBAF)

4.3.1 Experimental Set-Up

The background for this proposal is the CD-A experiment in Mt. Terri in which cyclic humidity variations cause measurable fluctuations in the crack mouth opening. However, it is less well known whether this leads to crack closure or progressive crack growth. From a modelling perspective, the intention is to investigate more deeply several coupling effects in a H^2M setting. Thus, the focus is on Experiment B in the description below (Fig. 4.23).

Therefore, the test is defined as follows:

- Three-point bending tests are performed on semi-cylindrical samples. the reason for the sample shape is simply an easier sample preparation compared to rectangular prismatic bars.
- In principle, anisotropy can be studied by using cores from appropriately aligned bore holes. Standard tests would allow studying orientation-dependent fracture toughness etc. (**Experiment A**)
- The CMOD and the displacement of the load point should be measured. If possible, the full displacement field on the sample's face can be measured by optical methods.
- For **Experiment B**, the same setup is used but the samples are put into a desiccator (or other humidity-controlled environment).
- A static load (dead weight) is applied to a fully saturated sample which is *subcritical*, i.e. this load does not lead to immediate failure. If rate-dependent effects

(creep) can be considered irrelevant, failure of the sample will not occur over time. This should be checked by maintaining one control sample at full saturation throughout by choosing the appropriate humidity in the chamber. If that breaks after a given time, the time-to-failure becomes another comparison metric.
- Subsequently (other samples), the humidity is decreased to a low value in order to initiate sample desaturation. An inhomogeneous saturation field will result.
- Cyclic effects, such as in the in-situ experiment, will not be considered here.

4.3.2 Model Approach

With the saturated sample serving as a reference for potential sub-critical crack growth, the models can now be used to shed light onto the relative relevance of the following phenomena:

- Desaturation causes shrinkage of the clay. Due to the desaturation front progressing from the sample's surface, tensile stresses will develop in the superficial zone of the sample depending on the drying rate. These may elevate local loads in the crack tip or even act as additional stress concentrators at the crack tip. Hence, desaturation may drive the sample to failure.
- A competing mechanism is brought about by the suction-induced strengthening. Desaturation is associated with the build-up of significant capillary pressures which act as suction stresses and increase the effective confinement, thus increasing strength. Thus, desaturation may via this mechanism act in a stabilizing manner. In that case, failure would not occur and the experimental task would be to observe whether F_{crit} increases with desaturation; thus, the sample would have to be loaded to failure. An extension of the test could entail loading each sample to failure after the desaturation experiment has finished, if it hasn't already failed.

WP1: Pathways Through Swelling/Shrinking Processes (Clay Rock)

4.4 Model-Experiment-Exercise MEX 1-1: Swelling of Red Salt Clay

Amir Shoarian Sattari (CAU), Keita Yoshioka (UFZ), Mathias Nest (IfG), and Christopher Rölke (IfG)

This exercise (MEX 1-1) investigates the closure or opening of the micro pathways by swelling process in red salt clay. Both numerical and experimental approaches are implemented to understand the governing factors, which result in materials behav-

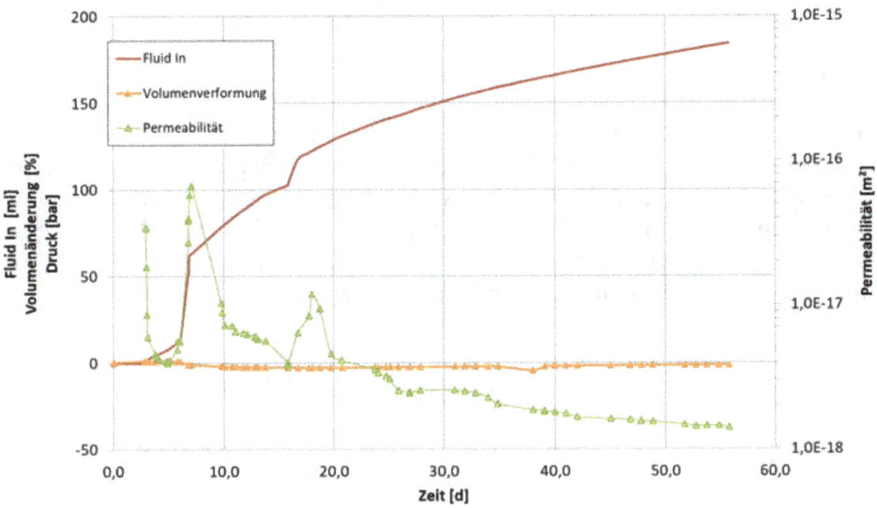

Fig. 4.24 Results of the 8 week swelling test of T4 clay

ior change during the swelling process. The experimental setup is described in Sect. 2.3.1.

4.4.1 *Experiment*

The experiment was conducted on a sample of dry red salt clay from the Aller series (T4) of the Zechstein in northern Germany. The experiment ran for 8 weeks, under isostatic stress conditions of 4 bar, and a brine pressure of 1 bar.

Figure 4.24 shows the amount of fluid that went into the sample (red), the change of the volume of the medium which applies the surrounding pressure (orange), and the permeability (green). We see that during the first three weeks the brine modifies the clay, and pathways are opened and closed. Apart from the wetting of the clay this may also be due to some salt going into solution and recrystallization. After this the permeability behaves more smoothly, and decreases with time. At all times the volume of the surrounding medium was below its initial value, indicating that swelling took place.

4.4.2 Model Approach

Lattice Element Model

With the application of the lattice model, the simulation of the swelling process in red salt clay sample is investigated. The experimental data provided by IfG Leipzig are used to determine the initial permeability and hydraulic aperture values. The effect of anisotropy is not investigated both in experimental and numerical approaches. The initial permeability value is assumed to be equal to $k_{per} = 1 \times 10^{-16}$ [m^2]. The initial hydraulic conductivity (k_h) is calculates based on,

$$k_h = \frac{k_{per}\rho_f g}{\nu_f} \tag{4.1}$$

where $\nu_f = 8.9 \times 10^{-4}$ [Pa s], $\rho_f = 998\,[\frac{\mathrm{kg}}{\mathrm{m}^3}]$ and $g = 9.8\,\frac{m}{s^2}$. The k_h is equal to 1.1×10^{-9} [m s^{-1}]. The initial hydraulic aperture or length of the interface element is calculated as,

$$a_h = \sqrt{\frac{12 k_h \nu_f^k}{g}} = 3.67 \times 10^{-8} \tag{4.2}$$

where $\nu_f^k = 1.004 \times 10^{-6}$ [m^2 s^{-1}]. The calculated parameters then are transformed into lattice model to simulate the swelling and change of the permeability with swelling process. The salt re-crystallization and closure of micro-pathways is not investigated here. The expansion of the elements based on the shrinkage and swelling model described in Sect. 3.2.2 is carried out. The expansion of the elements results in decrease of the hydraulic aperture and therefore lower permeability values. However, during the swelling process the micro fracking is also observed. The elements expansion lead to higher axial confinement stresses between the Voronoi cells, which is represented by interface elements. The domain is generated using the vectorizable lattice element (Fig. 4.25a). The dimension of the sample is given as 100 × 200 mm [DxL]. the fracking paths during the swelling process are shown in Fig. 4.25b. Figure 4.26 Depicts the change of hydraulic conductivity along different axis. The change of hydraulic conductivity along z-axis is the lowest due to the imposed boundary condition in this direction. Before initiation of the micro-fracking pathways, the hydraulic conductivity values are decreased. However, after initiation of micro-cracks the hydraulic conductivity values are increased as started at 0.025% axial strain of the elements. The embedded anisotropy in lattice model resulted in different hydraulic conductivity values in x and y-axis.

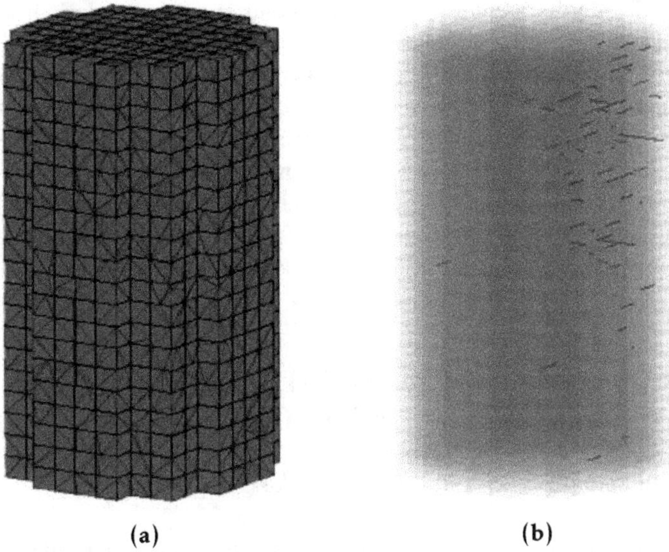

Fig. 4.25 The **a** generated domain for simulation of the swelling processes, and **b** fracking paths shown with red surfaces

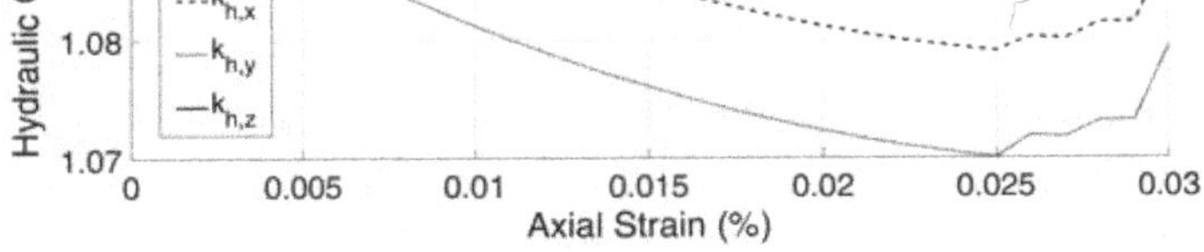

Fig. 4.26 The change of hydraulic conductivity along the axis for salt clay, swelling process

4.4.3 Results and Discussion

The swelling of the salt clay under isostatic stress conditions of 4 bar, and a brine pressure of 1 bar is investigated. The flow rate, change of the samples volume as well as the permeability change during the swelling process is measured. The experimental results indicate the sudden increase in the flow volume after 8 d of the test procedure. Initially, the salt minerals re-crystallization may result in opening and closure of the pathways as indicated from experimental data. Eventually, the closure of the pathways in salt clay sample is observed, which resulted in decrease of the permeability. The lattice model using the elements swelling strain is implemented to simulate the change of the permeability during the test procedure. The integrated interface element method is used here to assess the change of the hydraulic aperture. The anisotropy of salt clay is neglected in this study. Therefore, the initial permeability and hydraulic aperture along different axis are assumed to be equal. The numerical result indicate the decrease of the hydraulic conductivity before the micro-fracking paths formation. The results do not correspond to the experimental data, as in the experimental data, increase of the permeability values were not observed. Further quantitative investigation of the results and validation of the lattice model is required.

4.5 Model-Experiment-Exercise MEX 1-2: The Drying and Wetting Paths of Opalinus Clay

Amir Shoarian Sattari (CAU) and Keita Yoshioka (UFZ)

The simulation of the drying and wetting processes and evolution of the micro pathways in Opalinus Clay are the focus of the this model exercise (MEX 1-2). In this matter, the experimental data are used to model the shrinkage and swelling processes. Additionally, the change of hydraulic conductivity both in parallel and perpendicular to the embedded layering orientations is investigated. It is shown that due to the inherent anisotropy of claystone material, the materials linear strain in parallel and perpendicular orientations are different.

4.5.1 Experimental Set-Up

The prepared two cylindrical thin sections of sandy Opalinus Clay (Fig. 4.27a) the drying and wetting paths. The saturated salt solutions are used to apply different osmotic suctions. The considered salt solution and its induced suction and relative humidity values in the constant room temperature of 20 °C are listed in Table 4.4. The suction values range from 3.2 up to 367 MPa, which insures both drying and wetting paths. The fluctuation of the temperature is negligible. The first sample is used to determine the linear axial strains (Fig. 4.27b) along the embedded layers, which are arranged in a parallel and perpendicular orientations. The second sample is used to determine the water content change during the wetting and drying paths.

The considered salt solution and its induced suction and relative humidity values in the constant room temperature of 20 °C are listed in Table 4.4.

The samples dimension is 100×10 mm (DxL). The equilibrium inside the desiccator is reached when the change of the samples water content is equal to zero. Figure 4.28 depicts the change of suction and axial linear strains in parallel and perpendicular directions. Similarly, Fig. 4.28b shows the change of water content with applied suction using salt solutions. In drying path, the results indicate a higher

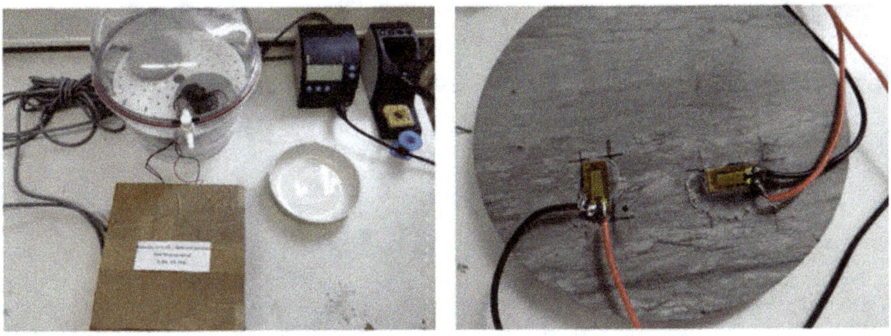

(a) (b)

Fig. 4.27 The **a** desiccator setup, and **b** arraignment of the strain gauge strips on Opalinus Clay surface

Table 4.4 The saturated salt solutions relative humidity and suction values at 20 °C

Salt solution	Relative humidity (%)	Suction (MPa)
K_2SO_4	97.6	3.2
KNO_3	94.6	7.5
KCl	85.1	21.8
NaCl	75.5	38
$Mg(NO_3)_2$	54	84
$MgCl_2$	33.1	149.5
LiCl	12	286.7
LiBr	6.6	367.5

strains for a strain perpendicular to the embedded layers. When the suction is higher than 150 MPa, the strains in perpendicular direction are almost 4.5 larger than parallel ones. Interestingly, in the wetting path, the differences between the strain gauges are much less. According to the water content data, the air-entry pressure for the sandy facies of a Opalinus Clay is around 25 MPa.

4.5.2 *Model Approaches*

The simulation results from both of the model methods, lattice element method (LEM) and variational phase field model (VPF), are described and the accuracy of numerical results for modeling the shrinkage process with change of linear or volumetric strains as well as the change of anisotropic hydraulic conductivity are investigated.

Lattice Element Model

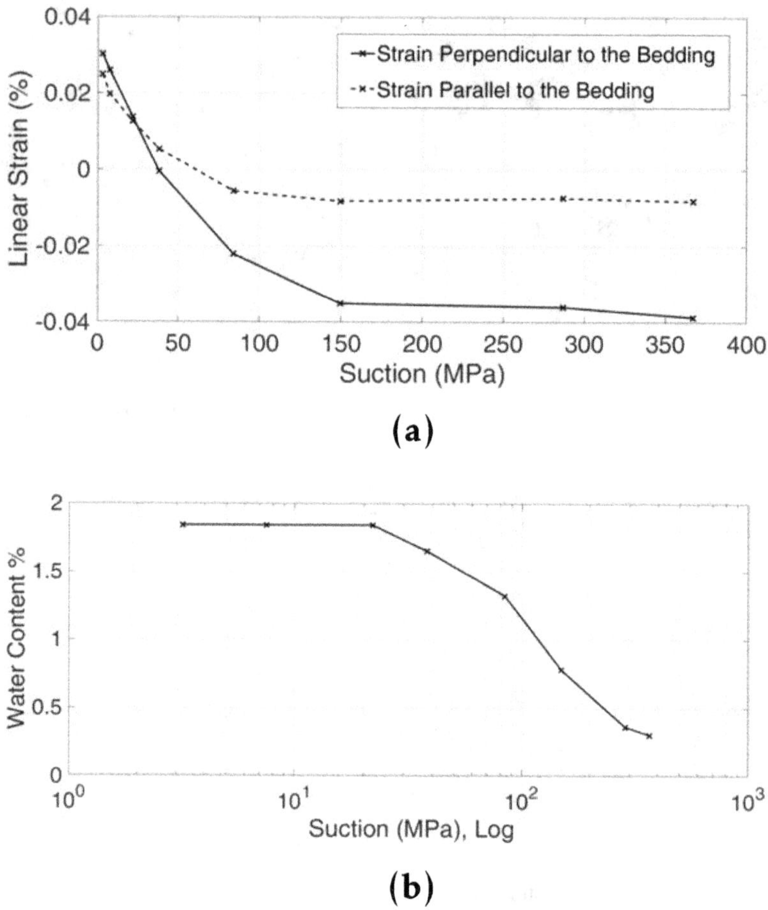

Fig. 4.28 The drying and wetting paths for Opalinus Clay **a** the suction versus linear strains, and **b** the suction versus the water content

With the application of the integrated interface element [5] the drying and wetting processes in the Opalinus Clay are simulated. The linear strain of the elements are calculated based on the experimental data. The initial hydraulic conductivity (k_h) values are approximated according to the technical Report, Mont Terri 2008-04 as,

$$k_{h,\parallel} = 2 \times 10^{-13} \quad , \quad k_{h,\perp} = 0.6 \times 10^{-13} \tag{4.3}$$

While considering the cubic law for flow transfer through the porous medium, the hydraulic aperture or length of the interface element is calculated as,

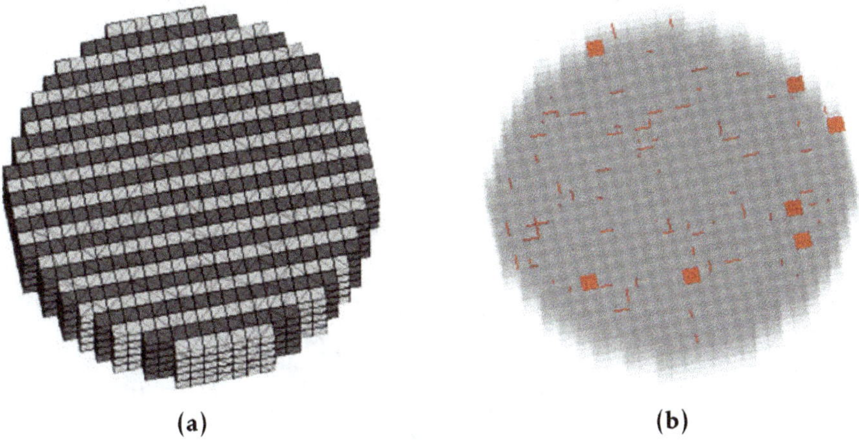

Fig. 4.29 The a generated domain for simulation of the drying and wetting processes, and b fracking paths shown with red surfaces

$$a_h = \sqrt{\frac{12k_h\nu_f}{g}} \quad , \quad a_{h,\parallel} = 4.95 \times 10^{-10} \quad , \quad a_{h,\perp} = 2.71 \times 10^{-10} \tag{4.4}$$

where $\nu_f = 1.004 \times 10^{-6}$ [$m^2\,s^{-1}$] and $g = 9.8$. The domain is generated using the vectorizable lattice element with defined layers as described in previous sections (Fig. 4.29a). The interface length is defined based on Eq. 4.4. The interface strength between two layers is assumed to be 5 times weaker than the bond between a same layer. Therefore, fracking along the layers in X and Z directions is expected, Which results in higher hydraulic conductivity values as the wetting and drying process continues (Fig. 4.29b). Figure 4.30 depicts the change of hydraulic conductivity along three axis of X, Y and Z. As expected, the change of the hydraulic conductivity during the drying process is the lowest along the Y axis, where the drying process is perpendicular to the embedded layering orientation.

Finite-Element-Method: Variational-Phase-Field (VPF)

While couplnng of unsaturated flow (Richards flow) and variational phase-field based formulation of fracture mechanics is underway in OGS, the contribution from VPF is being planned. Once the model becomes available, strain changes driven by shrinkage or swelling of the sample which is subject to the pore–pressure and the partial saturation changes is modeled utilizing the current Richard–Mechanics implementation in OGS. Then desiccation cracking will be simulated through the variational phase-field using the strain energy contributed from the shrinkage and swelling. A tentative FEM mesh prepared is shown in Fig. 4.31. The alternating hydraulic conductivities and the material strength will need to be assigned accordingly to the LEM simulation.

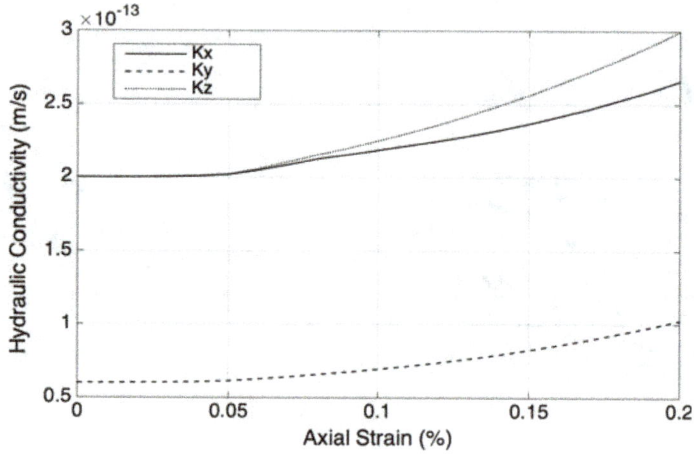

Fig. 4.30 The change of hydraulic conductivity along three axis for Opalinus Clay under drying process

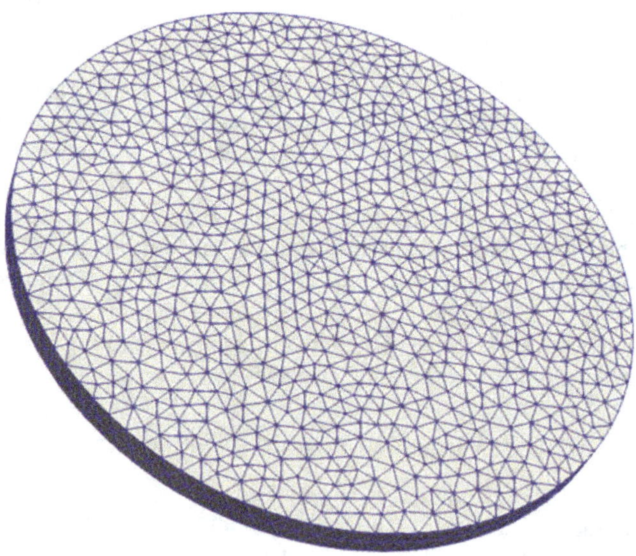

Fig. 4.31 The generated FEM mesh for shrinkage process

4.5.3 *Results and Discussion*

The experimental data assessed during the drying and wetting processes confirms the high anisotropy of Opalinus Clay. Using the saturated salt solutions to apply the osmotic suction, the change of water content as well as the micro deformations along two axis (parallel and perpendicular) are captured and recorded. Higher anisotropy factors during the drying path than in wetting path is observed (Fig. 4.28a). The initial water content of the sample was lower than expected in the literature, mainly due to the sample preparation process, where the sample was subjected to the different relative humidity. Figure 4.28b shows the air-entry pressure to be around 25 MPa. Similarly, the residual pressure corresponding to the residual water content is found to be 180 MPa (for more information regarding the soil-water retention curves see [6]. The numerical lattice model to simulate the drying and wetting paths is implemented and with defining the anisotropy, the change of the hydraulic conductivity along different axis is determined. The results show a similar behavior as seen from experimental data. However, more quantitative comparison of the data between experimental and numerical results is required.

4.6 Model Exercise 1–3: Desiccation Under In-Situ Conditions

Holger Steeb (UoS)

We focus on in-situ (image-based) characterization of desiccation processes of sandy-clay samples under confined conditions. The samples under investigation have to be prepared in sizes small enough for high-resolution X-Ray Computed Tomography (XRCT) and large enough to be representative.

- A cylindrical clay core with diameter $D = 30$ mm and height $H = 60$ mm (or smaller to get a higher spatial resolution) is embedded in a hollow PEEK (Polyether ether ketone) tube. The sample is prepared with a drill hole of $d = 5$ mm.
- Further, the sample is firstly sealed with a Teflon shrink tube. Secondly the gap between the sample and the PEEK tube is filled with an epoxy resin. Epoxy and PEEK have low X-Ray absorption properties. Through the embedding, radial deformations of the clay sample are preventing.
- The top and bottom parts of the samples are hydraulically sealed with end-caps corresponding to undrained conditions. The end-caps allow for axial deformations occurring during the shrinkage process. The drill hole is "open" allowing for water release/uptake.
- The prepared sample with in-situ humidity (partially saturation) is placed in an environmental chamber (Anton Paar, CDT 100, Peltier heated), cf. Fig. 4.32.

Temperature and humidity is controlled (ProUmid MHG 100 humidity controller, ProUmid).

- If necessary, a small van has to be placed in the enironmental chamber allowing for controlled humidity convection in the drill hole.
- Drying/desiccation will be image-based characterized within the open XRCT scanner in Stuttgart. Therefore, the drill hole is temporary sealed.
- We are aiming for a spatial resolution of $30/2500 = 12\ \mu$m/voxel. XRCT scans will be performed for equilibrium states at certain humidity conditions (at ambient temperature of $\Theta = 20\,°$C).

4.7 Model Exercise 1–4: CD/LP Experiment (Mont Terri)

Bernhard Vowinckel (BGR), Gesa Ziefle (BGR), Jobst, Maßmann (BGR)

4.7.1 Motivation

Various areas of research like disposal of nuclear radioactive waste, geothermal energy, and carbon capture and storage (CCS) among others deal with coupled hydraulic-mechanical as well as thermal and chemical processes in the underground. Finite element (FE) codes are a well established tool to strengthen the understanding of the related long-term effects. The German Federal Institute for Geosciences and Natural Resources (BGR) has gathered vast experience with the FE code Open-

Fig. 4.32 Anton Paar MCR 502 TwinDrive DMTA-rheometer with ProUmid MHG 100 humidity controller

GeoSys (OGS-5 as described in [7]) to compute coupled hydraulic-mechanical processes for example in the Mont Terri Rock Laboratory (RL) in Switzerland. Within the project GeomInt (Geomechanical integrity of host and barrier rocks), the focus is on the investigation of the development of discontinuities such as cracks and fissures in claystone, crystalline and rock salt. The Helmholtz Centre for Environmental Research (UFZ) is developing an enhanced version of the FE code OGS-6 (as described in [8]) which has an additional focus on the investigation of these effects. The present report summarizes the results of a comparison between two simulations of the seasonally induced hydraulic desiccation carried out using OGS-5 (version from November 13th, 2017) and OGS-6 (version from December 7th, 2018), respectively. The comparison focuses on the implementation of the two model approaches: (1) Unsaturated one phase flow, based on the Richards approximation [9] (Richards, RF) and (2) poro-elasticity coupled with Richards Flow following [10] (Richards Mechanics, RM). These models serve as a basis to extend OGS-6 further towards the dynamics of discontinuities.

4.7.2 Problem Statement

In the present study, we analyze an idealized niche that is located in the anticline of the Opalinus Clay formation of Mont Terri RL and has a width and height of 3.15 m and 3.3 m, respectively. The computational scenario is designed according to [11], who successfully reproduced the coupled hydraulic-mechanical behavior of a niche in the tunnel system of the Mont Terri RL. A cross-sectional view of the model of the niche is given in Fig. 4.33. The clayey rock in the niche is uncovered, so that seasonal changes in humidity cause cyclic desiccation and saturation of the rock. The porous medium is subdivided into two material groups. Far away from the niche (>2 m), the rock is undisturbed and reflects the properties of Opalinus Clay (ID 1). Close to the niche, an excavation damaged zone (EDZ) is defined (ID 2). For the present study, two relevant processes can be identified. First, owing to the exposure of the niche to seasonal change in humidity, there will be a saturation/desaturation of the porous rock. This process can be analyzed by considering hydraulic effects (RF) only. Second, the changes in capillary pressure induce a stress redistribution. To analyze this problem, the coupled hydro-mechanical process (RM) needs to be considered.

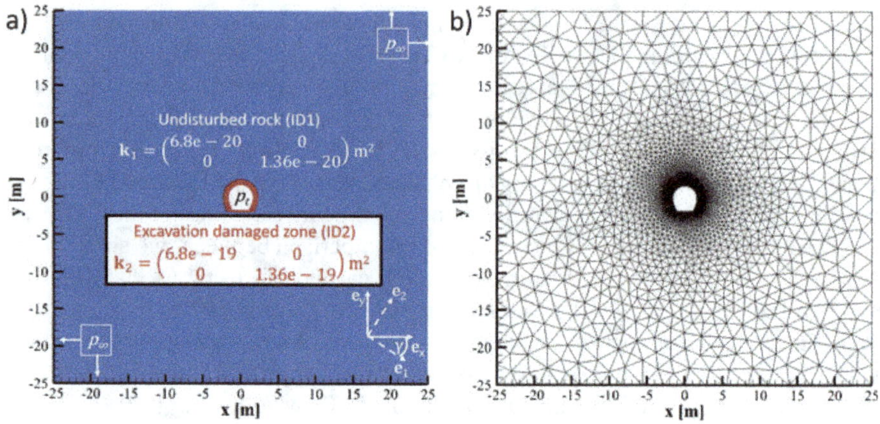

Fig. 4.33 **a** Computational setup and material parameters of the physical problem. **b** Triangulated computational grid comprising 5463 nodes

4.7.3 Unsaturated One-Phase Flow Using the Richards Approximation ("Richards Flow", RF)

4.7.3.1 Model Description

The model "Richards Flow" (RF) is based on the classical Richards equation for unsaturated conditions in porous media flows [9], which reads:

$$\phi \rho_f \frac{\partial S}{\partial p_c} \frac{\partial p_c}{\partial t} + \nabla \cdot \left(\rho_f \frac{k_{\text{rel}} \mathbf{k}}{\mu_f} (\nabla p_f - \rho_f \mathbf{g}) \right) = Q_f \quad , \tag{4.5}$$

where ϕ is the porosity, ρ_f the fluid density, S the fluid saturation, p_c the capillary pressure, t the time, k_{rel} the relative permeability, $\mathbf{k}$ the intrinsic permeability, μ_f the dynamic viscosity of the fluid, $p_f = -p_c$ the fluid pressure, $\mathbf{g}$ the vector of gravitational acceleration, and Q_f a source term. Note that this approximation neglects the change of gas pressure. Equation (4.5) is parametrized using the classical van Genuchten approach [12] for the capillary pressure

$$p_c = \max \left(p_d \left(S_{\text{eff}}^{-1/m} - 1 \right)^{1-m} ; p_{c,max} \right) \tag{4.6}$$

and the relative permeability

$$k_{\text{rel}} = S_{\text{eff}}^{1/2} \left[1 - \left(1 - S_{\text{eff}}^{1-\beta} \right)^{\beta} \right]^2 \quad , \tag{4.7}$$

respectively, where p_d is the air entry pressure, m and β are fitting parameters reflecting the pore size distribution, $p_{c,max}$ is the maximum capillary pressure and

$$S_{\text{eff}} = \frac{S - S_r}{S_{max} - S_r} \tag{4.8}$$

is the effective saturation. Here, S_r and S_{max} are the residual and maximum fluid saturation, respectively.

Parameters were taken from the simulations carried out by [11]:

Fluid density	$\rho_f = 1000\,\text{kg/m}^3$
Dynamic viscosity	$\mu_f = 1 \times 10^{-3}\,\text{Pa s}$
Porosity	$\phi = 0.16$
Air entry pressure	$p_d = 2 \times 10^7\,\text{Pa}$
Maximum capillary pressure	$p_{c,max} = 1 \times 10^9\,\text{Pa}$
Parameter for capillary pressure	$m = 0.41176$
Parameter for relative permeability	$\beta = 0.5$
Residual fluid saturation	$S_r = 0.0$
Maximum fluid saturation	$S_{\text{max}} = 1.0$
Gravitational acceleration	$\mathbf{g} = (0, 0)^T$
Source term	$Q_f = 0.0\,\text{kg/(m}^3\text{s)}$

The parameters ϕ, p_d, $p_{c,max}$ and m entering (4.6) and the values for β, S_r and S_{max} in (4.7) were reported for Opalinus Clay shale [13, 14], both of which were experimentally determined at the Mont Terri RL.

Consolidated clayrock typically has a bedding owing to the plate like shape of individual primary clay particles. As a result, clay rock becomes transverse isotropic. For the RF-process, this yields different permeabilities values perpendicular ($k_\perp$) and parallel ($k_\parallel$) to the bedding plane. In the present study, values were taken from the experimental report of [15]. For the EDZ, the permeability is assumed to be one order of magnitude higher than in the undisturbed rock:

Undisturbed rock (ID1)	$k_{1,\parallel} = 6.810^{-20}\,\text{m}^2$
EDZ (ID2)	$k_{1,\perp} = 1.36 \times 10^{-20}\,\text{m}^2$
	$k_{2,\parallel} = 6.8 \times 10^{-19}\,\text{m}^2$
	$k_{2,\perp} = 1.36 \times 10^{-19}\,\text{m}^2$

As mentioned above, the niche is located in the anticline of the Mont Terri RL. Hence, the transverse isotropic permeability needs to be transformed into the laboratory frame that corresponds to the bedding plane of the clay rock using

$$\mathbf{R}\,\mathbf{k}\,\mathbf{R}^T = \begin{pmatrix} \cos\gamma & -\sin\gamma \\ \sin\gamma & \cos\gamma \end{pmatrix} \begin{pmatrix} k_\perp & 0 \\ 0 & k_\parallel \end{pmatrix} \begin{pmatrix} \cos\gamma & \sin\gamma \\ -\sin\gamma & \cos\gamma \end{pmatrix} \quad , \tag{4.9}$$

where $\gamma = -32.96°$ is the angle of inclination. This transform yields a local coordinate system that forms an orthogonal basis with the set of vectors $e_1 = (0.8391, -0.5440)^T$ and $e_2 = (0.5440, 0.8391)^T$.

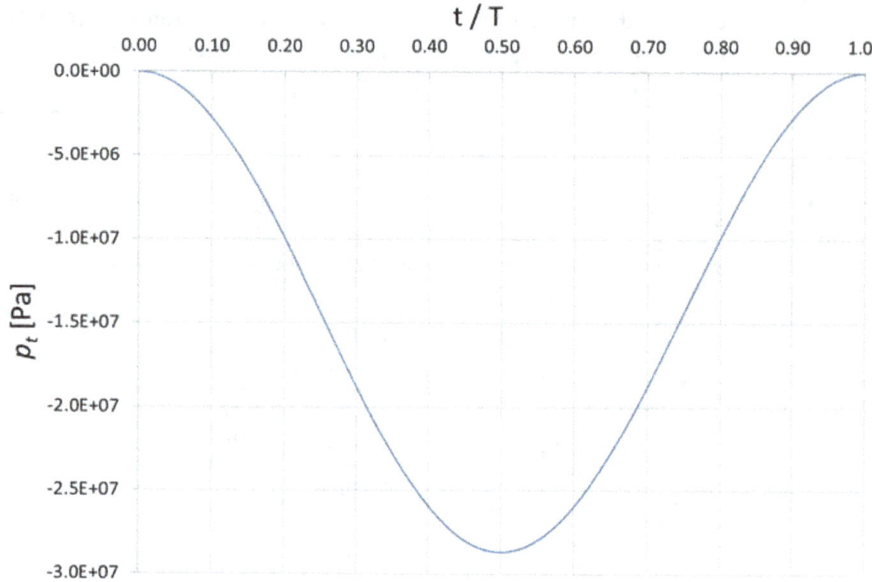

Fig. 4.34 Temporal evolution of the dynamic boundary condition at the wall of the niche for one seasonal period T. The total simulation time was $T_{\text{tot}} = 20T$

The primary variable entering the Richards equation as an unknown is the fluid pressure in the porous medium. This quantity is directly related to air humidity via the Kelvin equation [16]. The initial and boundary conditions for the simulation were set as follows:

Initial condition	$p_0 = -900$ Pa
Boundary condition	$p_\infty = -900$ Pa
	$p_t = \frac{1}{2} p_{\min} \left[\sin\left(2\pi\left(\frac{t}{T} + \frac{\phi_0}{360}\right)\right) - 1\right]$

where p_∞ is the pressure at the outer boundary for the undisturbed rock and p_t is the seasonal boundary condition at the walls of the niche that reflects the variation of humidity inside the niche (Fig. 4.34). Further, $p_{\min}$ is the minimum pressure reached for the saturation deficit at dry air conditions, $T = 365$ d is the duration of one seasonal period and $\phi_0 = 270°$ is the phase shift angle to start the simulations with a 'wet' boundary at the tunnel walls. For the present simulations, $p_{\min}$ was set to 2.87×10^7 Pa, which corresponds to $S_{\text{eff}} = 0.65$. The total simulation time is $T_{\text{tot}} = 20T$. To properly resolve the unsteady boundary condition p_t in time, a temporal discretization of $\Delta t = 21400$ s was used unless specified otherwise.

The simulation is set up as a two-dimensional problem. The total extent of the domain is 50 by 50 m and the domain is discretized by a total of 5463 nodes with elements of triangular shape (Fig. 4.34b) unless specified otherwise. This spatial discretization is the exact same mesh as used by [11]. A summary of the boundary conditions for the computational domain and the material parameters is given in

Table 4.5 Numerical parameters of the solvers used in OGS-5 and OGS-6, where $\mathbf{r}$ and $\mathbf{b}$ are the residual and the right-hand side vector, respectively, and ϵ the error tolerance. Further the abbreviations for the solvers and preconditioners mean Conjugate Gradient on the Normal Equations (CGNR), Biconjugate gradient Stabilized Method (BiCGSTAB), and Incomplete LU Factorization Technique (ILUT), respectively

Nonlinear solver			Linear solver		
Parameter	OGS-5	OGS-6	Parameter	OGS-5	OGS-6
Solver	Picard	Picard	Solver	CGNR	BiCGSTAB
Error	L2-norm	L2-norm	Error	$\|\mathbf{r}\| < \epsilon\|\mathbf{b}\|$	$\|\mathbf{r}\| < \epsilon\|\mathbf{b}\|$
Error tol. ϵ	1E–5	1E–12	Precon.	Jacobi	ILUT
Max. iter.	50	40	Error tol. ϵ	1E–12	1E–10
Relaxation	0.0	n/a	Max. iter.	20000	3000

Fig. 4.33a. The seasonal change in pressure boundary conditions at the surface of the niche is shown as a function of time in Fig. 4.34. The numerical parameters for the nonlinear and linear solver are listed in Table 4.5.

4.7.3.2 Well-Developed Stage

This well-developed stage is reached after 15 cycles of alternating wet and dry boundary conditions, i.e. 15 years of simulation time. The spatial distribution of p_f and S_{eff} is illustrated in Fig. 4.35 for the results generated by OGS-5. Figure 4.35a, b show the spatial distribution of the fluid pressure at a dry and a wet phase, respectively. Owing to the variations in pressure p_t at the niche, i.e. relative air humidity inside the gallery, the rock undergoes cyclic saturation and desaturation. As expected, this affects the saturation at the walls of the niche and in the vicinity of the walls. Farther away from the boundary, a stable pattern develops for p_f and S_{eff} that reflects the transverse isotropic permeabilities prescribed by (4.9). Even though the effect of desiccation becomes obvious in concentric ellipses surrounding the niche, the temporal evolution of the boundary condition at the wall of the niche is visible within the EDZ only. At this well-developed stage, the rest of the domain remains unaffected from these changes in p_t. The same can be observed for the effective saturation in Fig. 4.35c, d. Due to the nonlinear dependency of S_{eff} and p_f in (4.6), the area that is substantially influenced by the seasonal variation of p_t is even smaller in Fig. 4.35c, d compared to Fig. 4.35a, b.

4.7.3.3 Comparison of OGS-5 and OGS-6

A more detailed insight into the evolution of the the desaturated rock surrounding the niche is revealed by plotting p_f and S_{eff} probed over time at four different points around the niche. This analysis is shown in Fig. 4.36 together with the locations

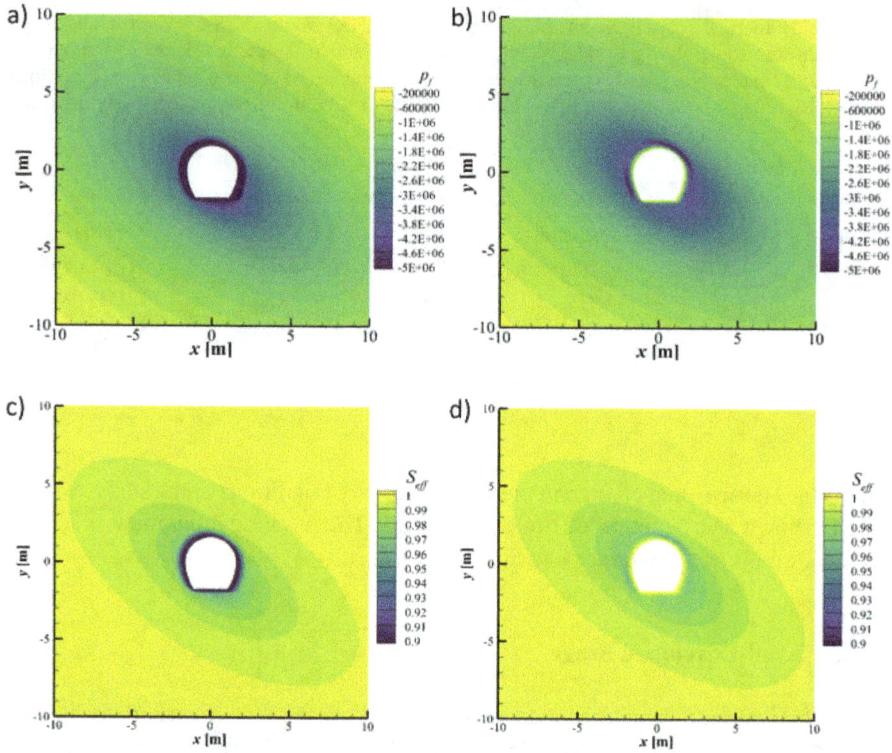

Fig. 4.35 Well-developed pattern of seasonally varying quantities fluid pressure (in Pa) and effective saturation. **a** fluid pressure at t = 19.5a, and **b** fluid pressure at t = 20a, **c** effective saturation at t = 19.5a and **d** effective saturation at t = 20a. The figures show a zoom into Fig. 4.34

of the sampling points as an inset. The seasonal variations are clearly visible, and the plot reaches a quasi-steady state after 15 years of simulation time. Hence, we conclude that the simulation ran long enough to reach a quasi-steady state for the zone surrounding the niche. Note, however, that these data were collected at locations 3 m away from the niche center in both, horizontal and vertical direction. For probing locations farther away from the niche, it takes longer to reach a steady state. This state, however, will be closer to fully saturated conditions the farther one moves away from the niche into the undisturbed rock.

It is now interesting to compare the results from both simulation codes, OGS-5 and OGS-6, to verify whether or not the results have changed after the new implementation of (4.5) in OGS-6. Hence, we show a direct comparison of the simulation results generated by OGS-5 and OGS-6 in the same figure (Fig. 4.36). As desired, all data curves collapse for OGS-5 and OGS-6 at the four different locations for the time simulated.

For a more quantitative comparison, we compute the relative deviation

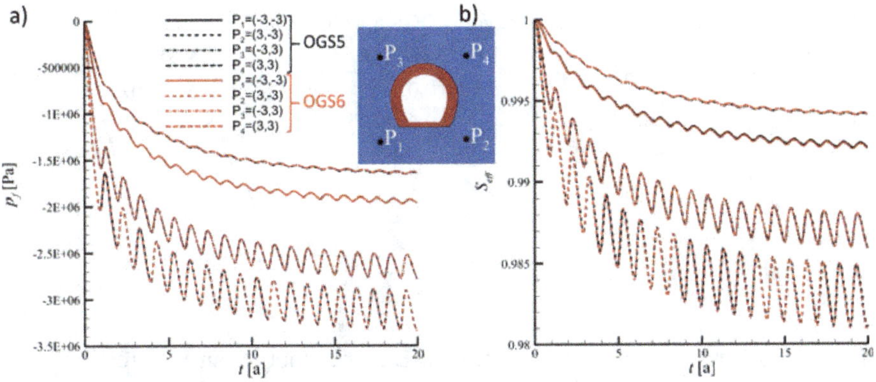

Fig. 4.36 Physical quantities probed over time at four different locations surrounding the niche for OGS-5 (black) and OGS-6 (red): **a** Fluid pressure and **b** effective saturation. The inset shows the location of the sampling points relative to the niche. The results collapse for OGS-5 and OGS-6

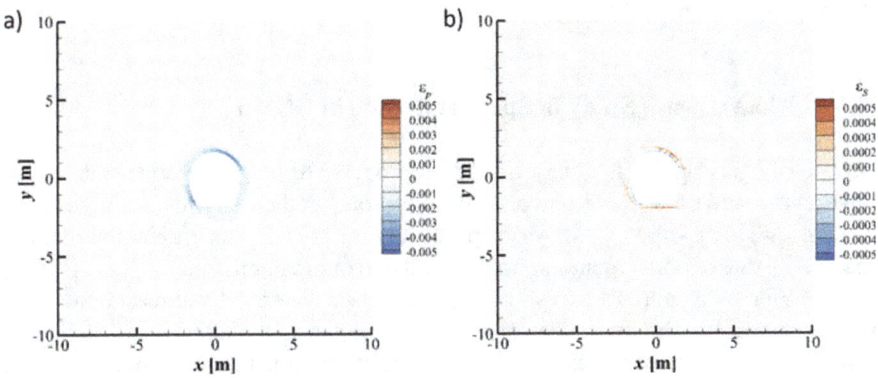

Fig. 4.37 Deviation between OGS-5 and OGS-6 after 20 years of simulation time. **a** Fluid pressure and **b** effective saturation

$$\epsilon_\theta = \frac{\theta^{OGS5} - \theta^{OGS6}}{\theta^{OGS5}} \tag{4.10}$$

where θ is a quantity of interest, e.g. p_f or S_{eff}. The two-dimensional plot of the error is shown in Fig. 4.37 for both of these quantities at time $t = 20$a. Far away from the wall of the niche, the error remains small (below 0.1%). However, we obtain that OGS-6 slightly underestimates the pressure in the EDZ compared to OGS-5 (Fig. 4.37a). This is most pronounced at the edges that are not pointing in the main direction of the anisotropy, i.e. at the top right and bottom left edges of the wall. Similarly, we see deviations for the saturation, but here, S_{eff} is underestimated at the wall and at the transition from the EDZ to the undisturbed rock, whereas it is overestimated in OGS-6 compared to OGS-5 within the rest of the EDZ (Fig. 4.37b).

Nevertheless, the maximum and minimum deviations as well as the RMSE

Table 4.6 Deviations observed when comparing OGS-5 with OGS-6 for the RF model

Quantity	min (ϵ)	max (ϵ)	RMSE
p_f	–0.0054	0.0	5.52×10^{-5}
S_{eff}	–0.0012	0.0078	6.77×10^{-4}

$$\langle|\overline{\epsilon}_\theta|\rangle = \text{RMSE} = \frac{1}{N_{\text{nodes}}}\sum_{i=1}^{N_{\text{nodes}}}\sqrt{\epsilon_\theta^2} \tag{4.11}$$

remain below 0.1% (Table 4.6). Here, RMSE is the Root Mean Square Error, i is the index of the nodes, and N_{nodes} is the total number of nodes of the computational mesh. These small deviations were deemed to be acceptable for the present scenario given that the simulation had run more than twenty cycles of desiccation/saturation. This is especially true considering the different convergence criteria used for the two different codes (Table 4.5).

4.7.3.4 Convergence Study in Space and Time in OGS-6

The rate of convergence is a very important property to judge the consistency of a numerical procedure. A consistent procedure approaches the analytical solution of the problem with increasing resolution in space and time [17]. For an efficient numerical scheme, the rate of convergence should be equal to or greater than unity, i.e. dividing the resolution by a factor of two should also decrease the error by at least a factor of two with respect to the reference solution. By comparing the numerical solution to some reference solution, one can also infer information about the minimum resolution required to obtain acceptable results. While the rate of convergence is often times determined under ideal conditions, i.e. on a regular Cartesian grid for an idealized problem, we apply the procedure to the present case of a niche surrounded by rock of two material groups, anisotropic permeability and cyclic boundary conditions for the fluid pressure at the walls of the niche to analyze a more realistic scenario. This is done in two consecutive steps: first we present order of convergence in time and then in space.

Convergence Study Time

To analyze the consistency in time, we use the exact same computational setup described in Sect. 4.7.3.1 above. While the results generated with this setup and presented in Sects. 4.7.3.2 and 4.7.3.3 were computed with $\Delta t = 21{,}400$ s, we now vary the time integration interval within the range $1350\,\text{s} \leq \Delta t \leq 345{,}600\,\text{s}$. We ran a total of nine simulations doubling Δt for each simulation. This yields an integration time interval ranging between 24 min and 4 d. The run with the smallest time integration interval, i.e. $\Delta t = 1350$ s, was chosen as a reference solution, since there is no analytical solution available for the current setup.

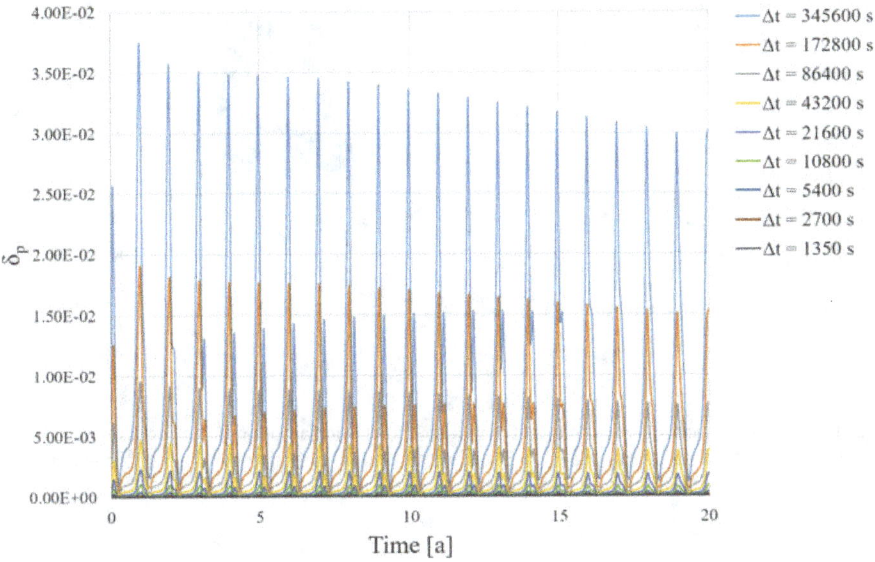

Fig. 4.38 Deviation between the simulations using different time discretizations Δt at sampling point $P_1 = (2.0, 0.0)$ (cf. inset in Fig. 4.39)

As a next step, we define sampling points, for which we probe data for fluid pressure and effective saturation over time. Since Fig. 4.37 suggested that deviations between OGS-5 and OGS-6 were largest at the transition zone from the undisturbed rock (material ID 1) to the EDZ (material ID 2), we chose four different locations inside the EDZ surrounding the niche at this critical distance, which is 2 m away from the domain center for P_1, P_2 and P_3 and 1.9 m for P_4, which is directly located underneath the niche. The coordinates are sketched in the inset in Fig. 4.39 and listed in Table 4.7. As a next step, we define the characteristic error

$$\delta_\theta(\Delta t) = \left| \frac{\theta(\Delta t) - \theta(\Delta t_{\text{ref}})}{\theta(\Delta t_{\text{ref}})} \right| \quad , \tag{4.12}$$

where again θ represents fluid pressure and effective saturation, respectively, and $\Delta t_{\text{ref}} = 1350$ s is the time integration interval of the reference run.

An example of the temporal evolution of δ_p for $P_1 = (2.0, 0.0)$ is given in Fig. 4.38. Similar to the observations reported for Fig. 4.34, there is a strong cyclic behavior of the error that depends on the evolution of p_t. A peak in p_t corresponds with a maximum deviation of the simulation result from the reference solution. This behavior was expected as a coarser discretization tends to smoothen out maximum values of a continuous function. For a given Δt, the deviations slightly decrease over time and decreasing Δt reduces the deviations from the reference solution substantially for all values of Δt investigated. Similar results were obtained for the other three sampling points.

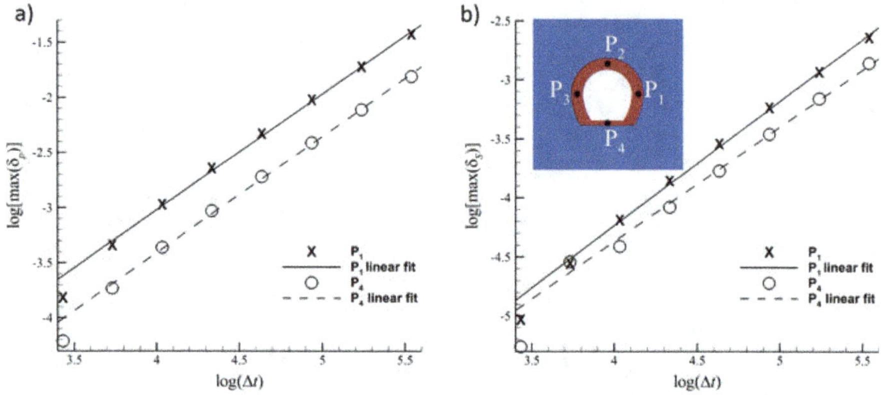

Fig. 4.39 Order of convergence in time for two characteristic sampling points on a log-log scale. The error increases with increasing Δt. **a** Fluid pressure and **b** effective saturation. The inset shows the locations of the analyzed sampling points

To quantify the deviations from the reference solution, we plot the maximum deviation max (δ_θ) against Δt in a log-log plot in Fig. 4.39. As already observed in Fig. 4.38, all data points follow the same trend of decreasing error with decreasing time step size. Note that we have only plotted results for P_1 and P_4, because the data from P_2 and P_3 collapsed with the data from P_1. The offset of P_4 from the rest of the sampling points is probably due to the slight difference in distance from the center of the domain and the flat bottom wall of the niche.

Finally, we determine the order of convergence by computing the linear fit of $y = ax + b$ for the log-log plot in Fig. 4.39. In fact, the linear fit is a very good approximation of the data presented in that figure, so that the fitting parameter a, which is the slope of the linear regression, serves very well to determine the order of convergence. Note that we have excluded the first data point (with the lowest value of Δt) from the fit, because it is slightly offset due to the fact that the reference case using Δt_{ref} does not represent an analytical solution as would have been desirable for a mathematically rigorous convergence analysis. The values of a and b for the fluid pressure and the effective saturation are summarized in Table 4.7 for all four sampling points investigated. For both physical quantities, the slope turns out to be 1.05 on average, which proves that the current time integration scheme of the "Richards Flow" model in OGS-6 is an efficient implementation of the *Backward Euler* scheme; a scheme that is known to be of first order [17].

Convergence Study Space

Similar to the convergence study in time described above, we use the computational setup described in Sect. 4.7.3.1 to repeat the simulations on different grids with different resolution. This time, the time integration interval was kept constant at $\Delta t = 21{,}400$ s, but the number of grid nodes N_{nodes} was varied within the range $3613 \leq N_{nodes} \leq 60{,}845$. Note that we switched from a triangulated domain to a

Table 4.7 Order of convergence in time extracted from the linear fit $y = ax + b$ as given in the log-log plot of Fig. 4.39 at the different sampling points shown in the same figure

	Pressure		Saturation	
Sampling point	a	b	a	b
$P_1 = (2.0, 0.0)$	1.051	–7.222	1.052	–8.441
$P_2 = (0.0, 2.0)$	1.050	–7.217	1.107	–8.715
$P_3 = (-2.0, 0.0)$	1.046	–7.192	1.054	–8.431
$P_4 = (0.0, -1.9)$	1.054	–7.622	0.968	–8.242

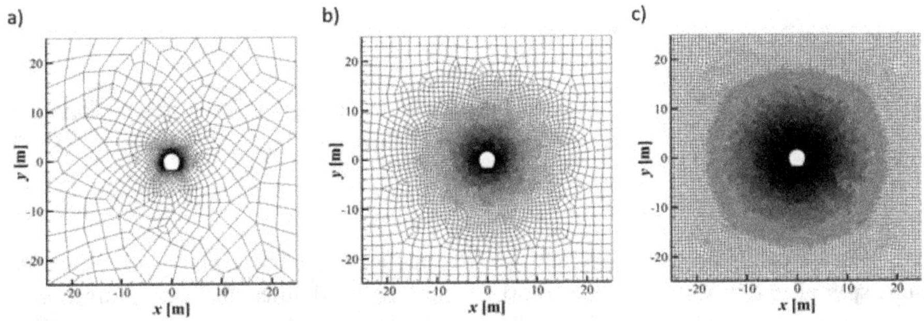

Fig. 4.40 Examples of the meshes used to conduct the grid convergence. **a** $N_{\text{nodes}} = 3613$, which was the coarsest grid employed, **b** $N_{\text{nodes}} = 11{,}047$, **c** $N_{\text{nodes}} = 60{,}845$, which was the finest grid employed

tetrahedral meshing due to its more regular behavior for grid refinements at the walls of the niche. We conducted a total of five simulations, where the simulation with $N_{\text{nodes}} = 3614$ used the coarsest grid (Fig. 4.40a), $N_{\text{nodes}} = 6500$ is comparable to the triangulated grid with 5463 nodes as shown in Fig. 4.33b, while we refined the grid to $N_{\text{nodes}} = 11{,}045$ (Fig. 4.40b) and $N_{\text{nodes}} = 31{,}582$, respectively. Again, the finest resolution with $N_{\text{nodes}} = N_{\text{ref}} = 60{,}845$ (Fig. 4.40c) serves as the reference run, since there is no analytical solution available.

It is important to note that the the mesh generator Gmsh [18] does not allow for the exact control over the number of nodes generated for the complex geometry in the present analysis. Hence, we adapt the number of grid cells by refining the spatial discretization at the outer boundary of the domain from 5 m ($N_{\text{nodes}} = 3613$) to 0.7 m ($N_{\text{nodes}} = 60{,}845$). This measure did not refine the grid at the walls of the niche in the same systematic manner, because the nodes have to coincide with the points defining the polygon course of the complex niche shape. In fact, we obtained the same spatial resolution $\Delta x = 0.0375$ m for the arch of the niche for all simulations, whereas the resolution at the more regular bottom was refined from 0.177 to 0.034 m.

We compute the error as

$$\delta_\theta(N_{\text{nodes}}) = \left| \frac{\theta(N_{\text{nodes}}) - \theta(N_{\text{ref}})}{\theta(N_{\text{ref}})} \right| , \tag{4.13}$$

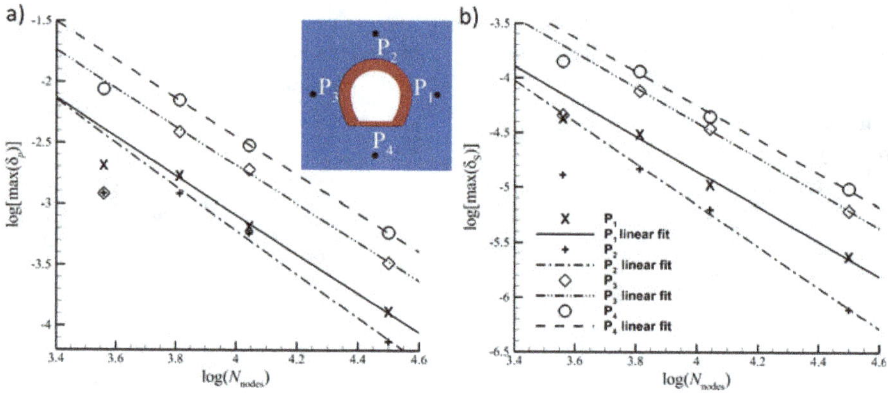

Fig. 4.41 Order of convergence in space for four characteristic sampling points on a log-log scale. The error decreases with increasing number of grid cells N_{nodes}. **a** Fluid pressure and **b** effective saturation. The inset shows the locations of the analyzed sampling points

to quantify the deviation of the results with respect to a given reference mesh, here the finest spatial resolution with $N_{\text{ref}} = 60{,}845$ nodes. We defined four sampling points that were moved farther into the undisturbed rock to a distance of 4 m away from the center of the niche (Fig. 4.41). This became necessary as the polyline defining the boundary between the two material IDs, which is about 2 m away from the center of the niche, predefines specific nodes. Sampling in this area would influence the convergence behavior of the simulations. Subsequently, we used a interpolation routine provided by the software TecPlot [19] to obtain data over time at these sampling points for all the simulations with different grid resolution. Due to the remeshing, however, the sampling points' location no longer coincide with the node locations. Hence, if the grid is too coarse, those sampling points can be closer or farther away from the next node location depending on the mesh. For these cases, the interpolation routine offered by the postprocessing software TecPlot will heavily influence the results and conclusions about the order of convergence will no longer be possible.

After probing the data at the different sampling points, we obtain results very similar to Fig. 4.38. The resulting convergence behavior with increasing grid resolution is shown in Fig. 4.41 for both fluid pressure and effective saturation. Overall, the error decreases with increasing grid resolution. This is true for both quantities recorded at all four sampling points, except for the values recorded for the coarsest mesh with $N_{\text{nodes}} = 3614$. Apparently, the interpolation routine of the postprocessing software has a big impact on the quality of the results of this run. It was therefore decided to exclude these data from the linear regression analysis and the grid with $N_{\text{nodes}} = 3613$ nodes was deemed to be too coarse to obtain meaningful results.

Nevertheless, after excluding two out of five simulations (with the coarsest being too coarse and the finest being the reference case), we are able to perform a linear regression analysis $y = ax + b$ with a very high correlation on the remaining

Table 4.8 Order of convergence in space extracted from the linear fit $y = ax + b$ as given in the log-log plot of Fig. 4.39 at the different sampling points shown in the same figure

Sampling point	Pressure		Saturation	
	a	b	a	b
$P_1 = (4.0, 0.0)$	−1.603	3.321	−1.593	1.521
$P_2 = (0.0, 4.0)$	−1.789	3.938	−1.883	2.374
$P_3 = (-4.0, 0.0)$	−1.574	3.611	−1.596	1.975
$P_4 = (0.0, -4.0)$	−1.571	3.836	−1.543	1.921

three data points for the four different sampling locations. The slope of this linear regression, which is the fitting parameter a, then yields the order of convergence. The results of this analysis is summarized in Tables 4.7 and 4.8. Note the change of sign in a as we obtain a decreasing error with an increase in the number of nodes. For both quantities recorded at all four points, the order of convergence in space averages out to be 1.64, which is well above 1. Hence, we conclude that improving the grid resolution can improve the simulation results drastically.

4.7.4 Unsaturated Single-Phase Coupled with Linear Elasticity ("Richards Mechanics", RM)

4.7.4.1 Model Description

The dynamic boundary conditions of p_t described for the simulations using the RF-model yields changes in local saturation. It is well known, however, that changes in fluid pressure will also have effects on the rock matrix of the Opalinus Clay [20]. To account for these processes, the hydraulic model described in Sect. 4.7.3.1 is extended to a coupled hydraulic-mechanical model [21]. In what follows, this model will be called "Richards Mechanics" (RM). The coupling between the fluid pressure and the local stress is based on the effective stress concept, which states that the total stress is equal to the sum of the pore pressure and the effective stress acting on the solid. This concept was initially developed for saturated porous soils [10, 22] and was later extended for unsaturated porous media flows [23].

The coupling of the fluid and the solid phase via the effective stress concept yields the balance of linear momentum:

$$\nabla \cdot \left(\boldsymbol{\sigma} - \alpha\chi(S)p_f\mathbf{I}\right) = 0 \quad , \tag{4.14}$$

where $\boldsymbol{\sigma}$ is the effective stress tensor, α is the Biot coefficient, $\chi(S) = S$ is a simple model for the Bishop's coefficient and $\mathbf{I}$ is the identity matrix. The mass balance for the fluid in a deformable porous medium, hence, becomes:

$$\phi \frac{\partial S}{\partial t} + S \left(\frac{\phi}{K_f} + \frac{\alpha - \phi}{K_s} \right) \frac{\partial p_f}{\partial t} + \nabla \cdot \mathbf{J}_f + S\alpha \nabla \cdot \frac{\partial \mathbf{u}}{\partial t} = 0 \tag{4.15}$$

where K_f and K_s are the bulk moduli for the fluid and the solid phase, respectively, and $\mathbf{u}$ is the displacement vector of the solid matrix. Furthermore

$$\mathbf{J}_f = \frac{k_{\text{rel}} \mathbf{k}}{\mu_f} (\nabla p_f - \rho_f \mathbf{g}) \tag{4.16}$$

is the fluid mass flux that already appeared in the RF-model (4.5). The constitutive relation for the effective stress tensor is the generalized Hooke's law

$$\boldsymbol{\sigma} = \mathbb{C} : \boldsymbol{\epsilon} \quad , \tag{4.17}$$

where we use the Voigt notation to write $\mathbb{C}$ the elasticity tensor, and $\boldsymbol{\epsilon} = \frac{1}{2}\left(\nabla \mathbf{u} + (\nabla \mathbf{u})^T\right)$ is the strain tensor.

4.7.4.2 Comparison of the Richards Equation Implementation for the Two Model Approaches "Richards Flow" and "Richards Mechanics" in OGS-6

The comparison made in Sect. 4.7.3.3 above is based on the process "Richards Flow" (RF). Since OGS-6 employs a monolithic scheme to solve coupled systems, this analysis does not allow for the conclusion that implementation of the Richards equation (4.5) is verified for all other coupled models as well. Hence, we ran the same simulation using "Richards Mechanics" (RM) to validate the implementation of the Richards equation in OGS-6 for this model. To focus on the Richards equation only, we turn off the hydraulically induced mechanical deformation by choosing a Biot coefficient of $\alpha = 0.0$ and set deformations at all outer boundaries to zero for all directions. Furthermore, we drive the storage term, which is the second term in (4.14) to zero by setting the bulk moduli to $K_f = K_s = 1 \times 10^{100}$ Pa.

The results of this comparison are shown in Figs. 4.42 and 4.43 as well as in Table 4.9. For two sampling locations, the evolution of pressure and saturation show very good agreement. Note that the sampling locations for the data shown in Fig. 4.42 were taken inside the EDZ, since Fig. 4.43 suggests that this is the area, where the strongest deviations can be expected. Indeed, while the cyclic saturation/desaturation is reproduced very well, there are differences in the maxima and minima between the two models for both quantities, p_f and S_{eff}.

Similar to the comparison made in Sect. 4.7.3.3, we define the simulation results reproduced by the RF model as the reference to compute the error

$$\epsilon_\theta = \frac{\theta^{RF} - \theta^{RM}}{\theta^{RF}} \tag{4.18}$$

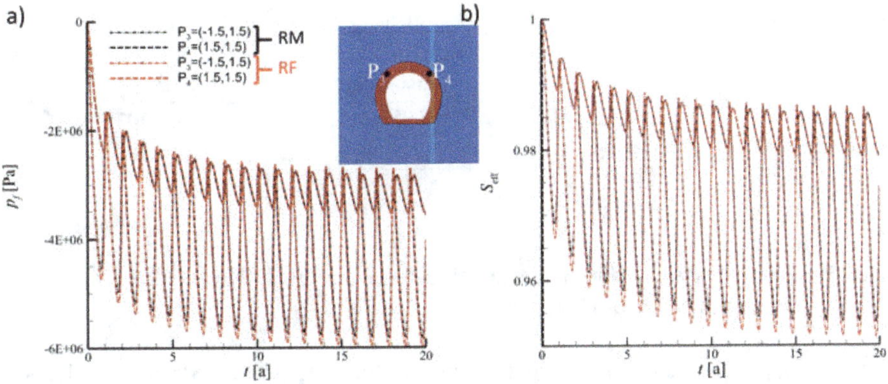

Fig. 4.42 Physical quantities probed over time at two characteristic locations for the model approaches 'Richards Flow' (RF, black) and 'Richards Mechanics' (RM, red). **a** Fluid pressure and **b** effective saturation. The inset shows the location of the sampling points relative to the niche

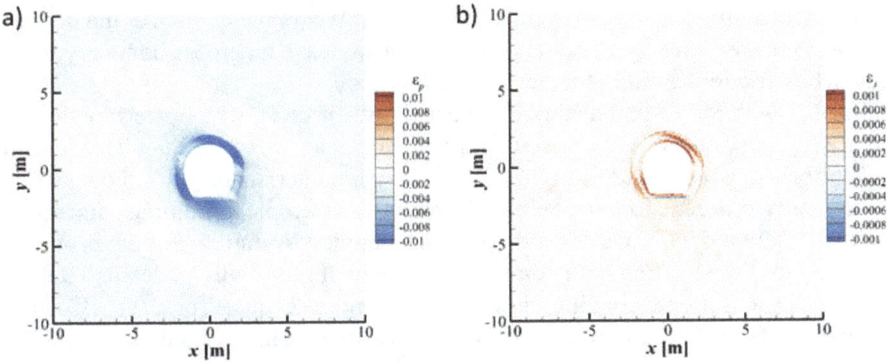

Fig. 4.43 Deviation between 'Richards Flow' and 'Richards Mechanics' after 20 years of simulation time. **a** Fluid pressure and **b** effective saturation

as a quantitative estimate of the differences in the results of the two models. Looking at the two-dimensional deviations (Fig. 4.43), it becomes immediately obvious that pressure is slightly underestimated by the RM model in comparison with the RF model within the EDZ (which yields that saturation is overestimated). The differences between the results computed by the algorithms of the two models is one order of magnitude larger compared to the analysis performed in Sect. 4.7.3.3. Especially the pressure shows higher deviations (Fig. 4.43a) at the transition between the EDZ and the undisturbed rock and at the walls of the niche. Nevertheless, the results computed by the two models are overall very similar. The RMSE of the simulation results remains far below 1% (Table 4.9). Hence, the implementation of the Richards equation into the model "Richards Mechanics" can be accepted as trustworthy for the present scenario.

Table 4.9 Deviations observed when comparing 'Richards Flow' with 'Richards Mechanics' in OGS-6

Quantity	Minimum	Maximum	RMSE
p_f	–0.0505	0.0055	8.56×10^{-4}
S_{eff}	–0.0405	0.0112	2.66×10^{-3}

4.7.4.3 Comparison of OGS-5 and OGS-6 for the Full "Richards Mechanics"-Model

For the analysis presented in this section, we use the exact same hydraulic model and the computational setup as described in Sect. 4.7.3.1. Furthermore, we parameterize the mechanical model, which was summarized in Sect. 4.7.4.1 by choosing $\alpha = 1.0$, and keep $K_f = K_s = 1 \times 10^{100}$ Pa to drive the storage term to zero. We define $\mathbf{u}_0 = 0$ m, which is zero displacements for the initial conditions, and use roller boundary conditions at $x = -25$ m and $x = 25$ m. We fix the domain at the bottom ($y = -25$ m) by defining $u_y|_{y=-25\text{m}} = 0$ m, but we leave the upper boundary to be completely frictionless in both, the x and y direction.

Finally, we need to parameterize the elasticity tensor $\mathbb{C}$, which is a function of the Young's modulus E and Poissons' ratio ν, where, again, we distinguish between undisturbed rock (E_1) and the EDZ (E_2). The parameterization of $\mathbb{C}$, however, is conceptually different for isotropic and transverse isotropic conditions, since one can use different definitions depending on the simplicity/symmetry of the problem. Since we are investigating both conditions, these definitions will be detailed below. The transverse isotropic condition is known to be more adequate for clayrock owing to the bedding of this particular rock and it was also used in the study of [11].

Isotropic Elasticity

As mentioned above, we use the Voigt notation to write the three-dimensional elasticity tensor. For isotropic conditions, $\mathbb{C}$ simplifies from the general form to

$$\mathbb{C} = \frac{E}{(1+\nu)(1-2\nu)} \begin{pmatrix} 1-\nu & \nu & \nu & 0 & 0 & 0 \\ \nu & 1-\nu & \nu & 0 & 0 & 0 \\ \nu & \nu & 1-\nu & 0 & 0 & 0 \\ 0 & 0 & 0 & \frac{1-2\nu}{2} & 0 & 0 \\ 0 & 0 & 0 & 0 & \frac{1-2\nu}{2} & 0 \\ 0 & 0 & 0 & 0 & 0 & \frac{1-2\nu}{2} \end{pmatrix} \tag{4.19}$$

Hence, $\mathbb{C}$ is fully parameterized with $\nu = 0.18$ for both material groups, $E_1 = 3.6 \times 10^9$ Pa for the undisturbed rock and $E_2 = 1.8 \times 10^9$ Pa for the EDZ.

While the simulation results of the RM-model yield very similar spatial distributions of p_f and S_{eff} that were reported in Fig. 4.35, we know obtain values for the deformation vector $\mathbf{u}$. The results for the two components after a simulation time of 20 years is shown in Fig. 4.44. The overall impact of the saturation/desaturation on the

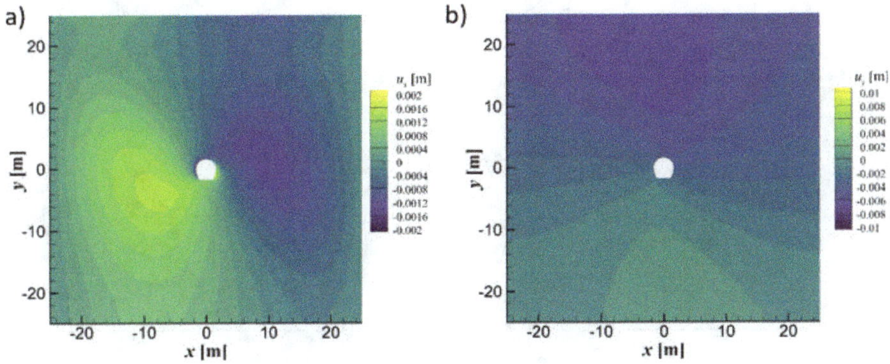

Fig. 4.44 Spatial distribution of the rock deformation **u** for isotropic conditions after 20 years of simulation time. **a** horizontal deformation u_x and **b** vertical deformation u_y

mechanical behavior is such that the tunnel is compressed in its height, whereas the lateral distance between the walls increases. This deformation leads to a compression in the undisturbed rock farther away from the niche.

To compare the simulation results of the two FE codes OGS-5 and OGS-6, we probe the generated data of p_f, S_{eff}, u_x and u_y over time at the sampling points illustrated in the inset of Fig. 4.36. Unlike the very good agreement obtained in that figure, we now find significant deviations between the simulation results of OGS-5 and OGS-6. This is true for all physical quantities at all sampling points. We obtain a systematically higher pressure and, hence, a higher saturation for OGS-6. This difference also causes different mechanical deformations of the rock, but these quantities do not differ as systematic as the pressure does. The difference can be explained by the different coupling schemes of the two codes. While OGS-5 uses a staggered scheme with a weak coupling of the two processes, OGS-6 provides a monolithic scheme that allows for a strong coupling of the hydraulic and mechanical processes by solving for p_f and **u** simultaneously. The strongly coupled monolithic scheme is known to yield better results than the weakly coupled staggered scheme [24, 25]. Hence, the simulation results generated by OGS-6 potentially provide more realistic results than OGS-5 does. However, a detailed comparison to analytic reference solutions or experimental benchmark data will be needed to verify this hypothesis.

Despite the differences of the results reported in Fig. 4.45, both simulations yield qualitatively the same results, albeit with different magnitudes in deformation. This becomes evident in the spatial distribution of the error (4.10), which is shown in Fig. 4.46. Note that values of $\epsilon_{u,x}$ and $\epsilon_{u,y}$ were normalized by the width of the niche, which is 3.2 m, rather than ϵ^{OGS5}, as normalizing by the small deformation values is prone to indicate unreasonably large errors. Even though there is a direct dependency of effective saturation on fluid pressure, the spatial distributions of the relative error for these two quantities look very differently (Fig. 4.46a, b). The rather large error for fluid pressure especially far away from the niche can be attributed to low values of p_f that are used to normalize (4.18). Closer to the niche, the error

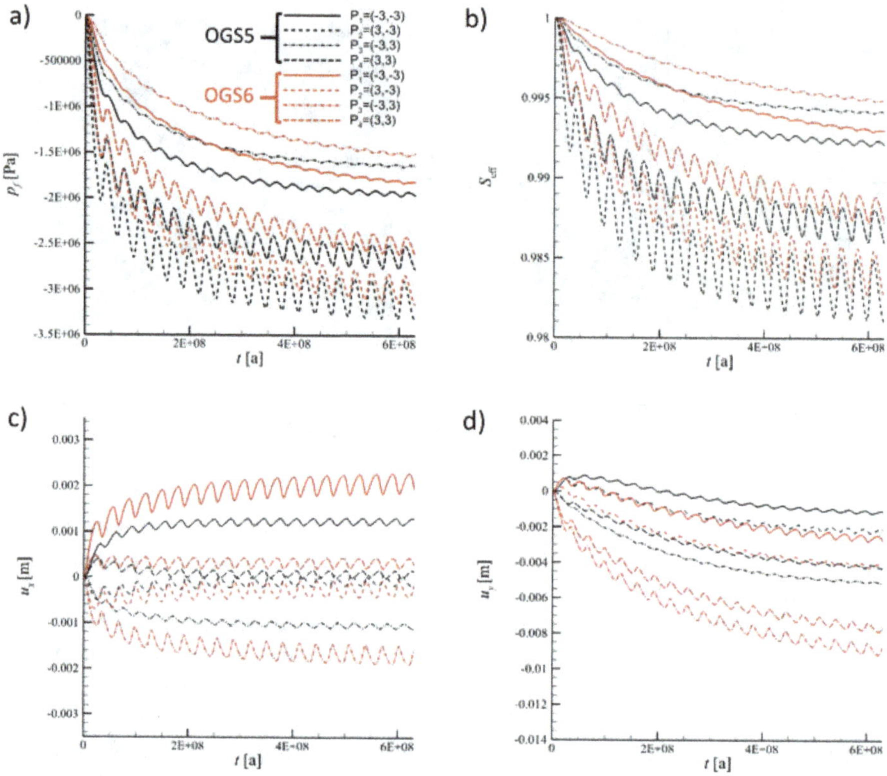

Fig. 4.45 Physical quantities computed by OGS-5 and OGS-6 for isotropic conditions probed over time at the same four locations depicted in the inset of Fig. 4.36. **a** Fluid pressure, **b** effective saturation, **c** horizontal deformation and **d** vertical deformation

decays to smaller values. Looking at the spatial distribution of the error for all other quantities, however, we can conclude that even though the coupling scheme of OGS-6 is more powerful than the scheme of OGS-5, both simulations yield similar results. This fact is also reflect in the maxima, minima and RMSE (4.11). Even though the deviations in pressure are rather large, the overall change in deformation with respect to the characteristic length of the problem, i.e. the width of the niche, remains small. Nevertheless, unlike the comparison presented in Sect. 4.7.3.1, we observe substantial deviations for the simulation results from OGS-6 compared to those generated using OGS-5. Further research will be needed to clarify these differences.

Anisotropic Elasticity

We consider the transverse isotropic behavior of clayrock described in Sect. 4.7.3.1 by defining a higher stiffness parallel to the bedding E_i and a lower stiffness normal to the bedding E_a. Here, the subscripts i and a indicate directions parallel and perpendicular to the bedding plane, respectively. Similarly, the Poisson's ratio has to be defined with respect to the anisotropy direction that is normal to the bedding ν_{ia} and parallel

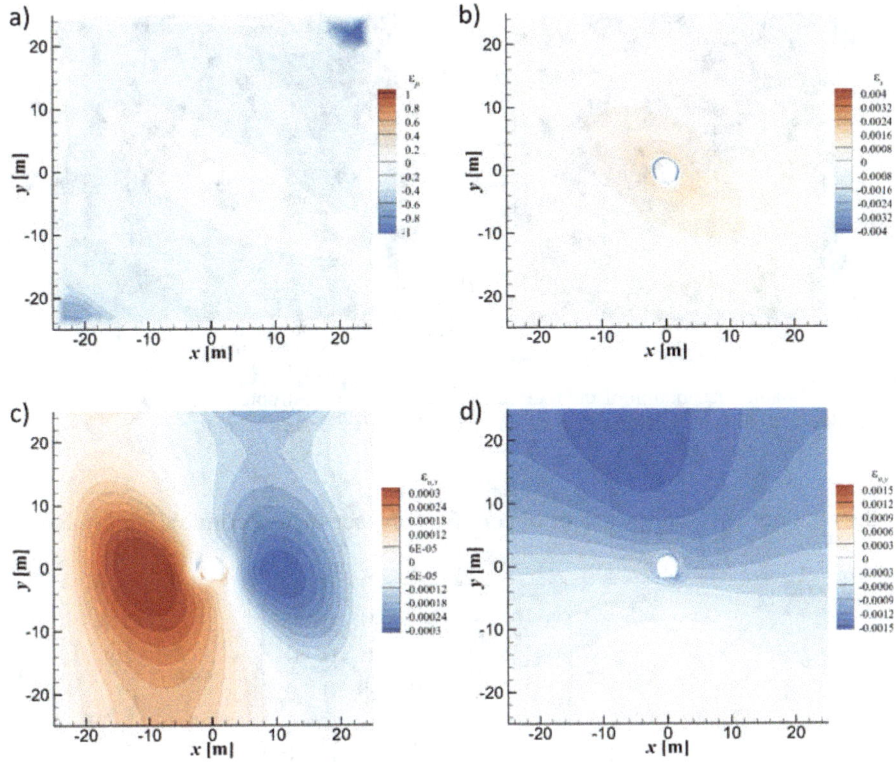

Fig. 4.46 Deviations between OGS-5 and OGS-6 for isotropic conditions after 20 years of simulation time. **a** Fluid pressure, **b** effective saturation, **c** horizontal deformation and **d** vertical deformation

Table 4.10 Deviations observed when comparing OGS-5 with OGS-6 for the RM model with isotropic conditions

Quantity	min (ϵ)	max (ϵ)	RMSE
p_f	–1.1514	0.1478	6.81×10^{-2}
S_{eff}	–0.0045	0.0017	1.12×10^{-3}
u_x	–0.0003	0.0004	8.97×10^{-5}
u_y	–0.0016	3×10^{-5}	8.44×10^{-4}

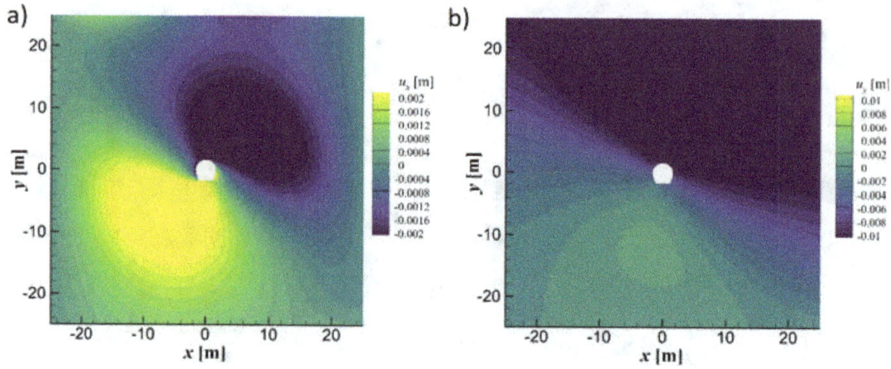

Fig. 4.47 Spatial distribution of the rock deformation **u** for anisotropic conditions after 20 years of simulation time. **a** horizontal deformation u_x and **b** vertical deformation u_y

to the bedding ν_{ii}. The inverse of the elasticity tensor then becomes (Table 4.10).

$$\mathbb{C}^{-1} = \begin{pmatrix} \frac{1}{E_i} & -\frac{\nu_{ai}}{E_a} & -\frac{\nu_{ii}}{E_i} & 0 & 0 & 0 \\ -\frac{\nu_{ia}}{E_i} & \frac{1}{E_a} & -\frac{\nu_{ia}}{E_a} & 0 & 0 & 0 \\ -\frac{\nu_{ii}}{E_i} & -\frac{\nu_{ai}}{E_a} & \frac{1}{E_i} & 0 & 0 & 0 \\ 0 & 0 & 0 & \frac{1}{G_a} & 0 & 0 \\ 0 & 0 & 0 & 0 & \frac{1}{G_i} & 0 \\ 0 & 0 & 0 & 0 & 0 & \frac{1}{G_a} \end{pmatrix} \tag{4.20}$$

where $G_i = E_i/(2 + 2\nu_{ii})$ and G_a are the shear moduli in isotropic and anisotropic direction, respectively. Note that the elasticity tensor is not symmetric. Instead, we need to rescale $\nu_{ai} = \nu_{ia}\frac{E_a}{E_i}$. Here, we have used the same values as [11], which were $E_i = 3.6 \times 10^9$ Pa, $E_a = 1.1 \times 10^9$ Pa, $\nu_{ia} = 0.16$ and $\nu_{ii} = 0.18$ and $G_a = 1.2 \times 10^9$ Pa. We rotate the elasticity tensor by the same angle of inclination $\gamma = -32.96°$ that was applied to the permeability tensor (4.9).

To illustrate the well-developed stage and to provide a comparison to the isotropic elasticity shown in Fig. 4.44, we plot the spatial distribution of the two components of the deformation in Fig. 4.47. As desired,the maximum and minimum horizontal deformations have rotated for the anisotropic case according to the angle of inclination. Other than that, the simulation results of the isotropic and anisotropic case remain very similar.

The similarity between the isotropic and the anisotropic behavior yield the same results for the pointwise sampling of physical quantities over time. Values for fluid pressure and effective saturation are systematically higher for OGS-6. However, the deviations of the deformation between OGS-5 and OGS-6 seem to be smaller for the anisotropic case. This becomes evident in the two-dimensional plot of the error (Fig. 4.49) and the RMSE-values listed in Table 4.11. Further research will be needed

Table 4.11 Deviations observed when comparing OGS-5 with OGS-6 for the RM process with anisotropic conditions

Quantity	min (ϵ)	max (ϵ)	RMSE
p_f	–1.8785	0.4051	1.27×10^{-1}
S_{eff}	–0.0120	0.0090	2.51×10^{-3}
u_x	–0.0002	0.0002	9.48×10^{-5}
u_y	–0.0004	0.0006	2.36×10^{-4}

Table 4.12 Runtime for the different simulations presented in Sects. 4.7.3 and 4.7.4. Both systems are run by the BGR and employ Intel Xeon E5-2690 processors. Cluster `B21x06` uses version two with a clock frequency of 3.0 GHz, wheras cluster `OGS02` uses version 3 with a clock frequency of 2.6 GHz

Analysis	Software	System	Runtime [h]
Section 4.7.3	OGS-5	`OGS02`	10.9
	OGS-6	`OGS02`	1.25
Section 4.7.4.2	OGS-6	`B21x06`	78.3
Section 4.7.4.3—isotropic	OGS-5	`OGS02`	81
	OGS-6	`B21x06`	91.6
Section 4.7.4.3—anisotropic	OGS-5	`OGS02`	153.5
	OGS-6	`B21x06`	113.6

to clarify whether or not these effects are significant for other processes, such as the dynamics of discontinuities.

4.7.5 Code Performance

To compare the performance of the two simulation codes, we measure the wall-clock time for the different simulations runs, which were carried out on a single core on the local `B21x06` and `OGS02` at the BGR. Both systems are equipped with Intel Xeon E5-2690 processors. The processors on the older cluster `B21x06` are the version 2 of this type of processor with a clock frequency of 3.0 GHz, whereas the newer cluster `OGS02` uses version 3 with a clock frequency of 2.6 GHz. Hence, we expect a similar performance for the two systems. For the analysis that follows, we only consider those simulations with a computational mesh of a total of 5463 nodes and 29,200 time steps for the entire simulation time.

All runtimes recorded for the simulations presented in Sects. 4.7.3 and 4.7.4 are summarized in Table 4.12. The simulation carried out with OGS-5 in Sect. 4.7.3 took 10 h and 54 min, whereas OGS-6 needed 1 h and 15 min for the same problem.

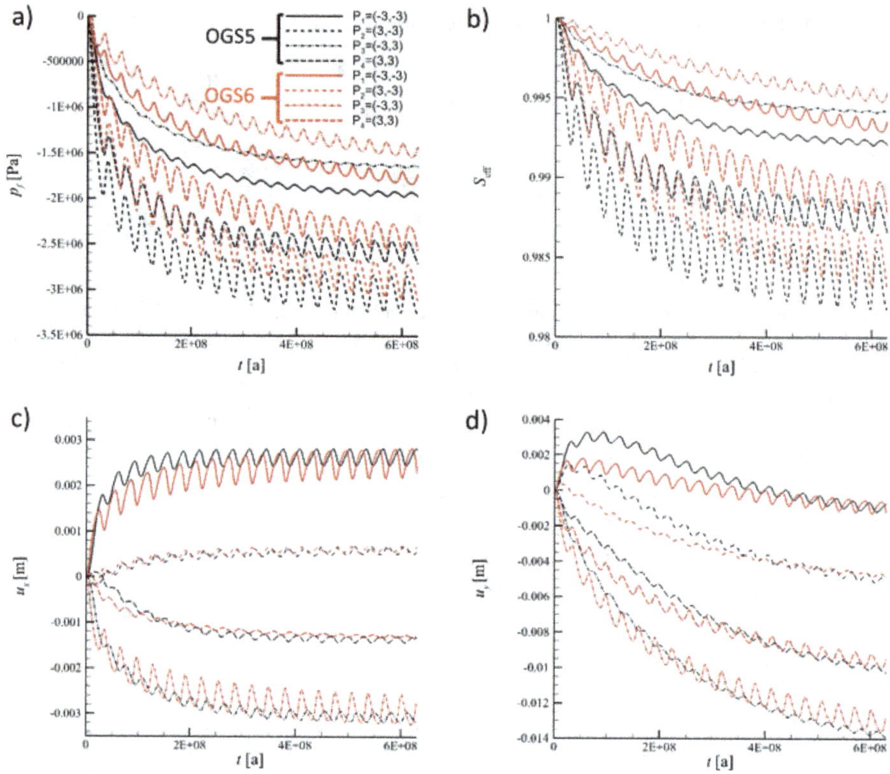

Fig. 4.48 Physical quantities computed by OGS-5 and OGS-6 for anisotropic conditions probed over time at the same four locations depicted in the inset of Fig. 4.36. **a** Fluid pressure, **b** effective saturation, **c** horizontal deformation and **d** vertical deformation

Hence, the computational time of OGS-6 excels over OGS-5 by a factor of 8.6 for the "Richards Flow" model. While the simulation using "Richards Flow" was completed after 1 h 15 min, the simulation using OGS-5 with the model "Richards Mechanics" (Sect. 4.7.4.2) took 78 h and 20 min on the local cluster `B21x06` at BGR, which is more than 3 d. This could be caused by the increased complexity of including mechanic deformation in the integration scheme (Eqs. (4.14)–(4.17)) (Fig. 4.48).

On the other hand, the computation for the setup described in Sect. 4.7.4.3 took 81 h (isotropic conditions) and 153.5 h (anisotropic conditions) on the (faster and newer) `OGS02` cluster, while OGS-6 needed 91.6 h (isotropic conditions) and 113.6 h (anisotropic conditions) on the (slower and older) cluster `B21x06` to complete the task. This shows that for the computational time becomes comparable for the codes when the full "Richards Mechanics" process is considered. Nevertheless, the fact that OGS-6 uses a stronger coupling suggests that the newly developed OGS-6 is superior to its predecessor version. A more rigorous speed test on identical hardware would be desirable to gain full insight into the enhanced capabilities of OGS-6.

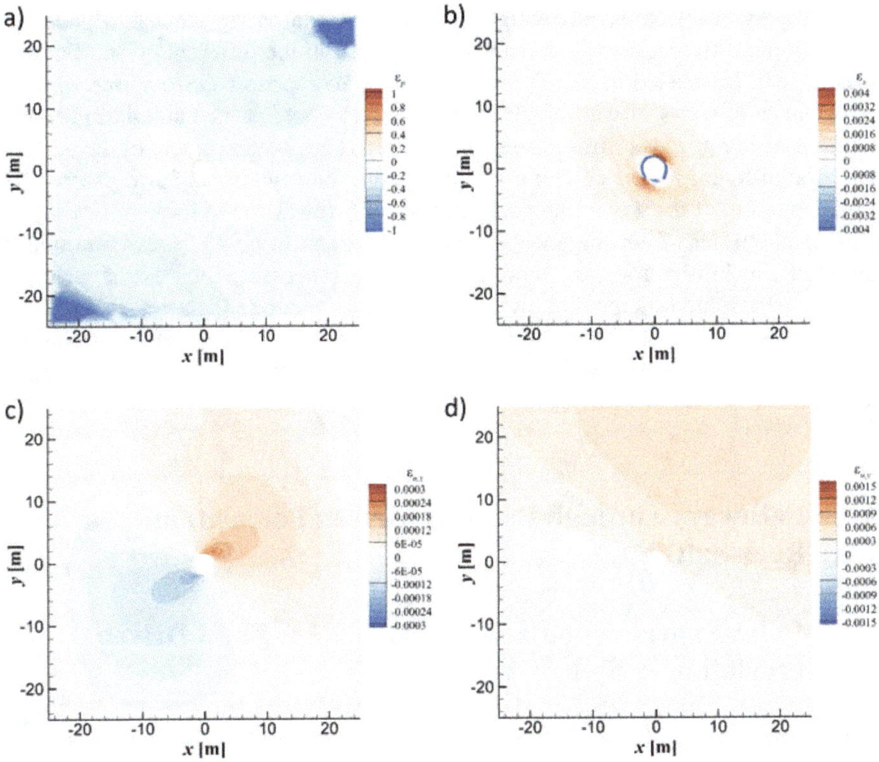

Fig. 4.49 Deviations between OGS-5 and OGS-6 for anisotropic conditions after 20 years of simulation time. **a** Fluid pressure, **b** effective saturation, **c** horizontal deformation and **d** vertical deformation

4.7.6 Conclusions

We have conducted a detailed investigation of a characteristic coupled hydro-mechanical problem of a saturating/desaturating niche in a rock laboratory to compare the accuracy of the newly developed FE-code OGS-6 to its predecessor version OGS-5. We employed characteristic properties of clayrock and apply realistic boundary conditions such as the seasonal change of air humidity at the wall of the niche and the tranverse isotropic behavior introduced by the bedding of the clayrock in a certain angle of inclination. The aim was to reproduce the results by [11], who successfully computed the hydro-mechanical behavior of a niche in the Mont Terri Rock Laboratory. We built the tests with increasing complexity starting from uncoupled hydraulic effects to transverse isotropic conditions of the fully coupled hydro-mechanical process.

As desired, we have found a high degree of agreement between the two codes for the uncoupled hydraulic process "Richards Flow". However, we found differences in the fully hydro-mechanically coupled problem "Richards Mechanics" which may be due to the different coupling schemes employed in the different codes. Further research will be needed to clarify this issue. The new code OGS-6 shows a good performance in terms of wall clock time needed to complete the simulation tasks, although more rigorous testing would be needed to get a better picture of the performance on different systems under controlled operation conditions. Furthermore, the implementation of the governing equations in OGS-6 was found to be of first order in time. The order of convergence in space was found to be 1.6. OGS-6 provides enhanced capabilities for the coupling of the two processes, such as the strongly coupled monolithic scheme that, in general, is more accurate than the weakly coupled staggered scheme. These developments provide a promising starting point for further simulations that allow for high fidelity investigations of coupled processes in clayrock.

WP2: Pathways Through Pressure-Driven Percolation (Clay/Rock Salt)

4.8 Model-Experiment-Exercise MEX 2-1a: Fluid Driven Percolation in Salt

Amir Shoarian Sattari (CAU), Mathias Nest (IfG), and Keita Yoshioka (UFZ)

Model Exercise 2 (ME 2) investigates the fluid driven percolation (hydraulic fracking) in salt [26] and clay stone samples under anisotropic confining stresses. The pressurized oil is injected through the pre-drilled cavity until the sudden pressure drop or increase in flow rate is observed. The main goal of this test is to determine the stress dependent fracking path and the required fracking pressure which should be higher than the minimum principle stress applied to the system. For the simulation of the fluid driven percolation, the continuum and discreet methods are considered.

4.8.1 Experimental Set-Up

The experimental tests on rock salt samples have been conducted at the IfG Leipzig [27]. The cubic samples with a side length dimension of 100 mm are prepared and placed in the true tri-axial apparatus (Fig. 4.50). The cubic samples are prepared with a drilled cavity with a length of 40 mm and a diameter of 16 mm. The pressurized fluid is injected through a tubing from the top and the fracking and change of flow rate is tracked. The applied two anisotropic stress configurations and the fracking paths are shown in Fig. 4.51a, b. Figure 4.52a, b depict the change of pressure inside of the borehole with time under the constant flow rate sequels.

4.8.2 Model Approaches

Different numerical methods (DEM, LEM, PFM) are implemented to investigate the applicability of each model to simulate the fracking path, fracking pressure and flow rate changes in the saltstone samples. Below, the description of each model and the discussion of the results are provided. The stress dependent fracking path due to anisotropic stress configurations are captured by both continuum and discrete models.

Fig. 4.50 ME2 setup configuration of saltstone placed inside the true tri-axial apparatus [27]

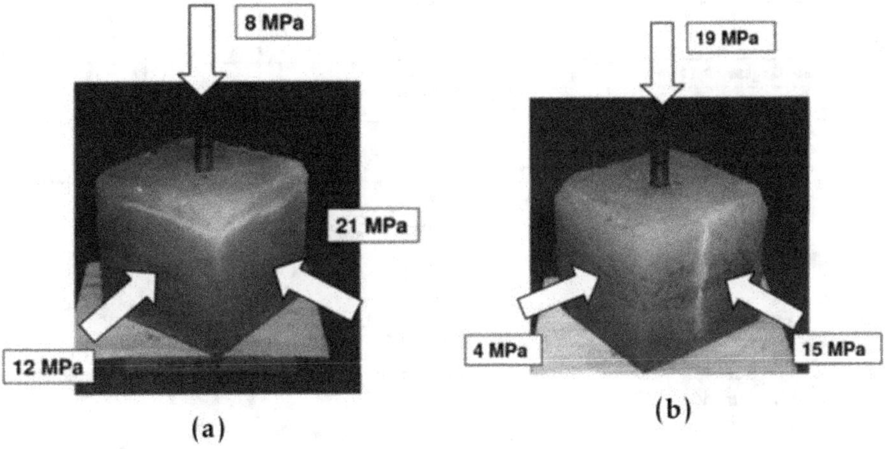

Fig. 4.51 The confining **a** 1st stress, and **b** 2nd stress configuration setup in saltstone [27]

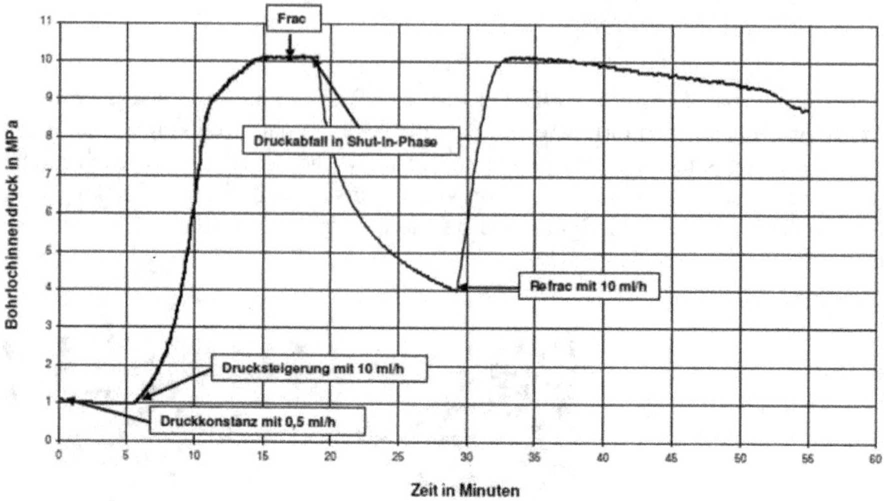

(a)

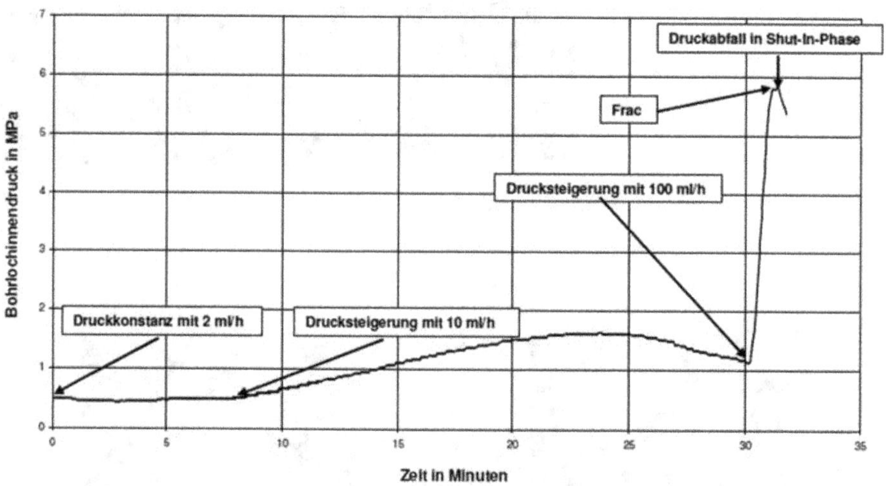

(b)

Fig. 4.52 The borehole pressure evolution under constant flow rate sequels for **a** 1st stress, and **b** 2nd stress configurations [27]

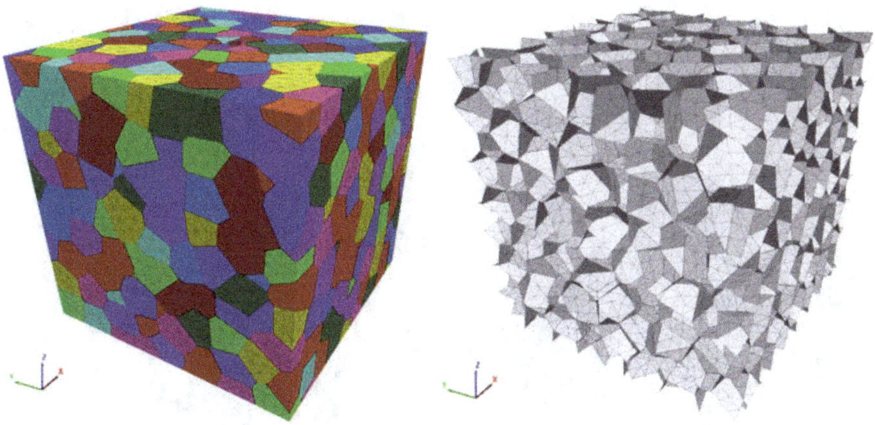

Fig. 4.53 ME2 DEM model set-up

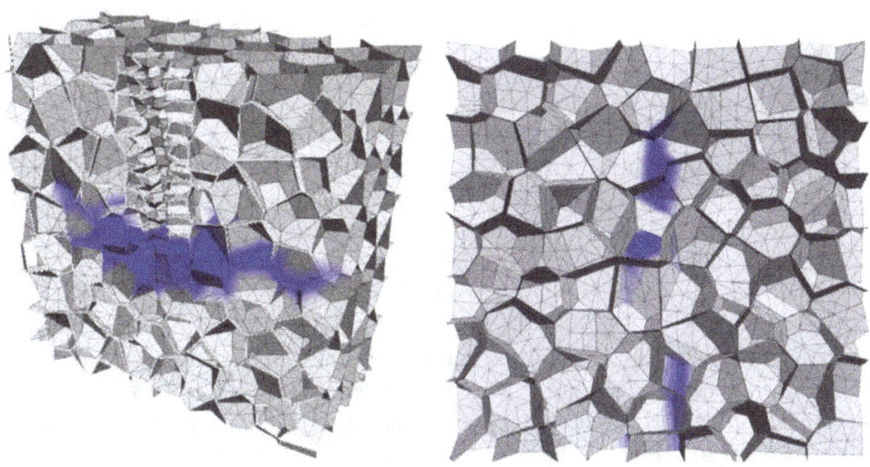

Fig. 4.54 ME2 DEM model results for the stress configuration 1

Discrete-Element-Model (DEM)

Figure 4.53 shows the elements and interfaces/grain boundaries used for the DEM simulation of this modeling exercise. For both of the stress states, the identical elements and interfaces were used. Simulation results from the first stress configuration are shown in Fig. 4.54.

Lattice-Element-Model (LEM)

The dual lattice model, described in the Sect. 3.2.2, is implemented to simulate the fluid driven percolation in the saltstone samples. The applied hydraulic pressures are transformed into the mechanical model using the weak coupling scheme and subsequently the elements failure and change of hydraulic aperture are determined

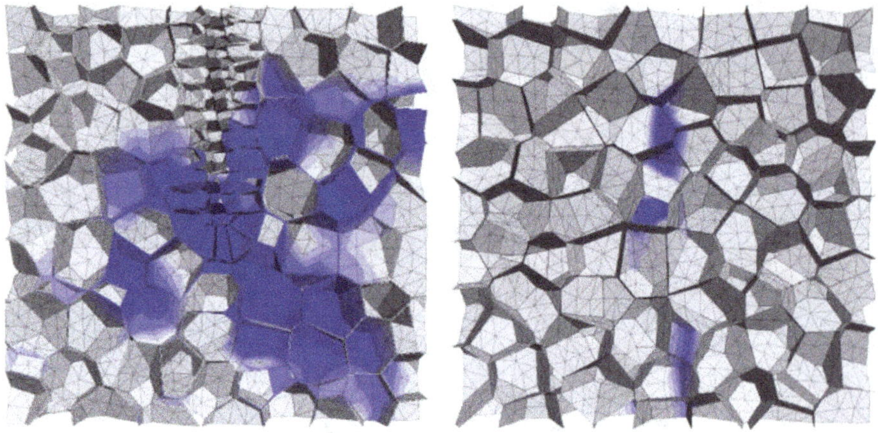

Fig. 4.55 ME2 DEM model results for the stress configuration 2

and transformed back to the hydro model. The considered mass conservation law results in the prediction of flow rate and change of reservoirs pressure as well as the flow and fracking paths, which are then compared to the experimental data. The total number of mechanical and conduct lattice elements are approximately 6000 and 45,000, respectively (Fig. 4.56). The experimental setup shown in Fig. 4.51a is simulated using the dual LEM and the developed fracture surfaces are shown in Fig. 4.57a. Similarly, the fracture surfaces under the second stress configuration (Fig. 4.51b) are illustrated in Fig. 4.57b. In these simulations the Young's modulus is assumed to be 30 GPa. While comparing the experimental and the numerical results, the frack propagation along the horizontal axis (visible on the surface frack path) for the 1st stress configuration (Fig. 4.57a) is observed. In Fig. 4.57b, the fracking path propagation in the vertical direction (visible on the surface frack path) similar to the experimental result (Fig. 4.51b) is observed (Fig. 4.55).

Finite-Element-Model: Variational-Phase-Field (VPF)

A computational domain for the variational phase-field model is depicted in Fig. 4.58. Relying on the symmetry of the domain, 1/4 of the domain was simulated. The whole domain was discretized with first-order tetrahedral elements. The total element count is 27,917,126 with 5,432,325 nodes. Two scenarios of boundary loading as in Fig. 4.51a, b were simulated.

4.8.3 Results and Discussion

The stress dependent fracking path in saltstone samples due to the defined anisotropic stress configurations is investigated. To do so, The experimental data are taken from the literature and the validation of the numerical models is performed. The VPF

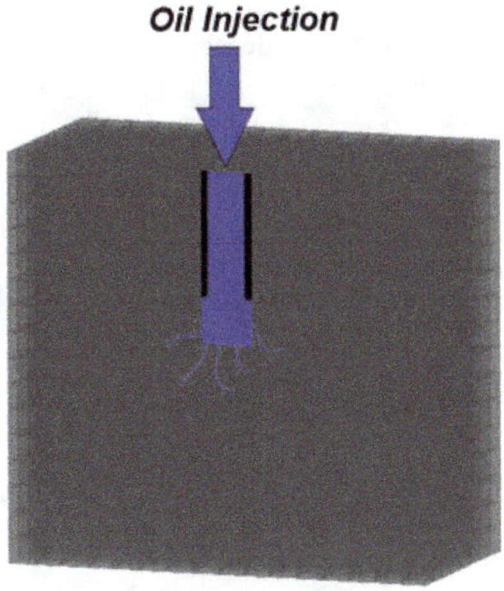

Fig. 4.56 The boundary condition in the lattice model, cross-section view

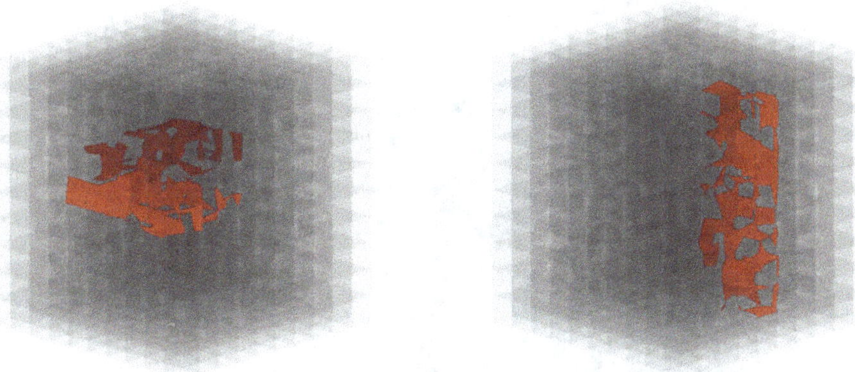

Fig. 4.57 The simulation of the fluid driven percolation and developed frack surfaces (red) for the **a** 1st, and **b** 2nd stress configurations

results of 1st and 2nd stress configurations are shown in Figs. 4.59 and 4.60, respectively. It should be noted that in the variational phase-field method, crack nucleation or propagation is not prescribed and they will be chosen as the result of the minimization of the energy. The implementation in this study adopts an anisotropic energy formulation—compression and tension split—to be able to distinguish the compressional contribution from the tensile one. Thus, even though the entire system in these experiments are under compression, the evolution of the phase-field (damage) is promoted where the material experiences tension.

For 1st case, with the vertical stress being the least stress, a more or less horizontal crack propagates towards the boundary of the sample. As the stress concentration is induced at the bottom of the injection borehole, the crack was initiated at that

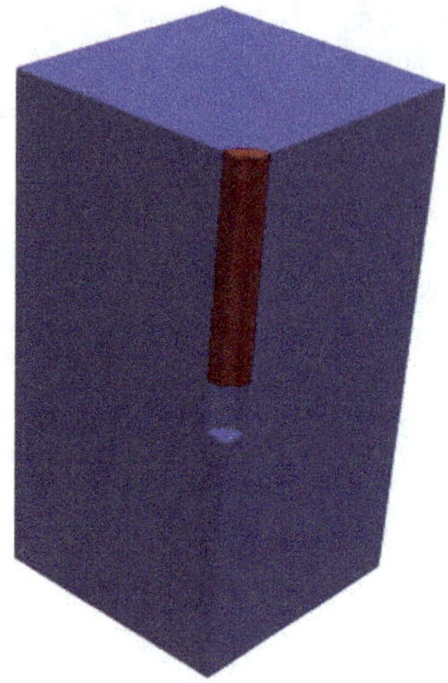

Fig. 4.58 A computational domain for the variational phase-field model

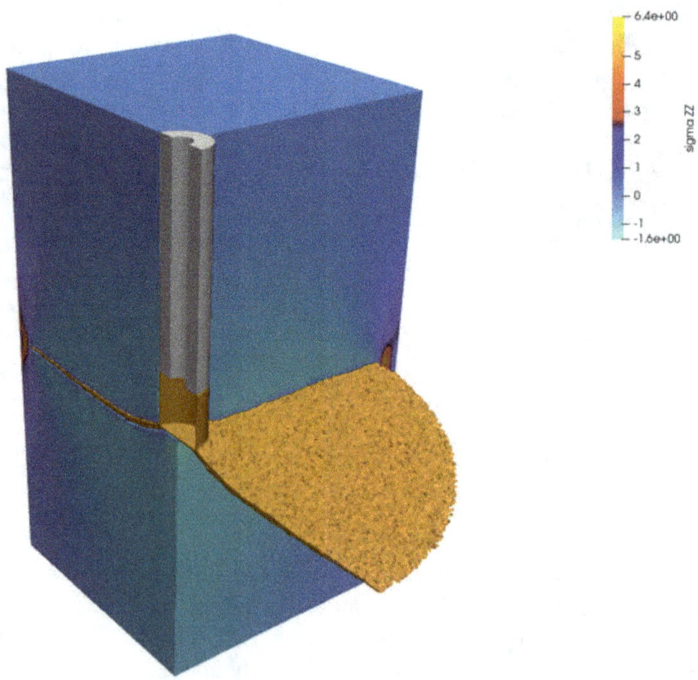

Fig. 4.59 The simulation of fluid driven percolation for case1. Created fracture and the vertical stress are shown

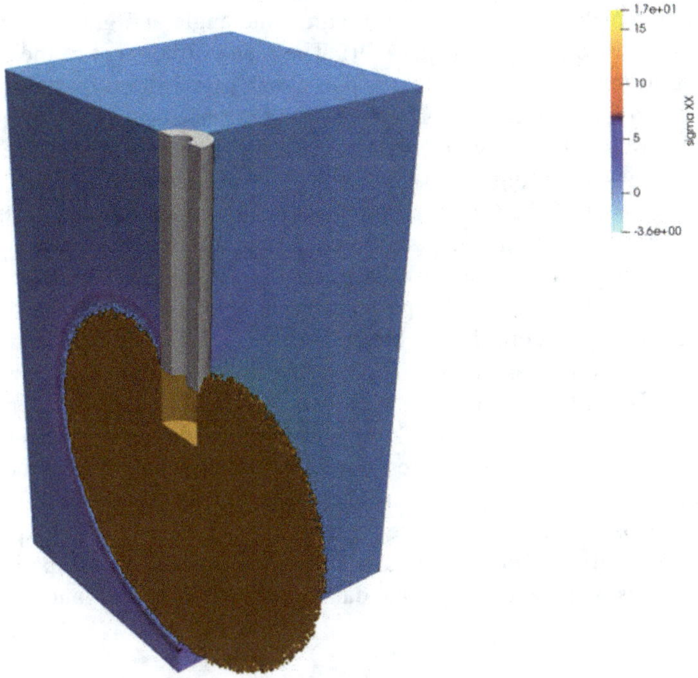

Fig. 4.60 The simulation of fluid driven percolation for case2. Created fracture and the horizontal stress are shown

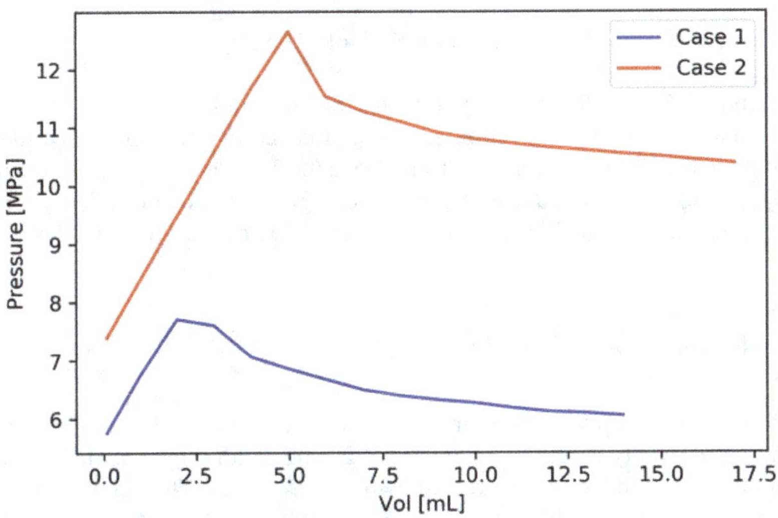

Fig. 4.61 Injection ressure responses from the cases against the injected volume

point (enough tensile strain is achieved) with some angle and gradually turned to align with the horizontal plane (Fig. 4.59). The plane orthogonal to the minimum stress direction for 2nd is the vertical plane. Crack propagation on that plane can be observed and its propagation is hinged by the presence of the casing tube in the borehole (Fig. 4.60). In both cases, similar pressure responses against the injected volume can be observed (Fig. 4.61). The peak pressures observed in these cases are 12.6 and 7.7 MPa and these values are proportional to the fracture toughness values input to the simulation. As the fracture toughness of the salt stone samples were not available in the original literature, 100 Pa m was used in the simulations.

The implemented lattice model depicts the fracking paths in Fig. 4.57a, b. For a 1st and 2nd stress configurations the fracking pressure is found to be 13.2 and 7.1 MPa, respectively. Both values are slightly higher than the given fracking pressure from the literature. However, the fracking pressures in both cases are higher than the minimum principle stresses. The fracking paths shown with red surfaces match the experimental results, where the 1st stress configurations results in horizontal fracking paths and the 2nd stress configurations results in vertical fracking paths. The discrete element method is also able to simulate the fracking path similar to the experimental results. The application of the numerical methods for quantitative comparison of the simulation results with the experimental data needs further investigation.

4.9 Model-Experiment-Exercise MEX 2-1b: Fluid Driven Percolation in Clay

Amir Shoarian Sattari (CAU) and Keita Yoshioka (UFZ)

In this model exercise, the fracking path and fracking pressure for a Opalinus Clay sample (sandy facies) under two different initial stress configurations are investigated. In addition to the stress configurations and due to the high anisotropy of the claystone samples, it is expected that due to the weak bond between the embedded layers, the fracking path and flow will be governed by orientation of the embedded layers.

4.9.1 Experimental Set-Up

The cubic Opalinus Clay samples are prepared in the side dimension of 43 mm and with the drilled cavity length and diameter of 20 and 8 mm (Fig. 2.35a). The applied mechanical stress configurations are as shown in Fig. 4.62a, b. The samples embedded layering orientations are shown in Fig. 4.63a, b, where in the first case the applied oil pressure is perpendicular to the layering orientations and in the second case it is parallel to the layering orientations.

The syringe pump is used to pump the pressurized hydraulic oil (up to 517 *Bar*) into the sample and with gradually increasing the pressure the fracking process is

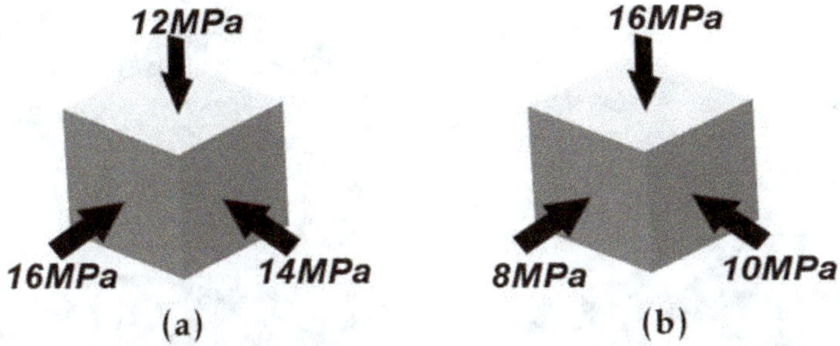

Fig. 4.62 The applied **a** 1st, and **b** the 2nd stress configurations

Fig. 4.63 The orientation of the embedded layers **a** perpendicular, and **b** parallel to the direction of the borehole pressure configuration

carried out. In the first setup, the hydraulic fracking is initiated at 23 MPa and the clear flow paths through the embedded layering surfaces is observed 4.64a. Similarly, for the second test setup, the hydraulic fracking is initiated at 10 MPa through the embedded layering surfaces as shown in Fig. 4.64b. Figure 4.65a, b illustrates the borehole pressure evolution with flow volume change obtained from the experimental data. The initial volume of the pump is around 265 mL.

4.9.2 Model Approaches

Lattice-Element-Model (LEM)

With the implemented lattice model, the effect of the Opalinus Clay anisotropy on the fracking paths and fracking pressure is investigated. The similar approach as described in Sect. 4.1 is taken to define the layering orientations in the medium. It is assumed that the interface elements bonding two different layers have 5 times

Fig. 4.64 The fracking paths through the Opalinus Clay **a** 1st stress configuration, and **b** 2nd stress configuration

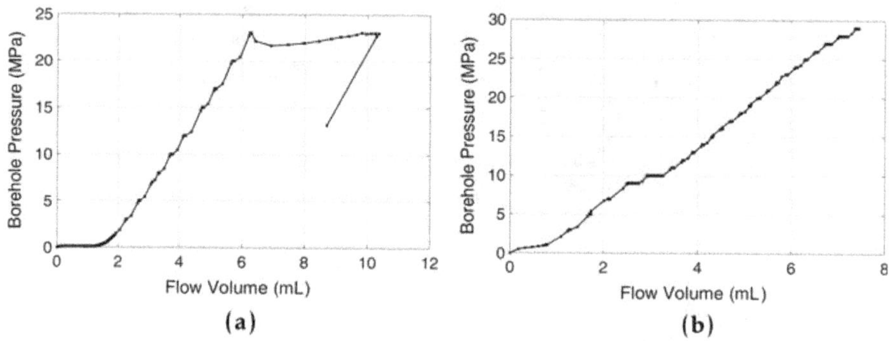

Fig. 4.65 The borehole pressure versus flow volume for a Opalinus Clay **a** 1st stress configuration, and **b** 2nd stress configuration

weaker strength than the same layers bond (found in Sect. 2.2.3). The generated 3D setup using the LEM is shown in Fig. 4.66a, b. The total number of mechanical and conduct lattice elements are approximately 28000 and 200000, respectively. The percolation tests under the two stress configurations and with different embedded layering orientations are simulated. Figure 4.67a, b depict the fracking surfaces (red) for the 1st and 2nd stress configurations, respectively. It is observed, similar to the experimental result, that the layering orientation controls the flow paths in Opalinus Clay samples.

Finite-Element-Model: Variational Phase-Field (VPF)

Using the variational phase-field model presented in this study, impacts of the anisotropy induced by the lamination parallel and orthogonal of the Opalinus Clay will be studied through 3D computational domains (Fig. 4.68).

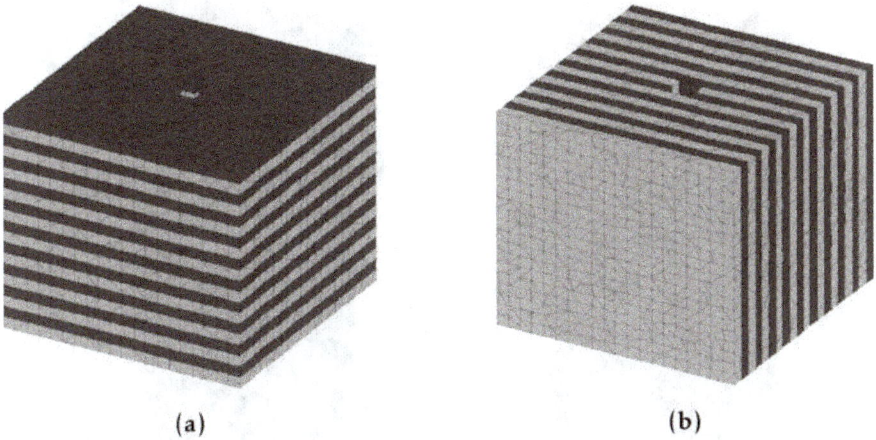

Fig. 4.66 The generated 3D domain in lattice model for the **a** 1st stress configuration, and **b** 2nd stress configuration

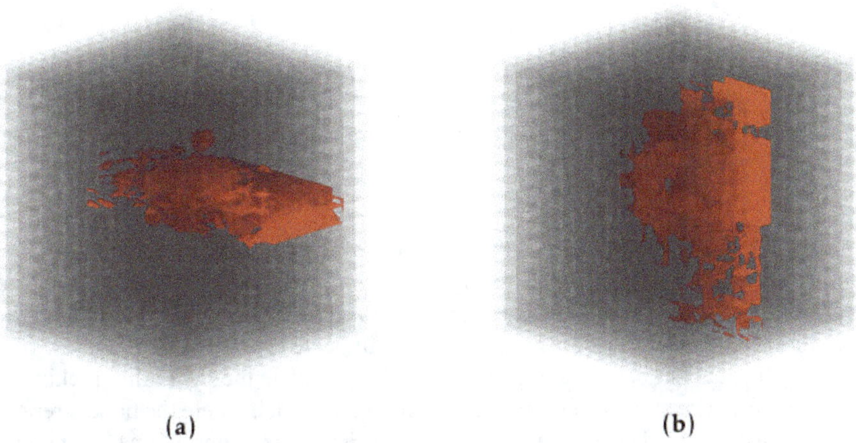

Fig. 4.67 The fracking surfaces (red) for the **a** 1st, and **b** 2nd stress configurations

4.9.3 Results and Discussion

The experimental results form the fluid driven percolation tests on Opalinus Clay show the effect of the anisotropy on fracking paths and stress distribution. The fracking and leakage through the weak bond along the embedded layers is observed. However, more experimental data under different principle stress configurations are required to detect and analyse the stress distribution and the frack paths dependency on the anisotropy. For both samples tested here the fracking stress was higher than the applied minimum principle stress. For a 1st stress configuration, the fracking pressure was much higher than applied maximum principle stress, which can be due

(a) (b)

Fig. 4.68 3D computation domain of VPF for **a** 1st stress configuration, and **b** 2nd stress configuration

to an error in the experimental results. The lack of the pre-defined notch during the sample preparation also can lead to higher fracking stresses. The fracking in 2nd stress configuration was subtle and no pressure drop is recorded. The increase in the flow volume at the borehole pressure of 10 MPa indicate the initiation of the fracking process. The dual lattice model is implemented to simulate the hydraulic fracking in Opalinus Clay. The anisotropy is implemented in the model, where the bond strength between the embedded layers is assumed to be 5 times weaker than a same layer bond. Due to this assumption which is found from Brazilian disk results (see Sect. 2.2.3), the fracking along the embedded layers is observed. The fracking pressures found to be higher than minimum principle stresses. However, more research for quantitative comparison of the results is needed.

4.10 Model-Experiment-Exercise MEX 2-2: Pressure Driven Percolation (Healing)

Mathias Nest (IfG) and Keita Yoshioka (UFZ)

4.10.1 Experimental Set-Up

One of the most important properties of barriers formed by salt or clay is their ability to close cracks and heal through creep. This means, that the materials can regenerate and regain their impermeability, after damage due to, e.g. excavation activities or pressure driven percolation because of a temporary violation of the minimal stress criterion.

In this study we focus on the sealing of pathways and healing of cracks in salt and clay by measuring and simulating their gas permeability after damage has been done.

For rock salt, we prepared cylindrical samples with a height of 20 cm and a diameter of 10 cm, with a borehole to the center, so that a gas pressure could be applied to a small area at the center. Initially all samples were placed under isostatic stress of 50 MPa for one day, to consolidate them. For the actual experiments the isostatic stresses were then changed to 10 MPa, 30 MPa, and 50 MPa, respectively. Before the gas pressure was applied, the plates which applied the axial stress were lock in place ("displacement boundary condition"). The gas pressure was increased in small steps, until a gas flow was detected. In all three cases a gas flow was detected before the percolation threshold was reached, which indicates that the samples suffered micro-fractures during preparation. After the last increase of the pressure, the flow rate was monitored for 24–70 h under constant conditions (Figs. 4.69 and 4.70).

The flow rate can be converted into permeabilities, which are shown in Fig. 4.71. For all three cases a reduction of the permeability is found, showing that the cracks are narrowing and slowly closing. The abrupt and discontinuous manner of the change of the flow rate makes it difficult to identify time-scales of the processes involved. In the case of 10 MPa confining stress one can distinguish two phases. Between 3 and 24 h after the start of the experiment, one can fit (admittedly only very crude) a time-scale of about 4.5 h. Some larger pathways seem to close abruptly. After that, the pathways narrow only very slowly, presumably by creep processes, on a time-scale of about 1000 h. In the cases of 30 MPa and 50 MPa we find time-scales of 30 h and 11 h, respectively. At least here, faster creep under higher stress is found as expected for the general trend.

The Opalinus Clay samples were prepared slightly differently, see Fig. 4.72. The cylinders had a height of 160 mm and a diameter of 80 mm. Two boreholes were used, to avoid that under isostatic stress conditions the gas flows in a radial direction. Again it turned out, that the sample preparation had induced dilatant damage, so that a significant gas flow was already detected before the gas pressure reached the value of the mechanical stress. The subsequent results were obtained at a pressure of 0.5 bar.

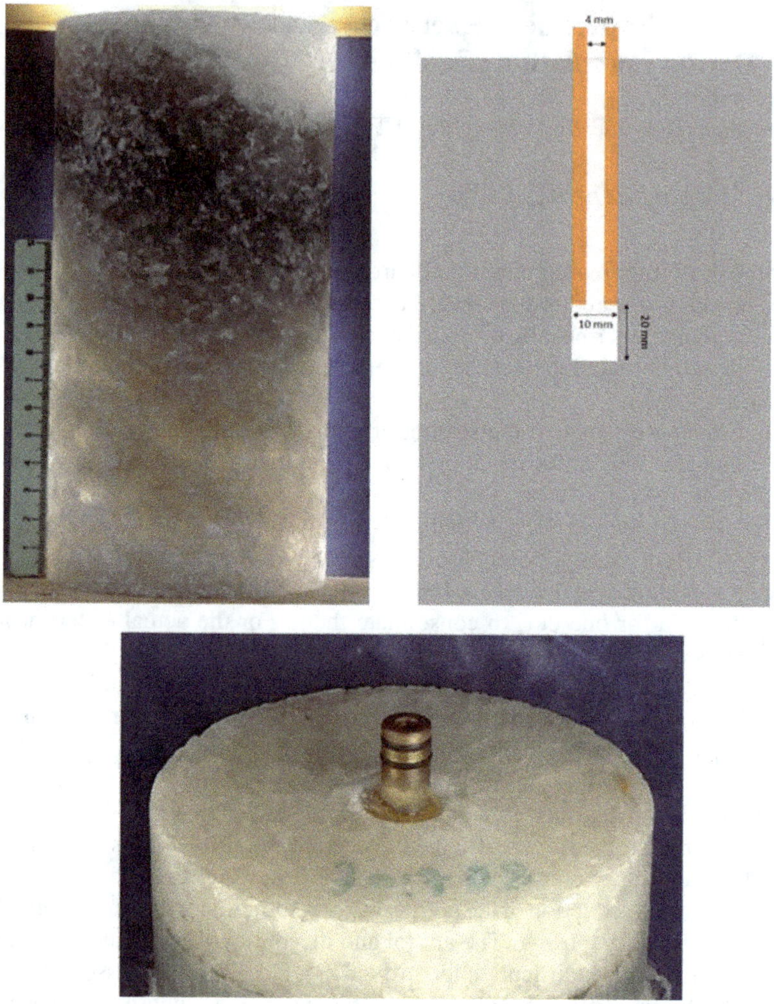

Fig. 4.69 Rock salt sample, borehole geometry, detail of metal pipe

The results are shown in Fig. 4.73. Initially, a stress of 7 MPa was applied, and the gas flow decreased from 38 to 24 ml/min in the course of about 100 h. Then the stress was reduced to 5 MPa, which lead to an increased flow of 26 ml/min. Again, a decreasing rate was observed, reaching 6 ml/min after 400 h. Finally the stress was increased to 9 MPa, and the rate dropped from 4.4 to 1.8 ml/min. So in all three cases the microfractures from the sample preperation were closing slowly, as a first step towards healing. The difference to the previous experiment with rock salt is stark. The clay reacted much more smoothly, and good fits to exponential decays could be obtained. The reason is the much smaller grain size in the clay, which permits much

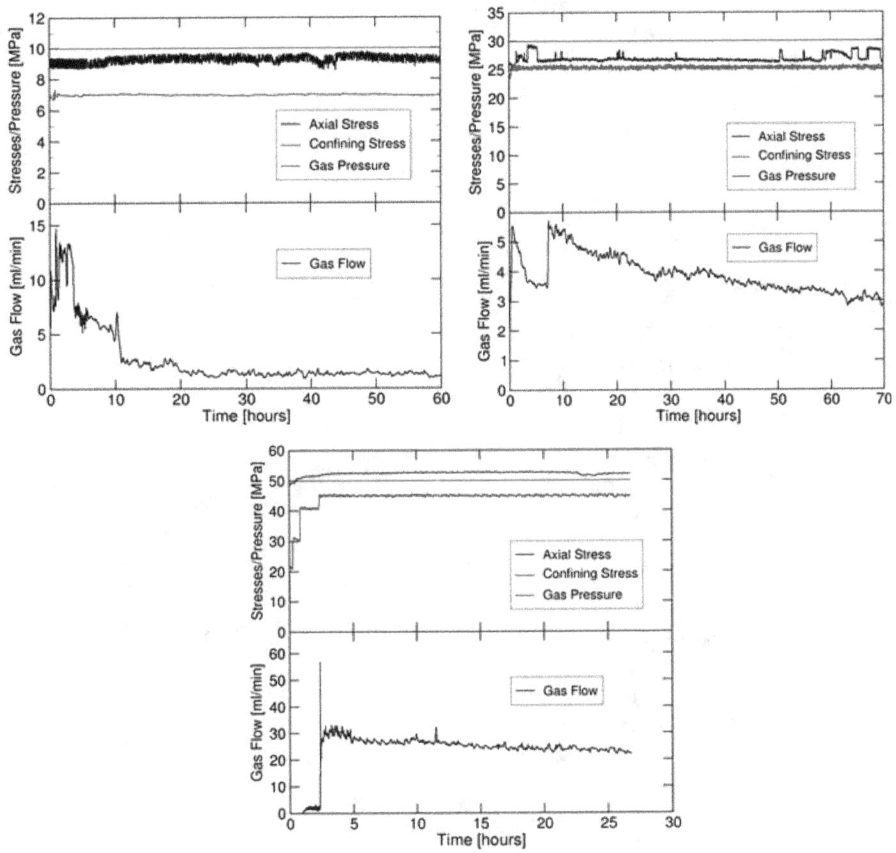

Fig. 4.70 Axial and confining stresses, gas pressures, and observed flow rates

smaller displacements. In addition, the general trend, that the time scales decrease as the stresses are increased, is observed as expected.

4.10.2 Model Approaches

Discrete-Element-Model (DEM)

The closure of cracks and the healing of rock salt after plastic deformation has been rarely looked upon and studied in detail. The purpose of this exercise is therefore to develop a phenomenological model, which can be incorporated into the 3DEC software. The mechanistic idea behind the process is shown in Fig. 4.74 Cracks have a rough surface, with high deviatoric stresses where they are in contact. This

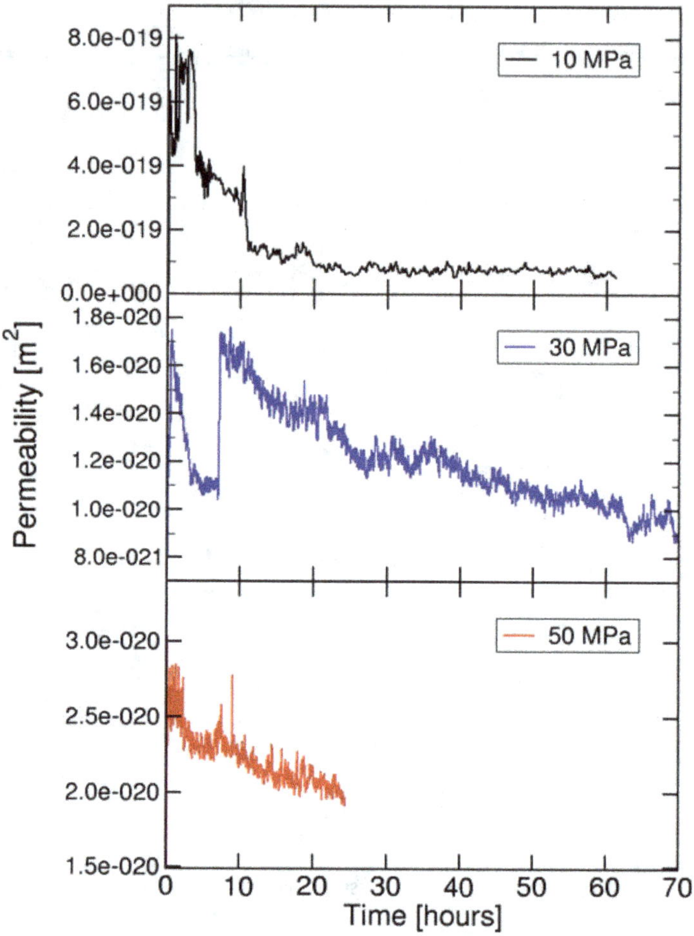

Fig. 4.71 Reduction of the permeability under different hydrostatic stress conditions

concentration of stress can lead to the abrupt abrasions observed above, and also to an increased creep rate.

In 3DEC the fluid flow is modelled using fluid knots, where apertures between two grains are defined. The size of the aperture depends on an elastic spring constant, the pressure, and two cut-off values, see Fig. 4.75. There, a_{max} and a_{res} are the maximally and minimally allowed fluid apertures, and a_0 is the aperture when the normal stress σ_N is equal to the fluid pressure p. In the following simulation we used $a_{res} = a_0 = 10^{-7}$ m. The narrowing of the crack is then modelled by decreasing a_{max} with time.

The net effect of this approach is hard to predict, because the total fluid flow through the sample depends on a complex network of fluid knots with different orientations and normal stresses. Two approaches were tried:

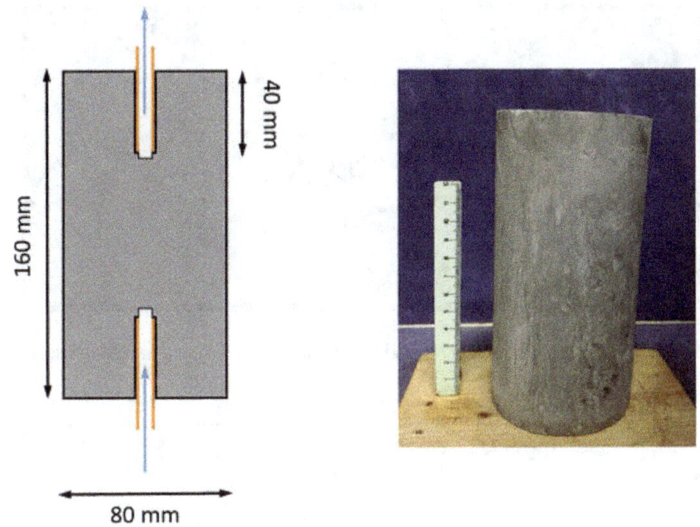

Fig. 4.72 Left: Setup of the fracture closing experiment with Opalinus Clay. Right: One of the samples

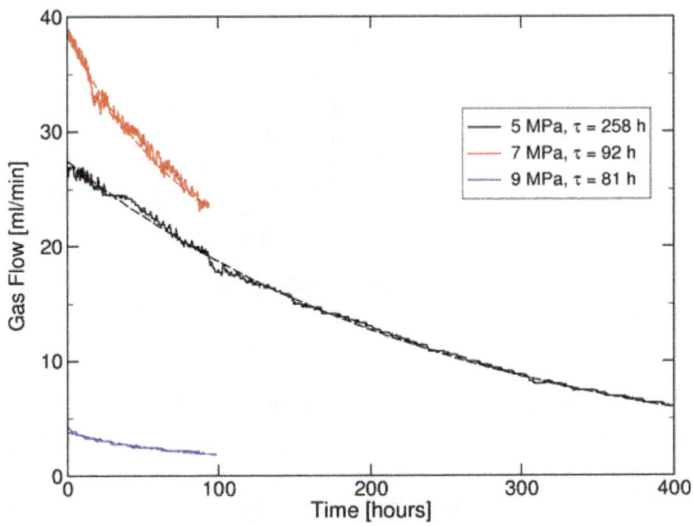

Fig. 4.73 Evolution of gas flow rates under different isostatic stresses

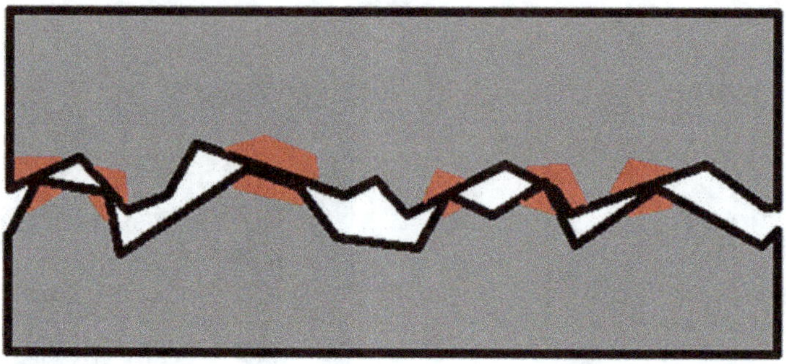

Fig. 4.74 Locations of high deviatoric stress in a crack. The deviatoric stress drives creep, which leads to a closing of the crack

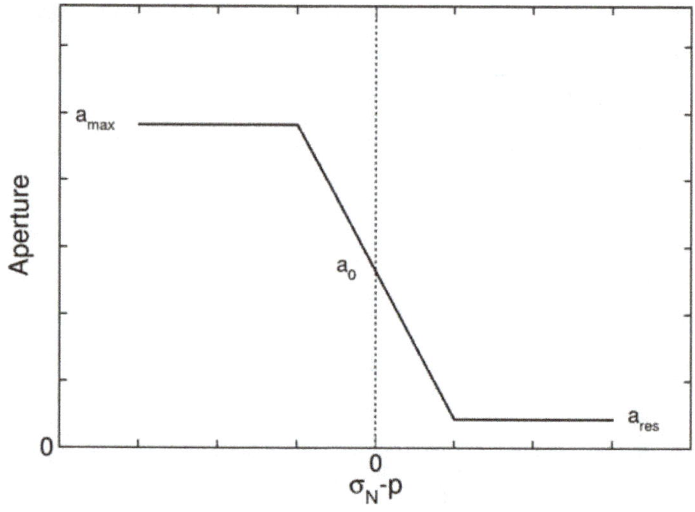

Fig. 4.75 Schematic representation of the relation between aperture on grain boundaries and effective stress in 3DEC

$$a_{max,n+1} = c \cdot a_{max,n} \quad \text{with } 0 < c < 1 \tag{4.21}$$

and

$$a_{max,n+1} = a_{max,n} - \Delta a \quad . \tag{4.22}$$

The model setup used for this simulation was the same as in Sect. 4.8. The first two phases of the simulation agree with the simulation of the pressure driven percolation. A fluid pressure was applied to the internal cavity, and after a while liquid started to leak out (see phase I in Fig. 4.76). After about 2.5 s the leak rate converged to a near constant level (phase II). From this point on, the simulation was run (phase III)

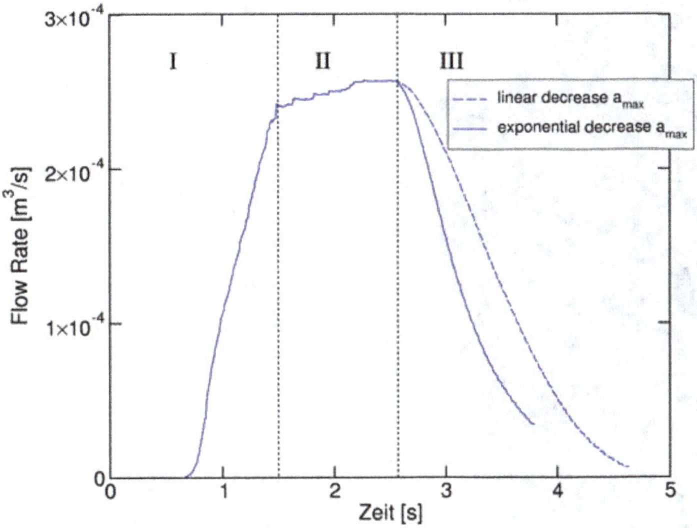

Fig. 4.76 Pressure driven percolation and subsequent crack closure

in two different versions, corresponding to the two different equations above for the decrease of a_{max}.

For the linear decrease of the maximum aperture we find a decrease of the flow rate that is not exponential as in the experiment (nor is it linear). For the exponential decrease we used $c = 0.9659$ and applied this value every 100,000th fluid timestep. (The full simulation required more than 10 million timesteps.) In this case a very good exponential fit with a time scale of 0.63 s could be obtained. It should be noted that in this simple setup, two time scales are actually mixed, which in reality are separated: In reality, creep (and thus the narrowing of the cracks) works much slower than the fluid dynamics. But as this exercise is a proof of concept, and serves to obtain a basic understanding of how how to model the phenomenon of crack closure, we accepted this approximation.

Finite-Element-Model: Variational Phase-Field (VPF)

Implementation of the crack healing through creep model within the OGS variational phase-field model is being planned. Crack opening (fracture width) is not a directly computed variable within the variational phase-field model, instead it is computed currently by taking a line integral [28] as:

$$[\![\mathbf{u} \cdot \mathbf{n}_\Gamma]\!] \approx \int_l \mathbf{u} \cdot \nabla v. \tag{4.23}$$

where l is the line that follows the normal direction to the crack Γ. With a creep model, the displacement will be decomposed into the elastic $\mathbf{u}^e$ and the creep part $\mathbf{u}^c$ as $\mathbf{u} = \mathbf{u}^e + \mathbf{u}^c$, and the crack opening can be computed then by

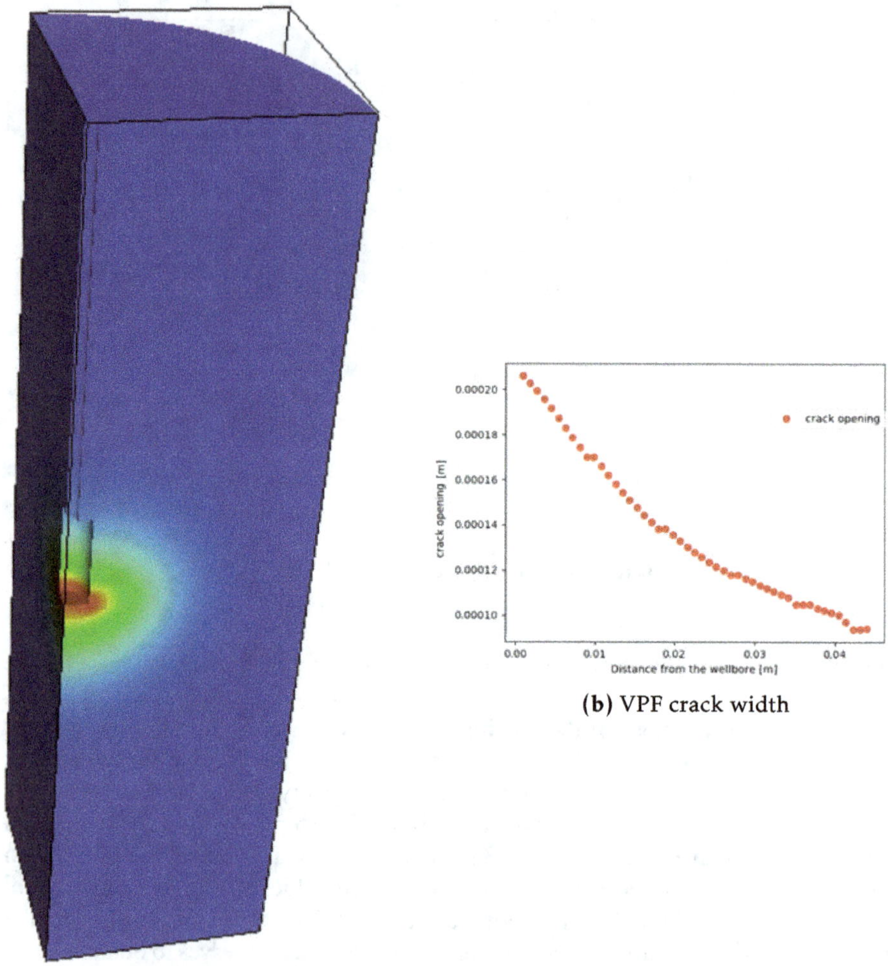

(**b**) VPF crack width

(**a**) VPF model set up for the model exercise 3

Fig. 4.77 VPF preliminary

$$[\![\mathbf{u} \cdot \mathbf{n}_\Gamma]\!] \approx \int_l \mathbf{u}^e \cdot \nabla v. \tag{4.24}$$

As $\mathbf{u}^c$ evolves with time, the crack opening will be subject to change and be able to simulate the aperture change observed in the experiments. A preliminary attempt to calculate the fracture aperture from the elastic response only is shown in Fig. 4.77.

4.11 Model-Experiment-Exercise 2–3: Effect of Compressibility on Pressure Driven Percolation

Mathias Nest (IfG), Amir Shoarian Sattari (CAU), and Keita Yoshioka (UFZ)

4.11.1 Model Set-Up

Salt barriers do not only protect mines against the intrusion of ground water, but they can also be important to keep substances securely isolated in underground repositories and reservoirs. Examples include caverns for gas and oil, as well as repositories for toxic or radioactive waste. In this model exercise we look at the difference between the pressure driven percolation of liquids and gases, which is not only due to their different viscosities, but also to their different response to a change in the available volume.

Liquids like brine react with a large change in pressure, because of their rather high bulk modulus K:

$$dp = -K\frac{dV}{V} \tag{4.25}$$

Gases, which can be described approximately with the ideal gas equation

$$pV = nRT \tag{4.26}$$

show a far smaller change in pressure. In this model exercise we explore this difference in a setup similar to the one in MEX 2-1a, under the assumption, that there is only a finite amount of substance available to spread into the rock salt. Another way to put this is, that a lab-scale model of the in-situ experiment MEX 2-4 is set up.

The basic reasoning behind the exercise is as follows: The total mass of the brine or gas is conserved, and consists of the mass in the reservoir M_R, the mass on the interfaces of the grains M_I, and the mass M_L that has leaked out of the sample.

$$M = M_R + M_I + M_L \quad = \text{conserved} \tag{4.27}$$

M_I is part of the model, M_L can be calculated using absorbing boundaries at the surface of the sample, and from the knowledge of the total amount M that is present initially, the mass that remains in the reservoir M_R can be calculated. This way, it is not necessary to model the reservoir (gas cylinder or brine container) explicitly. From the changing M_R under the assumption of a fixed reservoir volume the change of the pressure in the reservoir, which drives the pressure driven percolation, can be calculated. Initially, at time $t = 0$, $M_I = M_L = 0$, and the pressure in the reservoir is high enough to open pathways between grains. Then, pressures and the extent of the distribution of the substances is monitored.

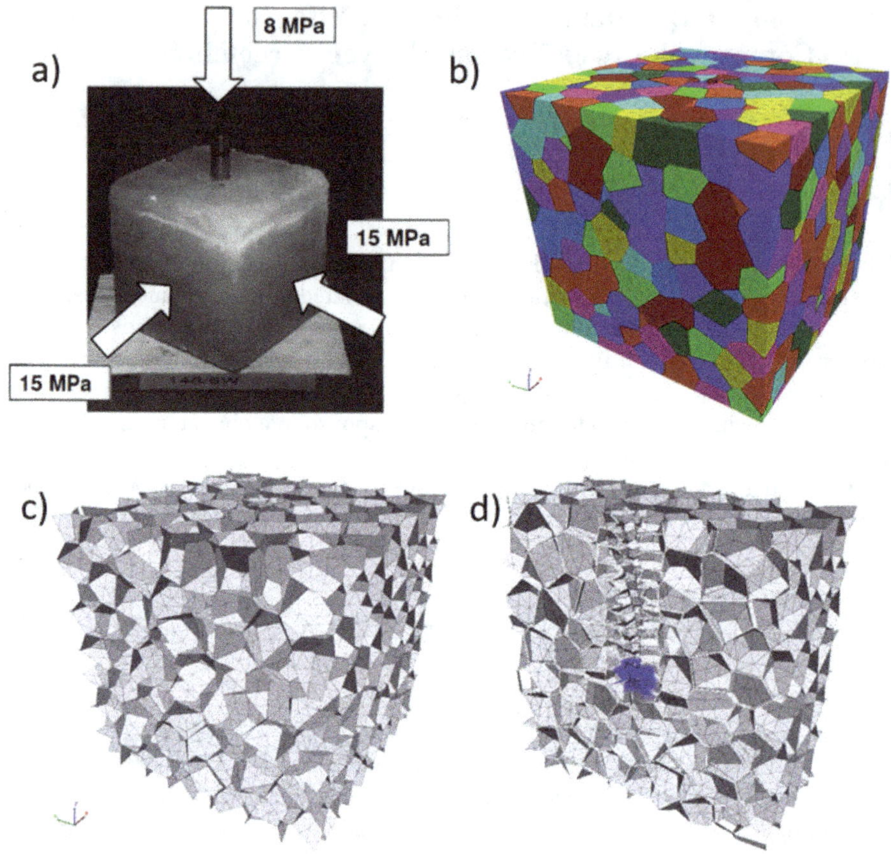

Fig. 4.78 **a** Stress boundary condition **b** discrete grain structure of 3DEC model **c** grain boundaries/interfaces **d** initial pressure on grain boundaries

4.11.2 Model Approaches

Discrete-Element-Method (DEM)

For the numerical model in 3DEC the same setup was used as in MEX 2, see Fig. 4.78. An initial reservoir pressure p_R of 12 MPa, between principal stress σ_2 and σ_3, was chosen, to allow percolation. The bulk modulus of the brine was set to 2 GPa. Various reservoir volumes were tested, and V_R = 13 litres was found to give the clearest difference between gas and liquid. For the simulation the masses M_I and M_L were calculated every 25 time steps. From this, the mass in the reservoir was calculated, and converted into a pressure by the constitutive relations above. The new reservoir pressure was then applied as an updated boundary condition as shown in Fig. 4.78d).

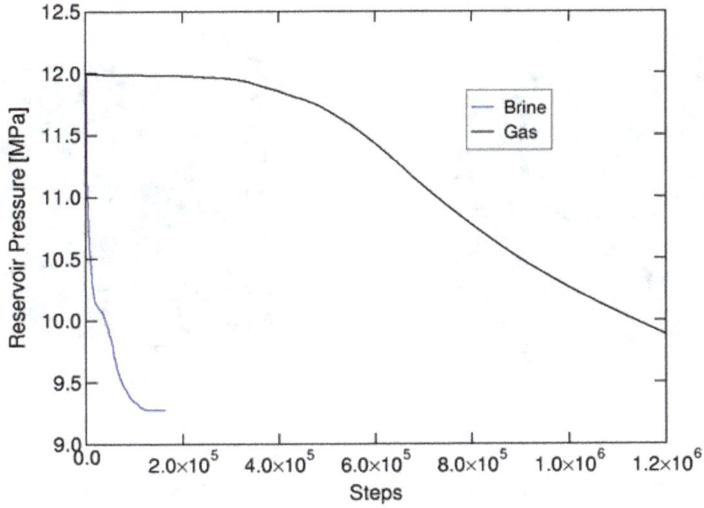

Fig. 4.79 Pressure evolution in a finite size gas/liquid reservoir

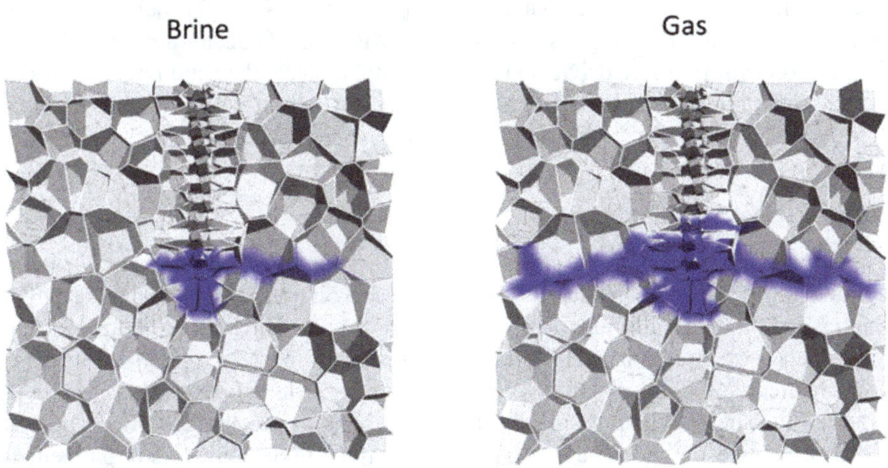

Fig. 4.80 Final pressure distributions. The brine (left) gets stuck, the gas (right) reaches the boundaries and leaks out

Figure 4.79 shows that there is a stark difference. The high bulk modulus of the brine leads to a very fast drop in pressure, while the gas pressure stays almost constant for quite a while. It should be mentioned that the time steps do not correspond to a physical timescale, because the rate of the fluid flow between adjacent fluid knots in the model was artificially increased, and not derived from the aperture between two grains. Still, the qualitative difference persists. The brine pressure drops so fast, that the liquid becomes stuck inside the cube (Fig. 4.80 left), while the gas pressure decreases only after a significant amount leaked from the sample (Fig. 4.80 right).

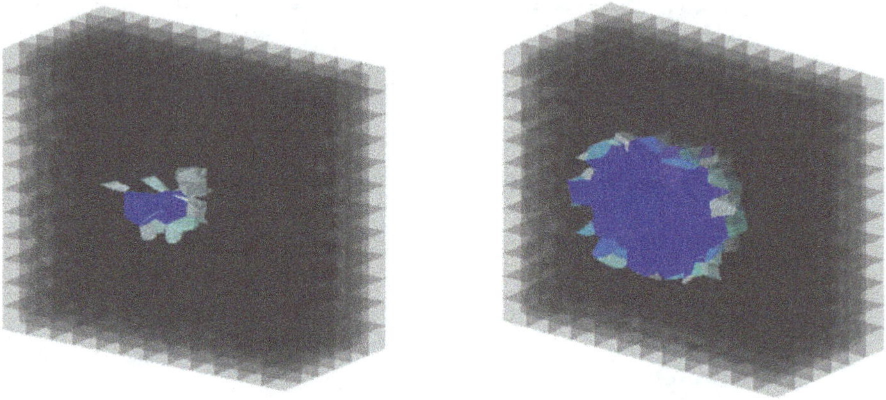

(a) (b)

Fig. 4.81 The flow potential for a **a** brine, and **b** gas reservoirs (cross-section view)

In addition, it should be noted that the brine pressure does not fall below 9.3 MPa, although the minimal principal stress is only 8 MPa. This is because of two effects: First, in order for pressure driven percolation to take place, the fluid pressure has to overcome the normal stress on the grain boundaries, which is a composition of all three external stresses due to the tilted orientations. Second, the fluid has to overcome the tensile strength, set to 1 MPa, of the joints, too.

Lattice-Element-Method (LEM)

In order to simulate the compressibility effect with the lattice model, the similar setup as shown in Fig. 4.78 is considered. The initial reservoir pressure is set to be 12 MPa, which is higher than minimum principal stress of 8 MPa. The bulk modulus of brine is assumed to be 2 GPa and the volume of reservoir to be 13 L. The total number of mechanical and conduct lattice elements are approximately 10,000 and 80,000, respectively. The rate of pressure drop in the reservoirs storing the brine and gas (methane) is investigated. The hydro model explained in Sect. 3.2.2 is implemented to grant the mass conservation in the domain. The boundary reservoir pressure is not constant and with the fluid/gas transport in the medium it gradually drops. Due to the high bulk modulus of brine it is expected that the pressure drop in the brine reservoir to be higher than gas reservoir. However, the gas is leaked in higher rate where the kinematic viscosity is higher. Figure 4.81a, b depict the flow potential in the conduct surfaces (blue) for a brine and gas reservoirs, respectively. Although minor cracks are observed, the borehole pressure was not enough to cause major fracking pathways. The amount of gas transport in the domain is higher than brine. Figure 4.82 illustrates the time dependent pressure drop for gas and brine reservoirs. As expected, the rate of pressure drop in the brine reservoir is much higher than gas reservoir. However, the pressure drop in comparison to DEM model is more subtle.

Finite-Element-Method: Variational Phase-Field

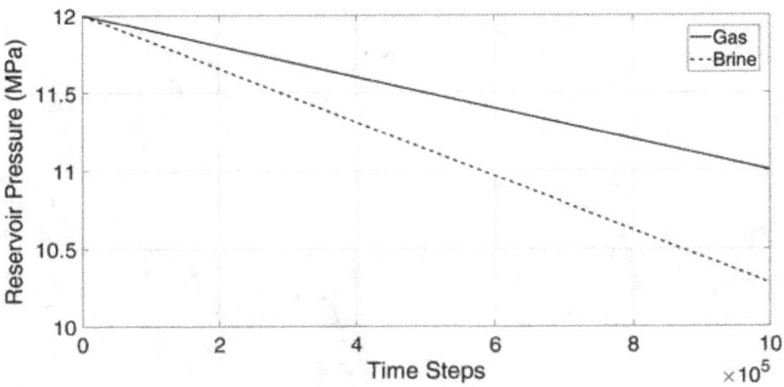

Fig. 4.82 The comparison between the pressure drop in gas and brine reservoirs

The simulation set-up will be identical to Fig. 4.58 except that the injection fluid will be compressible fluid (ideal gas).

4.11.3 Discussion (Preliminary)

The results above show, that at present at best a phenomenological description is possible with the available set of methods. A significant amount of implementation work, and increases of the speed and accuracy are required to capture the effect.

4.12 Model-Experiment-Exercise 2–4: Large Wellbore Test (Springen)

Mathias Nest (IfG)

The motivation for this experiment was two-fold. First, decommissioned caverns are usually filled with brine before they are closed and sealed. The lower density of the brine leads to a buoyant force, so that the pressure in the liquid can become larger than the lithostatic stress in that depth. This led to questions about the long-term tightness of abandoned caverns, and thus safety concerns [33]. Of central importance was the question, whether a large, long-range frac would be suddenly created, or a slow expansion between the salt crystals would take place.

Second, from a scientific point of view, it is interesting to see whether there is a difference in the pressure driven percolation between liquids and gases. To conduct experiments under realistic in situ conditions to answer these questions a large borehole with a volume of about 50 m^3 was excavated between two seams in the Springen mining field, see also Sect. 2.1.2.1 and Fig. 4.83. In 2011/2012 this borehole

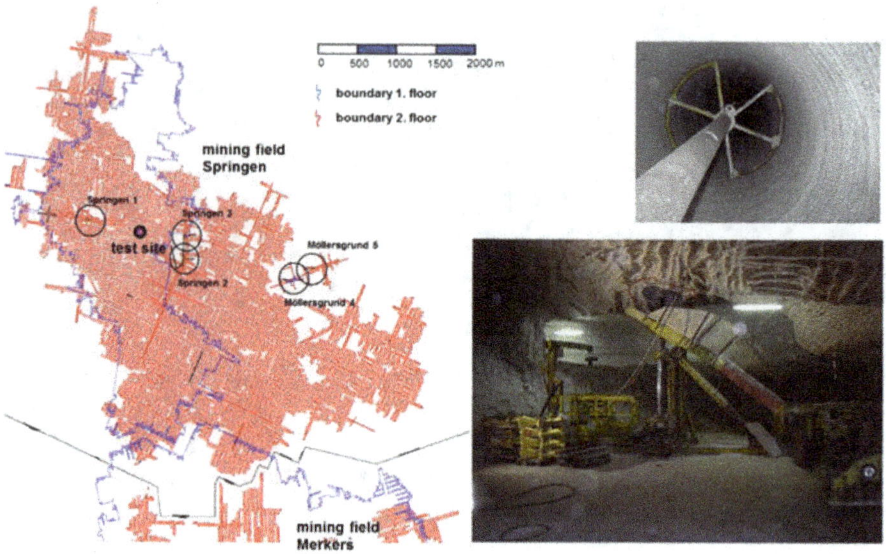

Fig. 4.83 Mining field Springen (left), borehole (right)

was used to conduct an experiment with pressurized air. Here, we report results of an experiment performed in 2018, where the borehole was loaded with brine. Starting in early 2018, the brine pressure was increased in small steps of around 5 bar, followed by periods of rest, where the surrounding rock was allowed to react. In the summer a pressure level was reached that was close to the minimal principal stress of about 50 bar. Figure 4.84 shows the last three pressure spikes over a span of 50 d. They were created by pumping brine with rates of 2 l/h, 3.3 l/h and 5.5 l/h, respectively. Each time, a rapid drop in pressure was observed consistent with the high bulk modulus of brine and the simulations in Sect. 4.11. The fact that more than 60 bar were required to induce cracks between salt crystals, i.e. more than minimal principal stress, is due to the tensile/adhesive strength of rock salt.

The central part of the figure shows, that the decay of the pressure still continues 6 weeks after the pressure spike, showing the slow expansion of the brine into the surrounding area. The final spike was an attempt to determine the highest pressure obtainable with available equipment (however, with the previously damaged rock salt). The experiment ended August 14, 2018. Acoustic emission data has been collected, but requires further analysis. Further pressure measurements in the spring of 2019 provided values of around 50 bar, consistent with the minimal stress criterion.

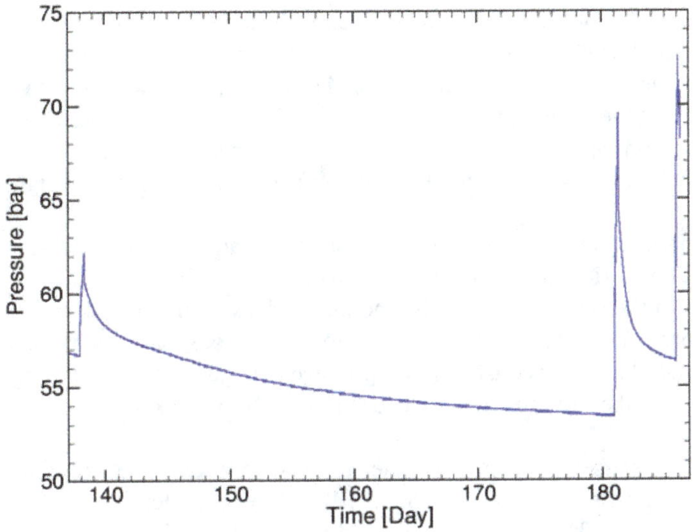

Fig. 4.84 Sequence of three pressure applications over a span of 50 d

WP3: Pathways Through Stress Redistribution (Crystalline Rock)

4.13 Model-Experiment-Exercise MEX 3-1: Constant Normal Load (CNL) Direct Shear Test

Daniel Pötschke (TUBAF) and Thomas Frühwirt (TUBAF)

4.13.1 Experimental Set-Up

Direct shear tests are conducted in rock mechanical laboratories to investigate the shear characteristics of rock fractures/joints. The samples are blocks or cylinders which are separated in two parts by a fracture. This fracture can be a natural one or an artificial one. Artificial fractures can be created by shearing of the intact sample or by a Brazilian test. A schematic illustration of direct shear test can be seen in Fig. 4.85. During a direct shear test a normal force acts on the rock fracture. One part of the sample is fixed and the other one is sheared against it. The shear forces are measured while the two parts are separated against each other at a specific shear rate.

In this model exercise a constant normal load (CNL) boundary condition has been applied. This means the normal stress acting on the fracture remains constant during the shear displacement. An analogy in the nature would be a rock boulder at a slope or a dam where in both cases the self weight of the top part creates the normal stress at the contact region/fracture. A rock fracture in nature is usually rough at some degree of detail. In the direct shear tests the two parts of the sample fit more or less perfectly at the initial position. The fracture is closed. When they are sheared against each other the unevenness, the roughness, causes an opening of the fracture and an uplift of the upper part. This so called normal displacement or dilatation is free like the rock boulder on the slope could easily uplift as only his constant weight is acting on it.

The experimental procedure involves a surface scan of the rock fracture as a first step. This geometric information are used to determine the roughness of the surface. The output of the scanning device is a point cloud, as seen in Fig. 4.86.

Other basic laboratory tests are done to obtain rock parameters of the intact rock material like elastic constants, compressive strength, cohesion or friction angle. The values for the granite which was used in this model exercise can be found in Table 2.1

The shear test itself has been conducted for this model exercise using four different normal stress levels of $\sigma_n = 1, 2.5, 5, 7.5$ MPa. After completing one level the parts were rearranged to their initial position and the next normal stress applied.

The task in this model exercise is to use the rock parameters from Table 2.1, the geometry data of the point cloud and the boundary conditions of the direct shear test to back calculate the lab results. The whole data sets are provided as a download, see Sect. 5.3.11. All research groups are invited to use this data to improve the quality

Fig. 4.85 Constant normal load (CNL) direct shear test. (*From* [29])

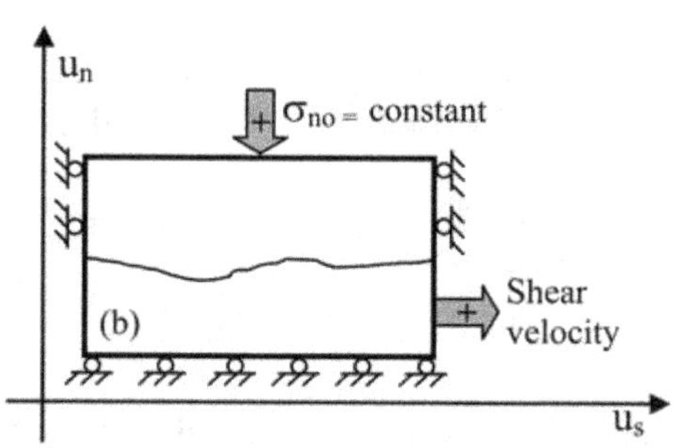

Fig. 4.86 Point cloud representing the surface of a granite sample from Saxony. The size is 65 mm by 170 mm and the cloud contains approx. 98000 points. The shear direction is parallel to the long edge in a way that the shown surface remains in its position and its top counterpart is moved to the right

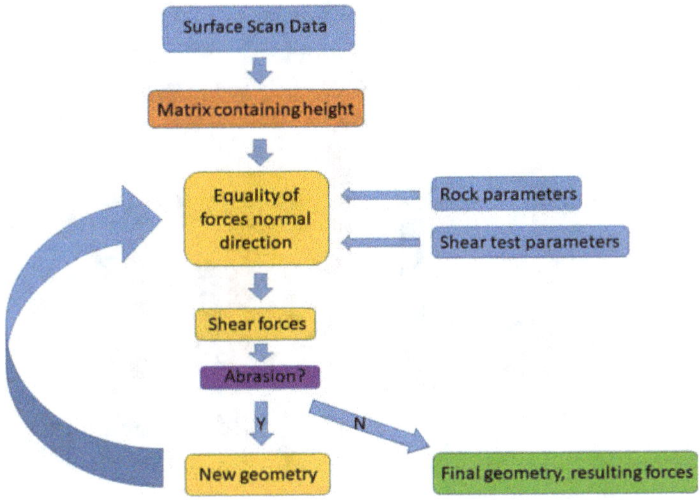

Fig. 4.87 The functionality of the new model

of the calculations. The shown results will be provided as well to make it easy to compare different methodologies.

4.13.2 Model Approach

A detailed description of the code can be found in the appendix. The modelling of the direct shear tests is done by a self developed code using MATLAB. This includes the usage of matrices to represent the geometry of a rough surface. This is a simplification which speeds up the numerical process. A function set does all the necessary steps from input through the calculations itself to the output. For comparison a couple of existing models where implemented as a reference.

- Barton-Bandis model [30], which is widely used and due to its simplicity easy to understand.
- Xia model [31], which uses quantitative roughness parameters with the drawback of complexity.
- Casagrande algorithm [32], a semi analytical approach which was used for the algorithm development within the GeomInt project.

A new algorithm was developed which uses the equality of normal forces as an approach to choose the surface elements which are in contact. The motivation is to make the calculation physically consistent. The scheme can be seen in Fig. 4.87. The shear forces are calculated using the approach of [32].

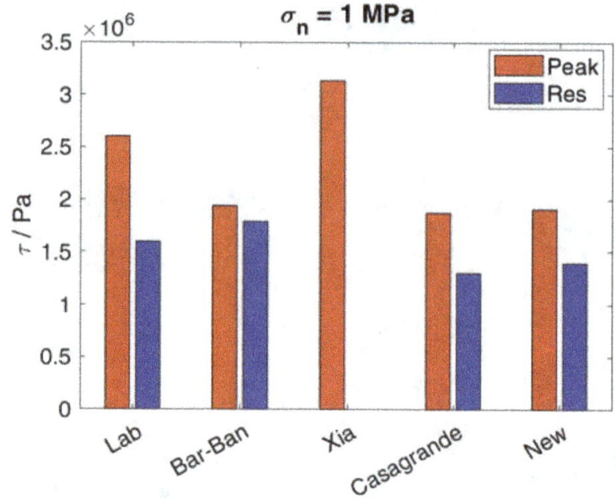

Fig. 4.88 Comparison of peak and residual shear strength for the first step of the shear test using different shear laws

4.13.3 Results and Discussion

The shear stress—shear displacement curves which result in lab testing have two values of main interest: The peak shear stress and the residual shear stress. A peak was just observed for the first shear stress level of $\sigma_n = 1$ MPa. In Fig. 4.88 the results are shown.

At higher stress levels and due to the repeated shearing which causes wearing of the rock surface the curve shows no peak but a plateau which is the residual shear stress. The residual shear stress results for all stress levels are shown in Fig. 4.89.

The New shear law and the Barton-Bandis model allow to calculate complete shear curves. So for this models a comparison with the lab curves is possible and displayed in Fig. 4.90.

4.14 Model-Experiment-Exercise MEX 3-2: Constant Normal Stiffness (CNS) Direct Shear Test

Daniel Pötschke (TUBAF) and Thomas Frühwirt (TUBAF)

4.14.1 Experimental Set-Up

Constant normal stiffness (CNS) direct shear tests follow the same basic principles like CNL tests (Sect. 4.13). Additionally to the initial normal load an extra load is added when the rock joint opens. This situation is comparable to a rock joint in the

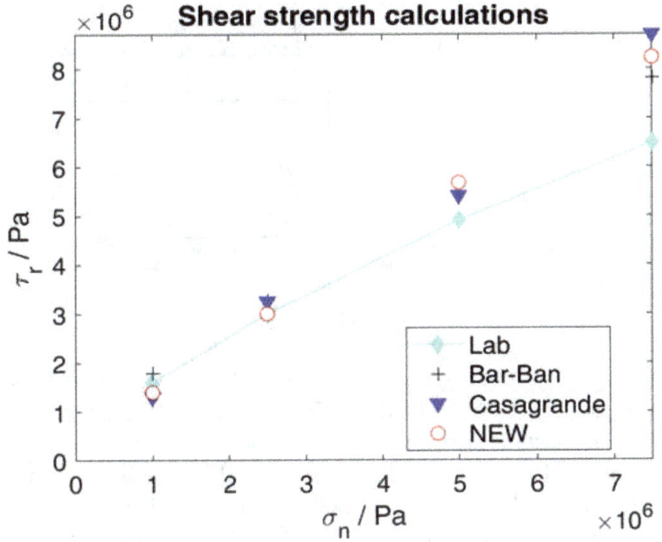

Fig. 4.89 Comparison of residual shear strength using different shear laws

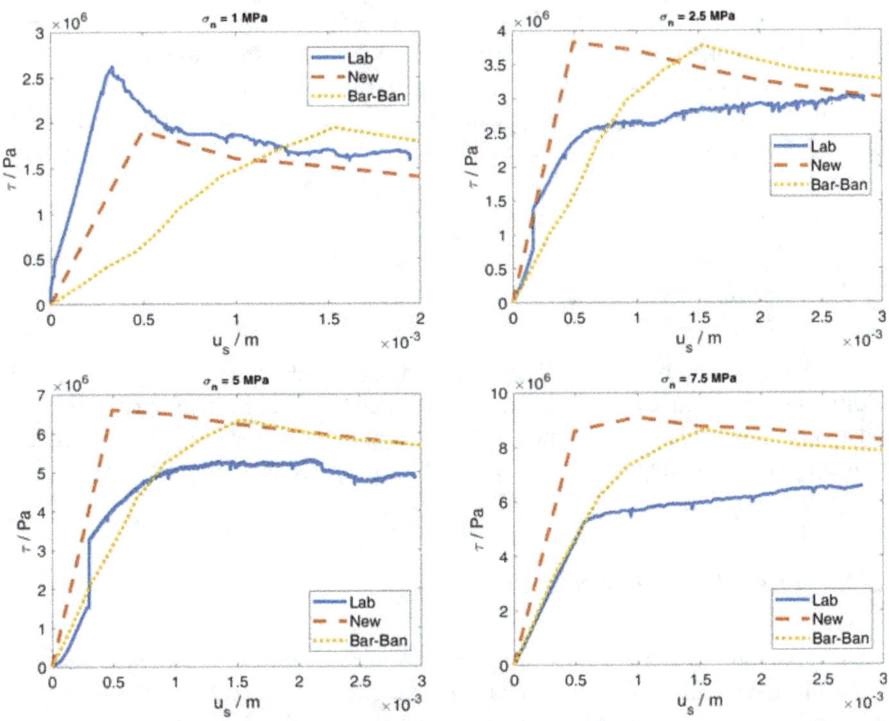

Fig. 4.90 Comparison of lab results with Barton-Bandis and New shear law for $\sigma_n =$ 1, 2.5, 5, 7.5 MPa using a grid constant of $a = 0.5$ mm

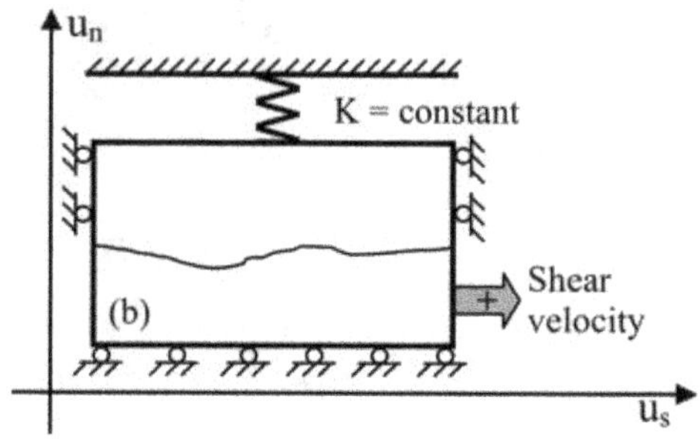

Fig. 4.91 Constant normal stiffness (CNS) direct shear test. (*From* [29])

underground where surrounding rock masses have to be deformed to open a joint. This deformation causes the extra loads. In Fig. 4.91 the stiffness term is represented as a spring with a constant stiffness. The difficulty is to estimate this stiffness. If the joint is surrounded by intact rock material it will be higher than for a jointed, weathered surrounding. A CNL test is a CNS test for which the stiffness is zero. This situation marks the lower boundary of the stiffness and is a valid boundary condition for rock joints close to the surface. The upper boundary of the stiffness is the stiffness of the intact rock. The real situation will be somewhere in between this extreme cases. The expected result is that for higher stiffness values the shear forces increase due to the increasing normal load when the joints starts to open.

The input data for this model exercise are explained in Sect. 5.3.12. The aim is to use this input to make good back calculations of the lab data. A good modelling approach for CNS tests can also be used for CNL tests by simply setting the stiffness to zero.

4.14.2 Model Approach

The model approach basically follows the approach of the CNL test, see Sect. 4.13. Additionally an estimation of the dilatation is needed. The normal stress has to be updated once the dilatation is known. The dilatation is modeled using a geometric approach where the dip angle of the worn surface is used as inclination angle.

4.14.3 Results and Discussion

The results show that in the lab testings an entire edge was destructed, see Fig. 4.92. Thus the simulation was done with 2 different scenarios: One with the original geometry and one with an adapted geometry where this edge was manually removed. The orientation of the edge in the left upper or right bottom corner is a result of different definitions using matrices and the plotting as surface or image.

Fig. 4.92 Visual comparison of the destructed areas of the rock joint. Notice the entirely destroyed edge

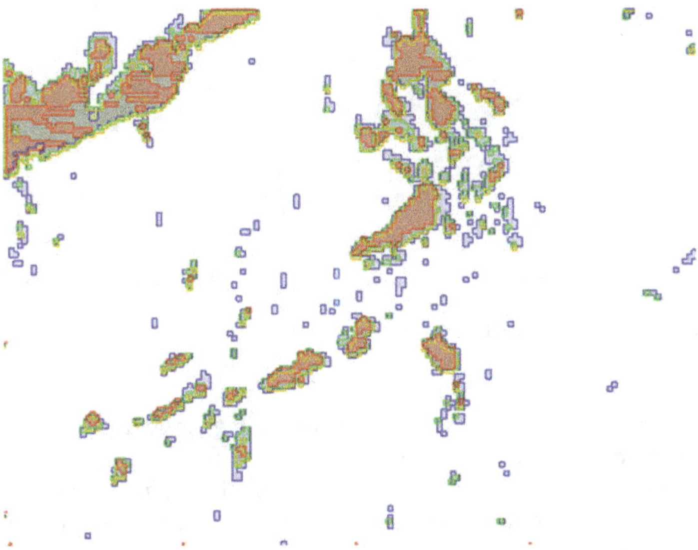

Fig. 4.93 The destructed area for different stiffness values. The area of destruction increases with the stiffness. *red*: 2 MPa, *yellow*: 4 MPa, *green*: 8 MPa, *blue*: 16 MPa

Figure 4.92 shows a visual comparison between the abrasion visible from the rock surface and the location of damage from the simulation. It can be concluded, that the used FFS approach (3.2.1) is able to roughly locate the occurrence of abrasion though it is not able to forecast the destruction of a bigger section.

In Fig. 4.93 a comparison between the simulated abrasion for different stiffness values during the CNS test series can be seen. As expected the area of damage

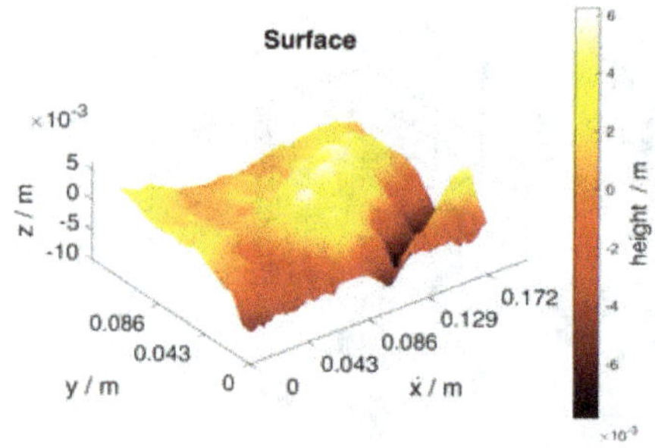

(a) Original geometry

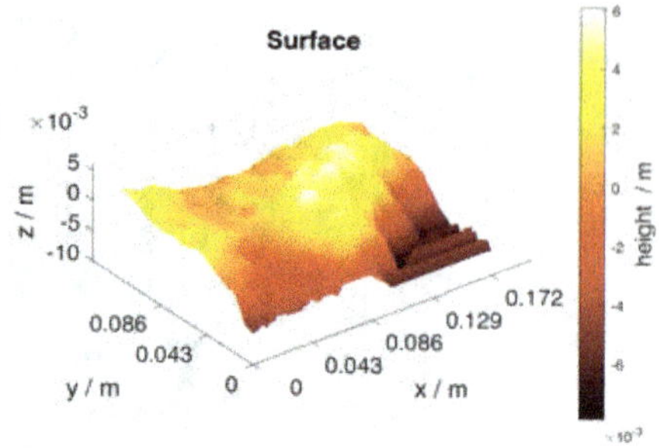

(b) Adapted geometry

Fig. 4.94 The used geometries for the CNS test

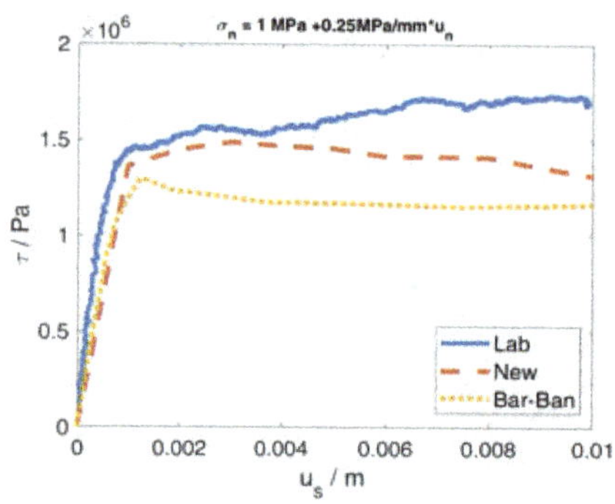

(a) Original geometry

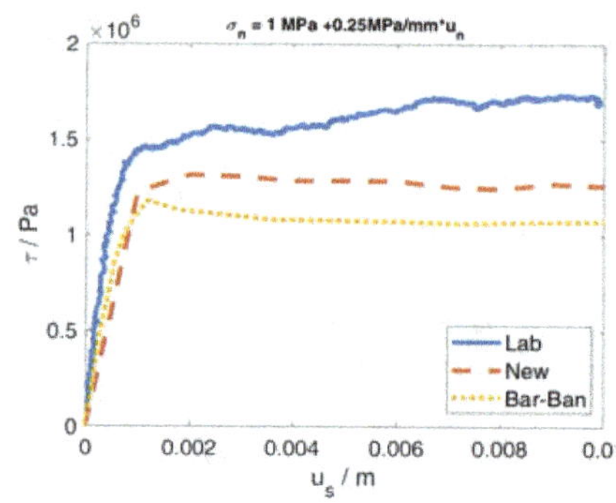

(b) Adapted geometry

Fig. 4.95 Shear curves for $K_n = 0.25\,\text{MPa}$

increases with increasing stiffness. The size of the bigger areas increases and new areas start to develop as well. This result acts as a verification of the code.

In Fig. 4.94 the used geometries for the the original morphology and the adapted one with the destructed edge can be seen. The results are shown for two different stiffness values and for comparison always the result for the original geometry and the adapted geometry are displayed.

In Figs. 4.95 and 4.96 the shear curves and the dilatation curves for a low stiffness value can be seen. The results show that the general trend of the shear curves is similar. On the other hand the dilatation curves are quite poor. Especially the dilatation curves for original and adapted geometry show some differences. In the shear curves this difference is minor because the low stiffness isn't dominating the overall normal stress and as a consequence the shear stress is not too strongly influenced.

In Figs. 4.97 and 4.98 the shear and dilatation curves for a higher stiffness can be seen. Now the stiffness dominates the normal stress and consequently the shear stress. That's why the curves for shear stress and dilatation have a similar shape. In this case the adapted geometry improves the quality of the results fairly though it is

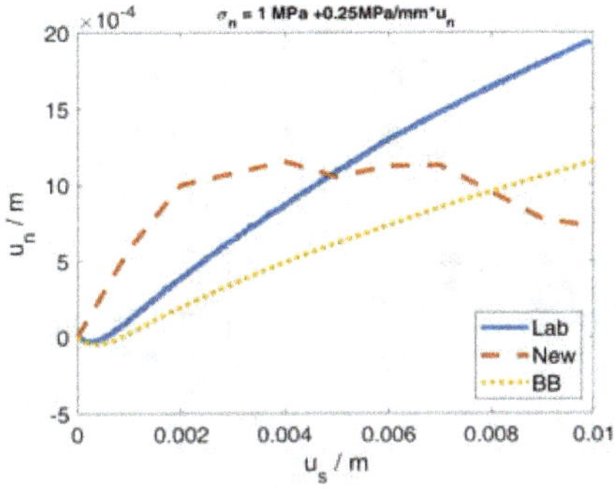

(a) Original geometry

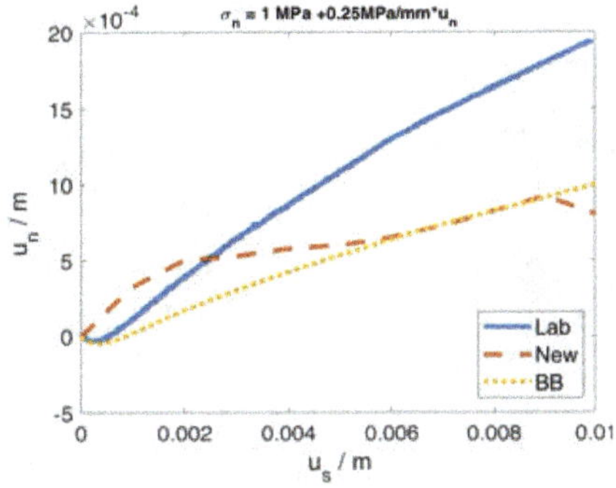

(b) Adapted geometry

Fig. 4.96 Dilatation curves for $K_n = 0.25\,\text{MPa}$

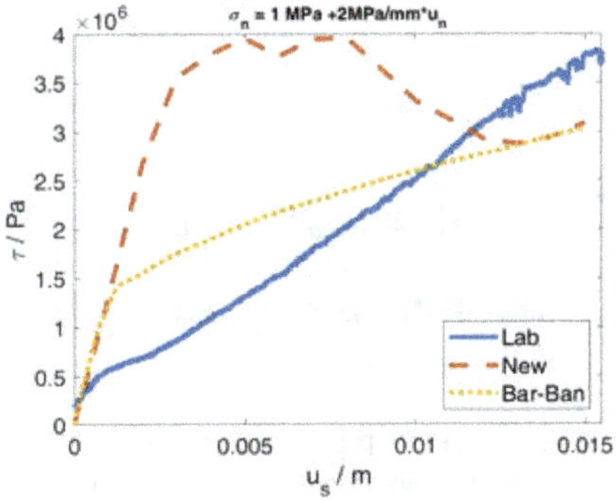

(a) Original geometry

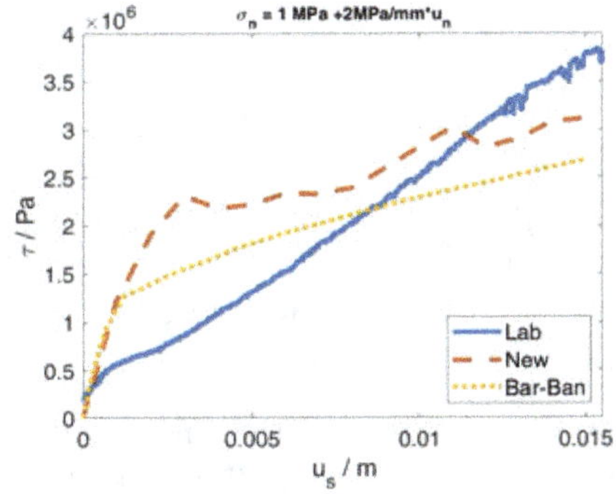

(b) Adapted geometry

Fig. 4.97 Shear curves for $K_n = 2.0\,\text{MPa}$

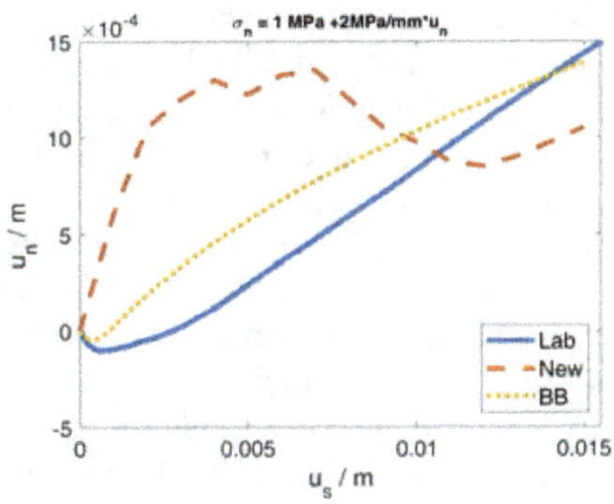

(a) Original geometry

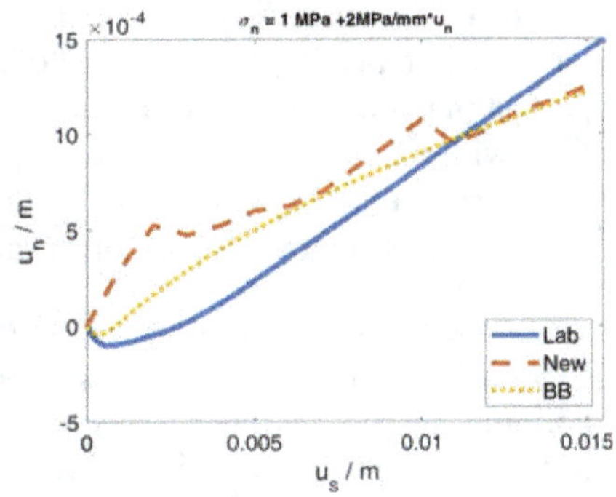

(b) Adapted geometry

Fig. 4.98 Dilatation curves for $K_n = 2.0\,\text{MPa}$

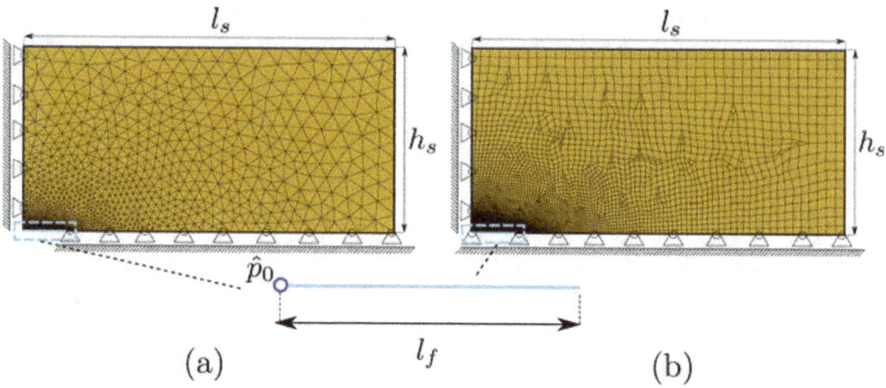

Fig. 4.99 Discretization and boundary conditions used for the weak coupling **a** and strong coupling scheme **b** throughout the validation

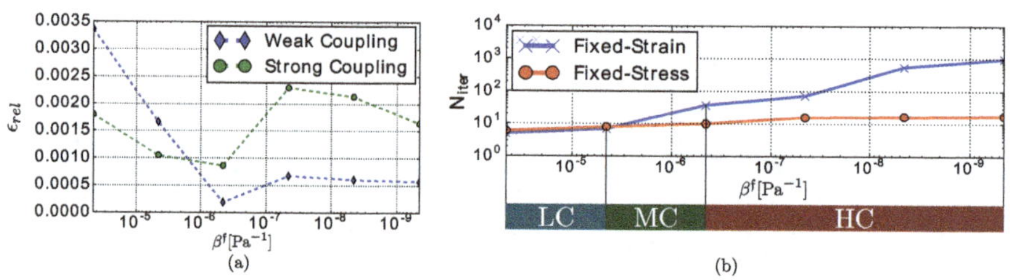

Fig. 4.100 Plot of the relative error ϵ_{rel} for the strong and weak coupling schemes (**a**) and recommended compressibility range of application for both methods based on the necessary number of iterations N_{iter} of the fixed-strain and fixed-stress scheme for **L**ow **C**ompressibilities, **M**oderate **C**ompressibilities and **H**igh **C**ompressibilities of the fluid (**b**)

not perfect. The main difference is that in the laboratory tests in the beginning of a shear test a compaction or settlement of the rock joint is observed. This is caused by imperfect matching of the two parts of the rock sample. The new code uses just one surface with a perfectly fitting counterpart. The result is the over prediction of the shear stress in the beginning stage of the shear test. For the original geometry this effect is more pronounced than for the adapted geometry (Figs. 4.99 and 4.100).

As a conclusion it can be said that the dilatation modelling is the key point to model CNS tests. With increasing stiffness values this dependency becomes more and more important. The pure geometric approach delivers results in the correct magnitude but it is still difficult to forecast how big the difference will be. By carefully inspecting all available information and using them the overall result quality could be increased significantly.

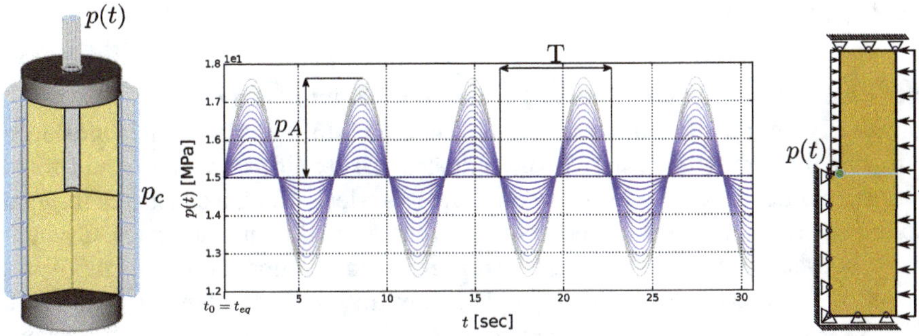

Fig. 4.101 Numerical set up of cylindrical probe under periodic hydraulic pressure $p(t)$ and confining pressure p_c

4.15 Model-Experiment-Exercise MEX 3-3: Cycling Loading Pressure Diffusion

Patrick Schmidt (UoS), Keita Yoshioka (UFZ), and Holger Steeb (UoS)

Fluid flow in high aspect ratio fractures results in perturbations of the equilibrium state of a fracture. Throughout harmonic testing transient fluid pressure changes induce small absolute fracture deformations which might lead to large relative changes of the hydraulic properties, such as the fracture's permeability. The variations within a testing period lead to a non-constant relationship between pressure and flow which can be investigated throughout numerical and experimental studies.

4.15.1 Experimental Set-Up

Shortly, a possible experimental setup for pressure diffusion measurements is described. We aim for a set-up which is close to the numerical investigation depicted in Fig. 4.101. Therefore, a cylindrical sample (diameter $D = 30$ mm, height $H = 75$ mm) of a crystalline rock (granite, marble, gneiss) is chosen and end faces are polished and prepared in parallel. The sample is cut in two identical parts and the two cutting areas are sandblasted. After sandblasting, a drill hole is established in one of the two pieces and a high-pressure fluid connector is pasted in. The roughness of the sandblasted surfaces should be well characterized by e.g. a profilometer. The drillhole is pre-saturated with a viscosity-adapted inert silicon oil (e.g. Fluorinert, 3M). Applying a sinusoidal total flux with a certain frequency through a syringe pump (Mid Pressure Syringe Pump neMESYS 1000N, Cetoni) while measuring the amplitude and phase shift of the pressure signal in the drill hole allows a comparison with numerical solutions depicted in Fig. 4.101.

4.15.2 Model Approach

Finite Element Approach: Hybrid-Dimensional Formulation

A physically sound flow model assuming compressible and viscous fluids in a deformable fracture setting is used to investigate pressure diffusion within fluid-filled fractures. In order to capture local and non-local transient effects a consistent implicit coupling of both domains is necessary to guarantee stability and accuracy. Calculations on non-conformal meshes and two different computational domains motivates a so-called weak coupling scheme implemented in the finite element framework FENICS. A robust numerically scheme is obtained by strongly coupled interface elements implemented in the DUNE framework.

Evaluation of Strong and Weak Coupling Scheme

Implementation of both schemes are firstly evaluated for a single fracture under linear conditions and compared with a reference solution obtained by Biot's formulation. The weak coupling scheme is implemented using a fixed-stress and fixed-strain algorithm providing a physical preconditioning and is mathematically similar to a Richardson type iteration. The system becomes numerically stiff for low fluid compressibilities on a local level since the physics are similar to volumetrically coupled problems. Strong and weak coupling schemes are tested on a single fracture with a length of 100 m under a constant fluid pressure of 20 kPa with regards of accuracy and stability (Table 4.13).

Non-linear Evaluation

Large deformations in high aspect ratios are reached for small absolute deformation values already and highly influences the fracture flow by local (permeability) and non-local (volumetric) deformation related perturbations. Time dependent pressure/flow stimulation requires a continuous transient change of fracture aperture and deformation dependent pressure diffusion. Validation of the non-linear formulation is using a cylindrical probe with a single fracture under a confining pressure σ_c and a harmonic pressure stimulation $p(t)$. In order to study the behaviour the amplitude of stimulation pressure is increased to enforce large deformation changes within a period.

Table 4.13 Collection of parameters used for validation of the weak and strong coupling schemes

Quantity	Value	Unit	Quantity	Value	Unit
Poroelastic domain $\mathcal{B}^{Pe}$					
Intrinsic permeability k^s	1.1×10^{-19}	[m^2]	Min fluid comp. β^f_{min}	4.5×10^{-10}	[1/Pa]
Max fluid comp. β^f_{max}	4.5×10^{-4}	[1/Pa]	Sample length l_{Pe}	1.0×10^{3}	[m]
Sample height h_{Pe}	5.0×10^{2}	[m]			
Fracture domain $\mathcal{B}^{Fr}$					
Min fluid comp. β^f_{min}	4.5×10^{-10}	[1/Pa]	Max fluid comp. β^f_{max}	4.5×10^{-4}	[1/Pa]
Fracture aperture δ_0	5.0×10^{-3}	[m]	Fracture length l^{Fr}	1.0×10^{2}	[m]
Pumping pressure p_0	2.0×10^{4}	[Pa]			
Numerical parameter					
Time step size Δt	1.0×10^{-2}	[s]	Fracture discret. in $\mathcal{B}^{Fr}$ Δx^{Fr}_{Fr}	1.0×10^{-1}	[m]
Fracture discret. in $\mathcal{B}^{Pe}$ Δx^{Fr}_{Pe}	1.0×10^{-1}	[m]	Evaluation time t_0	1.0×10^{1}	[s]
Evaluation position x_0	1.0×10^{1}	[m]	Error tolerance ϵ_{max}	1.0×10^{-6}	[–]
Number of DoF	1.4×10^{5}	[–]			

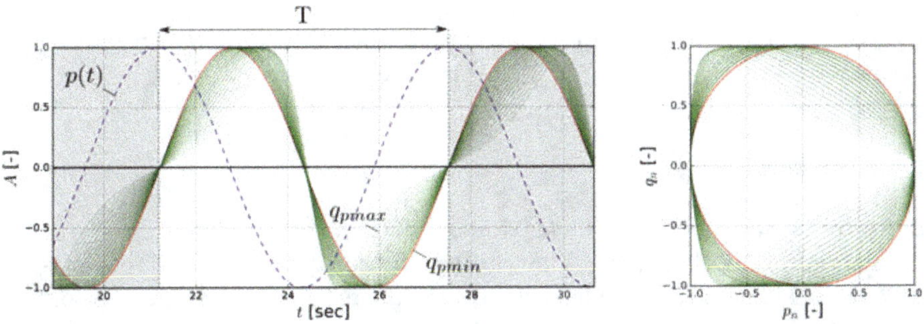

Fig. 4.102 Presentation of $p - q$ comparison via time and hysteresis plot. Flux solutions q_n are highlighted in green scales for varying pressure amplitudes normalized with respect to its maximum value, the normalized linear flux solution q_{lin} is highlighted in red and the normalized stimulation pressure $p_n(t)$ is highlighted in blue

Table 4.14 Collection of parameters used throughout numerical periodic hydraulic stimulation experiment on a sandstone probe

Quantity	Value	Unit	Quantity	Value	Unit
Rock matrix					
Dry frame bulk modulus K	2.2×10^{10}	[Pa]	Grain bulk modulus $K^{\mathfrak{s}}$	4.6×10^{10}	[Pa]
Shear modulus μ	1.77×10^{10}	[Pa]	Porosity ϕ	1.0×10^{-2}	[–]
Intrinsic permeability $k^{\mathfrak{s}}$	5.0×10^{-19}	[m²]	Fluid compressibility β^f	4.17×10^{-10}	[1/Pa]
Effective bulk modulus K_{eff}	3.98×10^{10}	[Pa]	Effective shear modulus μ_{eff}	1.77×10^{10}	[Pa]
Fracture domain					
Effective fluid viscosity η^{fR}	1.0×10^{-3}	[Pa s]	Initial fracture aperture δ_0	2.5×10^{-6}	[m]
Equilibrium fracture aperture δ_{eq}	4.61×10^{-6}	[m]	Fluid compressibility β^f	4.17×10^{-10}	[1/Pa]
Sample geometry and hydraulic stimulation					
Sample height h	7.5×10^{-2}	[m]	Sample diameter d	3.0×10^{-2}	[m]
Borehole diameter d_b	6.0×10^{-3}	[m]	Stimulation period T	$2.0 \times \pi$	[sec]
Equilibrium pressure p_{eq}	1.5×10^7	[Pa]	Reference Amplitude p_A^{lin}	1.0×10^3	[Pa]
Minimum Amplitude p_A^{min}	0.5×10^6	[Pa]	Maximum Amplitude p_A^{max}	3.0×10^6	[Pa]
Pressure increment Δp_A	0.5×10^6	[Pa]	Confining pressure p_c	2.0×10^7	[Pa]

Model by Variational Phase-Field Model

The hydromechanical coupling with varying fluid compressibility and rock mass in the variational phase-field implementation and the resulting crack opening computation in OGS is tested under the same condition. The hydromechanical coupling in OGS is achieved via a staggered approach with fixed stress. Crack opening in the variational phase-field mode is not direct variable but is a reconstruction from the diffused variables [28]. The axisymmetric model with the same settings is prepared as shown in Fig. 4.103.

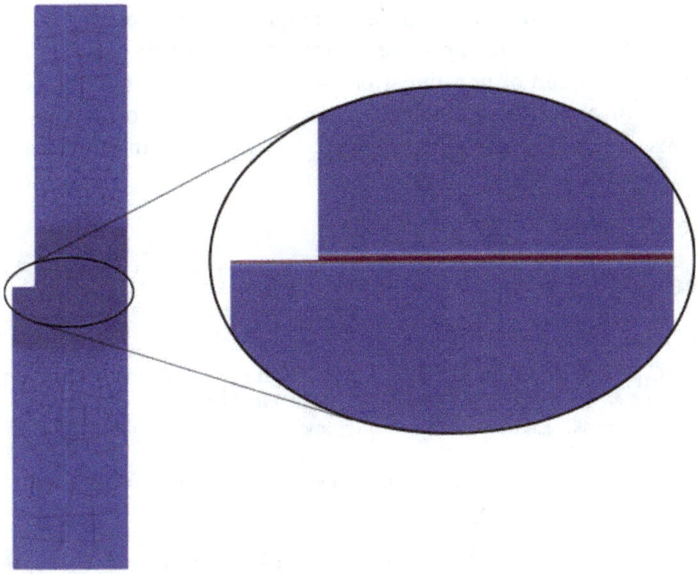

Fig. 4.103 Finite element mesh set up for variational phase-field simulation

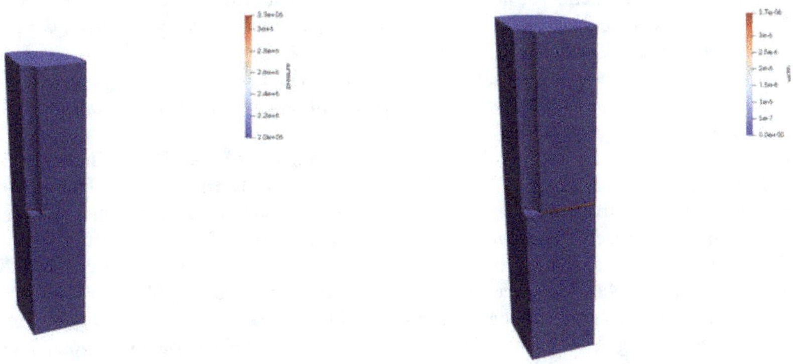

(**a**) Pressure simulation from the variational phase-field

(**b**) Crack opening from the variational phase-field

Fig. 4.104 Crystalline rock samples under investigation

4.15.3 Results and Discussion

Harmonic, fluid pressure induced perturbations of the fracture's equilibrium state result in a non-linear pressure-flow $(p - q)$ relationship as shown in Fig. 4.102. Injected fluid volume results in a volume change of the fracture leading to non-constant permeabilities throughout a testing period and changes of the characteristic diffusion time of the investigated system. The effect scales with the magnitude of the applied pressure amplitude since deformations of the fracture and changes in permeability are responses of the system to the non-constant acting pressure conditions. The phenomenon of non-linear pressure flow $(p - q)$ relations emphasizes the importance of characteristic changes due to induced deformations of investigated fractures and might serve as a tool to characterize the transient volume change (Fig. 4.104 and Table 4.14).

References

1. A. Tarokh, C.-S. Kao, A. Fakhimi, and J.F. Labuz. Spalling and brittleness in surface instability failure of rock. *Geotechnique*, 66(2):161–166, 2016. cited By 5.
2. M. Ostoja-Starzewski. Lattice models in micromechanics. *Applied mechanics*, 55(1):35–60, 2002.
3. R. Alessi, M. Ambati, T. Gerasimov, S. Vidoli, and L. De Lorenzis. *Comparison of Phase-Field Models of Fracture Coupled with Plasticity*, pages 1–21. Springer International Publishing, Cham, 2018.
4. R. Ince, A. Arslan, and B.L. Karihaloo. Lattice modelling of size effect in concrete strength. *Engineering Fracture Mechanics*, 70:2307–2320, 2003.
5. A. S. Sattari, H. B. Motra, Z. H. Rizvi, and F. Wuttke. A new lattice element method (lem) with integrated interface elements for determining the effective thermal conductivity of rock solids under thermo-mechanical processes. *International Symposium on Energy Geotechnics (SEG), Energy Geotechnics*, pages 266–275, 2019.
6. A. Sattari and et al. Thermo-mechanical fracture behaviour of opalinus clay - experimental and numercial analysis. *preparation for the Geo:N Topical Collection in Environmental Earth Sciences*, 2020.
7. Olaf Kolditz, Sebastian Bauer, Lars Bilke, Norbert Böttcher, Jens-Olaf Delfs, Thomas Fischer, Uwe J Görke, Thomas Kalbacher, Georg Kosakowski, CI McDermott, et al. Opengeosys: an open-source initiative for numerical simulation of thermo-hydro-mechanical/chemical (thm/c) processes in porous media. *Environmental Earth Sciences*, 67(2):589–599, 2012.
8. Dmitri Yu. Naumov, Lars Bilke, Thomas Fischer, Yonghui Huang, Christoph Lehmann, Xing-Yuan Miao, Thomas Nagel, Francesco Parisio, Karsten Rink, Haibing Shao, Wenqing Wang, Norihiro Watanabe, Tianyuan Zheng, and Olaf Kolditz. Appendix a: Opengeosys-6. In Olaf Kolditz, Thomas Nagel, Hua Shao, Wenqing Wang, and Sebastian Bauer, editors, *Thermo-Hydro-Mechanical-Chemical Processes in Fractured Porous Media: Modelling and Benchmarking*, pages 271–277. Springer International Publishing, 2018.
9. Lorenzo Adolph Richards. Capillary conduction of liquids through porous mediums. *Physics*, 1(5):318–333, 1931.
10. Maurice A Biot. General theory of three-dimensional consolidation. *Journal of applied physics*, 12(2):155–164, 1941.
11. Gesa Ziefle, Jean-Michel Matray, Jobst Maßmann, and Andreas Möri. Coupled hydraulic-mechanical simulation of seasonally induced processes in the mont terri rock laboratory (switzerland). In *Mont Terri Rock Laboratory, 20 Years*, pages 197–214. Springer, 2018.
12. M Th Van Genuchten. A closed-form equation for predicting the hydraulic conductivity of unsaturated soils 1. *Soil science society of America journal*, 44(5):892–898, 1980.
13. W.J. Xu, H. Shao, J. Hesser, W. Wang, K. Schuster, and O. Kolditz. Coupled multiphase flow and elasto-plastic modelling of in-situ gas injection experiments in saturated claystone (mont terri rock laboratory). *Engineering Geology*, 157:55–68, 2013.

14. Katrin M Wild, Linda P Wymann, Sebastian Zimmer, Reto Thoeny, and Florian Amann. Water retention characteristics and state-dependent mechanical and petro-physical properties of a clay shale. *Rock Mechanics and Rock Engineering*, 48(2):427–439, 2015.
15. H Bock. Updated review of the rock mechanics properties of the opalinus clay of the mont terri url based on laboratory and field testing. Technical report, Mont Terri Technical Report 2008-04, Q+ S Consult, Germany, 2009.
16. Alexander Bond, Alain Millard, Shigeo Nakama, Chengyuan Zhang, and Benoit Garritte. Approaches for representing hydro-mechanical coupling between sub-surface excavations and argillaceous porous media at the ventilation experiment, mont terri. *Journal of Rock Mechanics and Geotechnical Engineering*, 5(2):85–96, 2013.
17. Joel H Ferziger and Milovan Peric. *Computational methods for fluid dynamics*. Springer Science & Business Media, 2012.
18. Christophe Geuzaine and Jean-François Remacle. Gmsh: A 3-d finite element mesh generator with built-in pre-and post-processing facilities. *International journal for numerical methods in engineering*, 79(11):1309–1331, 2009.
19. Tecplot. Tecplot: Data visualization and CFD post-processing software (https://www.tecplot.com/), 2019.
20. Katrin M Wild, Patric Walter, and Florian Amann. The response of opalinus clay when exposed to cyclic relative humidity variations. *Solid Earth*, 8(2):351–360, 2017.
21. Roland Wynne Lewis and Bernard A Schrefler. *The finite element method in the static and dynamic deformation and consolidation of porous media*. John Wiley, 1998.
22. Karl Terzaghi. Theoretical soil mechanics. johnwiley & sons. *New York*, pages 11–15, 1943.
23. Alan W Bishop and GE Blight. Some aspects of effective stress in saturated and partly saturated soils. *Geotechnique*, 13(3):177–197, 1963.
24. Björn Christoph Hübner. *Simultane Analyse von Bauwerk-Wind-Wechselwirkungen*. Inst. für Statik, TU Braunschweig, Germany, 2003.
25. Charbel Farhat and Michael Lesoinne. Two efficient staggered algorithms for the serial and parallel solution of three-dimensional nonlinear transient aeroelastic problems. *Computer methods in applied mechanics and engineering*, 182(3-4):499–515, 2000.
26. Minkley, W., Brückner, D., Lüdeling, C. (2018): Percolation in Salt Rocks. *Mechanical Behavior of Salt IX, 2018, Proceedings* (ISBN 978-3-9814108-6-0), 29-46.
27. W.P. Kamlot. *Habilitationsschrift: Gebirgsmechanische Bewertung der geologischen Barrierefunktion des Hauptanhydrits in einem Salzbergwerk*. des Instituts für Geotechnik der Technischen Universität Bergakademie Freiberg, 2009.
28. Keita Yoshioka, Dmitri Naumov, and Olaf Kolditz. On Crack Opening Computation in Variational Phase-Field Models for Fracture. *Computer Methods in Applied Mechanics and Engineering*, under review.
29. Van-Manh Nguyen, Heinz Konietzky, and Thomas Frühwirt. New methodology to characterize shear behavior of joints by combination of direct shear box testing and numerical simulations. *Geotechnical and Geological Engineering*, 32(4):829–846, 2014.
30. N. Barton, S. Bandis, and K. Bakhtar. Strength, deformation and conductivity coupling of rock joints. *International Journal of Rock Mechanics and Mining Sciences and Geomechanics Abstracts*, 22(3):121–140, 1985.
31. Cai-Chu Xia, Zhi-Cheng Tang, Wei-Min Xiao, and Ying-Long Song. New peak shear strength criterion of rock joints based on quantified surface description. *Rock Mechanics and Rock Engineering*, 47(2):387–400, 2014.
32. Davide Casagrande, O Buzzi, Anna Giacomini, Cedric Lambert, and G Fenton. A new stochastic approach to predict peak and residual shear strength of natural rock discontinuities. *Rock Mechanics and Rock Engineering*, pages 69–99, 08 2017.
33. W. Minkley, M. Knauth, T. Fabig, N. Farag. Stability of salt caverns under consideration of hydro-mechanical loading. *Mechanical Behavior of Salt VIII*, pages 217–227, 2015. Rapid City, USA, 26–28 May 2015.

5

GeomInt and Data Management: Architectures, Practices and Procedures

Carolin Helbig, Uwe-Jens Görke, Mathias Nest, Daniel Pötschke, Amir Shoarian Sattari, Patrick Schmidt, Bernhard Vowinckel, Keita Yoshioka, and Olaf Kolditz

Data management includes the development and use of architectures, guidelines, practices and procedures for accurate managing of data during the entire data lifecycle of an institutional unit or a research project defined as different information units such as numbers, alphabetic characters, and symbols that are particularly formatted and can be processed by computer. The data in the project is provided by various actors which can be GeomInt partners, their legal representatives, employees, and external partners. GeomInt Data is provided at GeomInt data management portal (DMP).

In GeomInt project the partners work in very close cooperation. Project-owned and connected infrastructures are synergetic used (as illustrated in Fig. 5.1). In addition to the rock mechanics laboratories of the partners CAU, IfG and TUBAF with partly unique equipment data from ongoing experiments in the underground laboratories

C. Helbig (✉) · U.-J. Görke · K. Yoshioka
UFZ, Helmholtz Centre for Environmental Research, Leipzig, Germany
e-mail: carolin.helbig@ufz.de

M. Nest
IfG, Institut für Gebirgsmechanik Leipzig, Leipzig, Germany

D. Pötschke
TUBAF, Technische Universität Bergakademie Freiberg, Freiberg, Germany

A. S. Sattari
CAU, Christian-Albrechts-Universität zu Kiel, Kiel, Germany

P. Schmidt
UoS, University of Stuttgart, Stuttgart, Germany

B. Vowinckel
Leichtweiß-Institute for Hydraulic Engineering and Water Resources, Technische Universität Braunschweig, Braunschweig, Germany

O. Kolditz
UFZ, Helmholtz Centre for Environmental Research, Technische Universität Dresden, Dresden, Germany

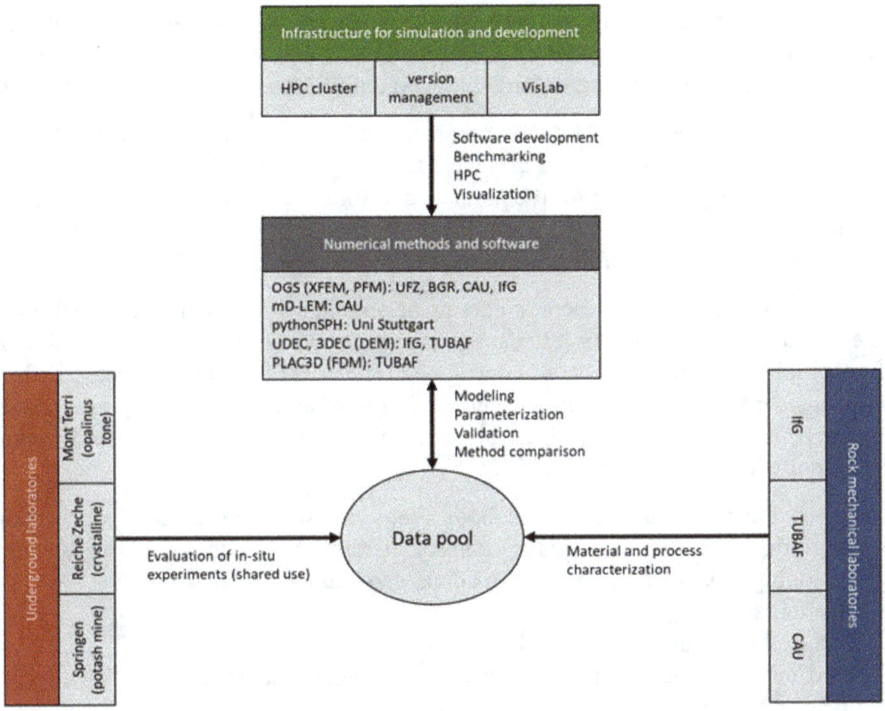

Fig. 5.1 Network diagram to illustrate synergies and dependencies in the GeomInt network in connection with the infrastructure elements and numerical methods in the project

are accessed. An essential element of the GeomInt project is the simulation and software development structures illustrated in Fig. 5.1. With regard to the use of the simulation platform OpenGeoSys, the development of which is coordinated by the UFZ and in whose further development the GeomInt partners BGR, CAU, IfG and TUBAF are involved, the simulation and development infrastructures located at the UFZ, including version management, is available to these partners.

The collaborative work requires data management structures and guidelines. Therefore, the first step was to set up a document that includes a user agreement and a data management plan which is the basis for data management in the project.

5.1 User Agreement and Data Management Plan

The GeomInt project partners agreed to set up a user agreement which includes specifications for data structures including metadata, data formats, access authorization for data, the possible publication of data, as well as the handling of the data after the end of the project and outside the project. A first version of this user agreement was created six months after the start of the project.

The user agreement includes guidelines and definitions for the following aspects

- Which data will be generated in the project and has to be managed?
- How will data be provided and exchanged?
- What are the rights of use for the partners and for third parties?
- How to cite data?
- How to supervise the compliance of the user agreement?

As part of the user agreement, a data management plan, which is a formal document that describes how project data is managed during the research period and after completion of the project, was developed. The goal of a data management plan is to consider the aspects of data management (metadata creation, data preservation and analysis) before the start of the project. Following points are discussed in the GeomInt data management plan:

1. Generation and management methods (data infrastructure, external data, data integration, data formats, quality control, user groups, data processing stages, versioning, documentation and meta data, geocoding)
2. Data Legal Management
3. Data exchange and provision, citation rules
4. Short-term storage and data management (storages, data transfer, backup, security)
5. Long-term storage (characteristics, metadata and documentation, responsibility)
6. Resources (organizational roles and responsibilities for data management).

5.2 GeomInt Data

The project results include specific data from laboratory and in-situ experiments, software components and data sets from numerical simulations (i.e. model and result files). An estimation of the extent of the data generated in GeomInt could not be made before the project. Therefore, the data management concept had to be flexible. This uncertainty was mainly due to the fact that the evaluation of test and calculation results may lead to a change in test and calculation planning and may even lead to additional experiments or simulation calculations.

The availability of experimental and numerical data generated in the project, including existing metadata, is realized on an internal area of the Geomint homepage. The UFZ is responsible for the project data and has many years of experience in data management regarding the cooperative development of open source software (OpenGeoSys) as well as the acquisition, storage and processing of data from experiments on different scales, exploration and monitoring campaigns, numerical simulations and scientific 3D visualizations.

The UFZ has sufficient capacities and modern data management systems for data storage, which are available as a central data infrastructure for the research network. Specifically, data sets are managed by means of an ORACLE database. Access is via a web portal, where each data record must be provided with metadata before uploading. The metadata standard used is compatible with the INSPIRE Directive 2007/2/EC and also regulates the rights for access, use and transfer of the data. A tape system is also available for the long-term storage of very large amounts of data. For the provision of exploration and monitoring data, geo-services mentioned in the GDI-DE are used as far as possible. Since such services for complex modelling and simulation data do not exist so far, the provision is done via a data research portal, where data can be found by means of stored metadata.

As software components are part of non-commercial, scientific program platforms and are open source products (e.g.. OpenGeoSys), they are hosted by the responsible partner via established source code hosting services (e.g. GitHub) and is publicly available. A possible public access to project data, which goes beyond the status quo as described in technical publications, as well as the handling of the data after the end of the project is regulated in the cooperation agreement or in the cooperation contract between the project partners.

The handling of data obtained from the in-situ experiments in the underground laboratories through synergies with other projects is also regulated separately (access authorisation for these data, storage location, publication, handling of the data after the end of the project). Such an approach is necessary because specific parts of these data can be used for the scientific purposes of GeomInt, but they are generated in other projects with partly other partners.

5.3 GeomInt DMP

In this section, exemplary data sets of every project partner are described. A table of these data sets including description and link are available only for project partners at the website (Fig. 5.2). Some data sets can be found on the UFZ data investigation portal https://www.ufz.de/drp/. These data sets are uploaded to the data management portal at UFZ (DMP@UFZ).

The GeomInt data management system (DMS) is organised in three sections (Fig. 5.3):

- Experimental data (Fig. 5.3 left),
- Simulation data (Fig. 5.3 right),
- and data connected to URLs (Fig. 5.3 bottom right).

Table 5.1 summarizes the MEX related data concerning experiments and simulations. A selection will be described in the following sections.

The following codes (and related input files) are used (see Chap. 7 for detailed code introductions).

Fig. 5.2 GeomInt DMS portal https://www.ufz.de/geomint/index.php?de=46799

Software Codes

- LEM: In-house developed MATLAB In-house code available in executable P-file format.
- DEM: Commercial code by Itasca Ltd.
- SPH: In-house code.
- OGS (OpenGeoSys): https://www.opengeosys.org/releases/.
- HDF: In-house code.

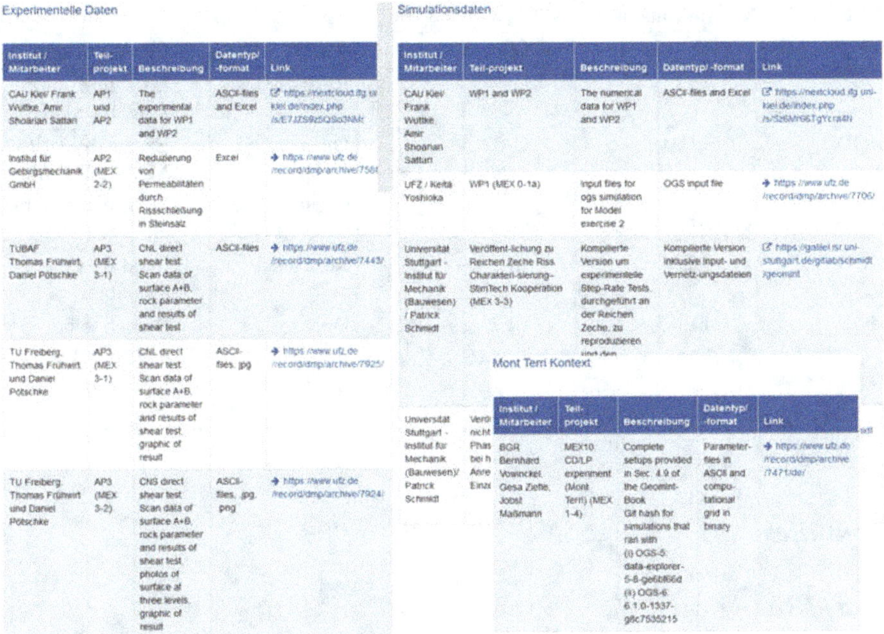

Fig. 5.3 GeomInt DMS portal: Data areas for experimental, simulation and URL related information

Table 5.1 MEX data management

MEX	TOP	EXP	MOD				
WP			LEM	DEM	FEM	HDF	FFS
0-1a	Bending fracture test	LIT	✓	✓	✓		
0-1b	Bending fracture test (aniso)						
0-2	Humidity controlled bending	Concept					
1-1a	Swelling of clay						
1-1b	Swelling of clay						
1-2	Shrinkage of clay						
1-3	Desiccation of clay						
1-4	CD/LP experiment						
2-1a	Pressure driven percolation						
2-1b	Pressure driven percolation						
2-2	Healing / closure						
2-3	Compressible fluids						
2-4	URL Springen						
3-1	CNL test						
3-2	CNS test						
3-3	Cyclic loading						

Table 5.2 MEX 0-1a: Data overview

Type	Spec.	Owner	Access	Comment
EXP	LIT	[1]	Restricted	Literature available via UFZ
MOD	LEM	CAU	Open	Executable MATLAB P-file
			Open	Input files will be uploaded
MOD	DEM	IfG	Restricted	Input files available
MOD	FEM	UFZ	Open	Branch to be merged into OGS
			Open	Input files available

Input files

Input files

- LEM: In-house documentation
- DEM: Itasca Ltd. (user's manuals)
- SPH/HDF: In-house documentation
- OGS (Benchmarks): https://www.opengeosys.org/docs/benchmarks/
- HDF: In-house documentation.

5.3.1 MEX 0-1a: Bending Fracture Test

Participating institutions of MEX 0.1a (see section 4.1): CAU, UFZ, IfG*[1] (Table 5.2).

[1] IfG as a user of commercial codes has specific data management rules.

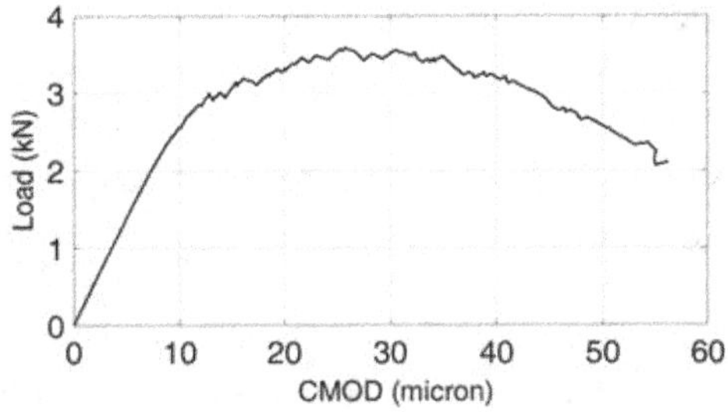

Fig. 5.4 The load versus CMOD from the lattice simulation

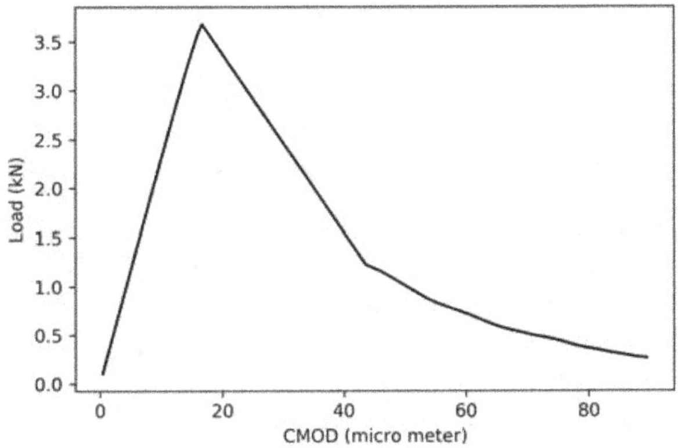

Fig. 5.5 The load versus CMOD from the VPF simulation results

CAU (LEM)

The required LEM code and the input variables of the three-Point fracture toughness test on the Rockville Granite samples are uploaded to the IfG (Kiel) NextCloud server. The data is accessible through the following link: https://nextcloud.ifg.uni-kiel.de/index.php/s/pRmBPJ9gK5Se6ci.

The uploaded protected MATLAB file in a *.p format requires a MATLAB version with a built-in Voronoi Tessellation and Delaunay Triangulation functions. The input variables are prepared in a single file for the simulation of the fracture toughness in Rockville Granite. Figure 5.4 shows the relation between the load versus CMOD as described in Sect. 4.1.

UFZ (FEM-VPF)

The source code can be found in OpenGeoSys project on github and the input files for the three point bending test performed on the Rockville Granite samples have been uploaded. The files include the unstructured finite element mesh files in `vtu` format and an OGS input file in `xml` format. As homogeneous properties such as Young's modulus are assigned int the computational domain, the spatially constant material properties are specified in the OGS input file rather than in the mesh file. The load and crack mouth opening displacment computed from the simulations are shown in Fig. 5.5 as described in Sect. 4.1.

Table 5.3 MEX 0-1a (CAU)

Data label	GeomInt I CAU I Bending fracture test, Granite
URL (Numerics)	https://nextcloud.ifg.uni-kiel.de/index.php/s/pRmBPJ9gK5Se6ci
Subject	LEM simulation of bending fracture test (Granite)
Type of data	Executable MATLAB P-file, input parameters
Data quality	Quality assured data
Status of data	Unprocessed data
Data format	txt, MATLAB executable P-file
Creators	Kiel University, Institute of Geomechanics and Geotechnics, Ludewig-Meyn-Straße 10, 24118, Kiel
Source/Origin	In-house code
Publisher	Kiel University, Institute of Geomechanics and Geotechnics, Ludewig-Meyn-Straße 10, 24118, Kiel
Rights holders	Kiel University, Institute of Geomechanics and Geotechnics, Ludewig-Meyn-Straße 10, 24118, Kiel
Contributors	Kiel University, Institute of Geomechanics and Geotechnics: Amir Shoarian Sattari, Frank Wuttke
Time/period of creation	2018–2019
Language of the content	English
Update policy	Stored data is final
Access permissions	Full access

MEX 0-1a (UFZ) will be also provided as an OGS benchmark case at: https://www.opengeosys.org/docs/benchmarks/phase-field/pf_tpb/.

Meta Data Overview (According to Dublin Core)
See Tables 5.3, 5.4 and 5.5.

5.3.2 *MEX 0-1b: Three-Point Fracture Toughness Test, Opalinus Clay*

Participating institutions of MEX 0-1b (see Sect. 4.2): CAU, UFZ (Table 5.6).

CAU Kiel

The experimental results of the three-point fracture toughness test on the Opalinus Clay samples are uploaded to the IfG (Kiel) NextCloud server. The data is accessible through the following link: https://nextcloud.ifg.uni-kiel.de/index.php/s/pJxp2eNEJb6PfiS.

The data set, which includes the time, applied force (N) and the displacement of the sample at the loading point (mm), is provided in a *.txt file. The crack mouth opening displacement (CMOD), which is determined from the image processing technique (Sect. 2.2.2), is given in a *.xlsx file. The data includes the time and the calculated CMOD (mm).

Table 5.4 MEX 0-1a (UFZ)

Data label	MEX 0-1a (UFZ)
URL	http://www.ufz.de/record/dmp/archive/7706
Subject	OGS-VPF simulation of bending fracture test (crystalline rock)
Type of data	Data set (structured data in a defined format)
Data quality	Quality assured data by benchmarking
Status of data	Processed data
Data format	OGS files
Creators	Yoshioka, Keita
Source/Origin	Open source
Publisher	Helmholtz Centre for Environmental Research UFZ
Rights holders	Helmholtz Centre for Environmental Research UFZ
Contributors	Yoshioka, Keita
Time/period of creation	2019–2020
Language of content	English
Update policy	To be merged to OGS benchmarks (see below)
Access permissions	Free access

Table 5.5 MEX 0-1a (IfG)

Data label	MEX 0-1a (IfG)
URL	
Subject	DEM simulation of bending fracture test (crystalline rock)
Type of data	Data set (structured data in a defined format)
Data quality	Quality assured data by benchmarking
Status of data	
Data format	
Creators	Nest, Mathias
Source/Origin	Commercial code
Publisher	Institut für Gebirgsmechanik IfG
Rights holders	Institut für Gebirgsmechanik IfG
Contributors	Nest, Mathias
Time/period of creation	2019–2020
Language of content	English
Update policy	
Access permissions	Limited access

Table 5.6 MEX 0-1b: Data overview

Type	Spec.	Owner	Access	Comment
EXP	Ben	CAU	Open	Output files are uploaded
MOD	LEM	CAU	Open	Executable MATLAB P-file
			Open	Input files available
MOD	FEM	UFZ	Open	Branch to be merged into OGS
			Open	Input files available

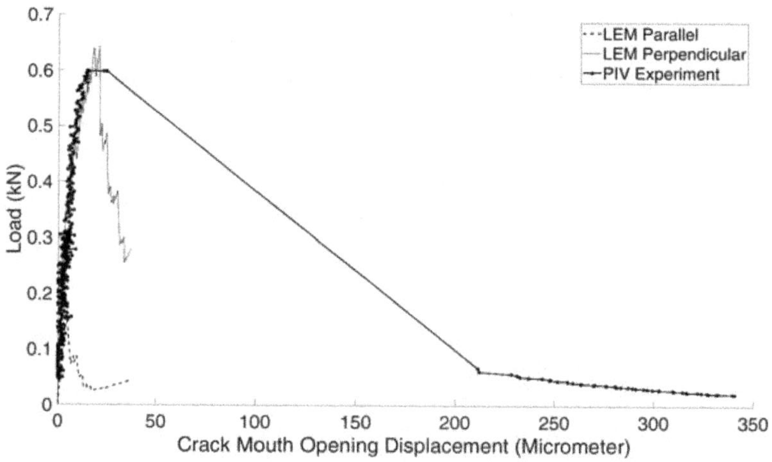

Fig. 5.6 The load versus crack mouth opening displacement (CMOD) response of the Opalinus Clay

The required LEM code and the input variables of the three-Point fracture toughness test on the Opalinus Clay samples are uploaded to the IfG (Kiel) NextCloud server. The data is accessible through the following link: https://nextcloud.ifg.uni-kiel.de/index.php/s/ZBFN2rSZ99kPY9M.

The uploaded protected MATLAB file in a *.p format requires a MATLAB version with a built-in Voronoi Tessellation and Delaunay Triangulation functions. The input variables are prepared in two different files for a parallel and perpendicular embedded layer orientations. Figure 5.6 shows the comparison between the experimental and numerical data as described in Sect. 4.2.

UFZ

The input files for OGS, which were used to simulate the three point bending test performed on the orthogonal and parallel laminations of Opalinus Clay samples, have been uploaded. The files include the unstructured finite element mesh files in vtu format and OGS input files in xml format. Also in the mesh files, the material properties are defined per element. Particularly for the orthogonal and the parallel Lamentations in the samples are represented through a contrast in the fracture tough-

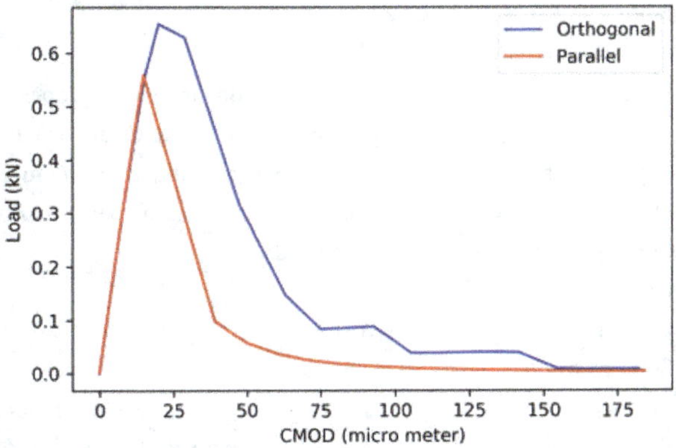

Fig. 5.7 The load versus crack mouth opening displacement (CMOD) response simulations for orthogonal and parallel lamination by VPF

ness in the samples and can be found in the mesh files. The load and crack mouth opening displacment computed from the simulations are shown in 5.7 as described in Sect. 4.2.

MEX 0-1b (UFZ) will be also provided as an OGS benchmark case at: https://www.opengeosys.org/docs/benchmarks/phase-field/pf_tpb_ani/.

Meta Data Overview (According to Dublin Core)

See (Tables 5.7 and 5.8)

Table 5.7 MEX 0-1b (CAU)

Data label	GeomInt, MEX 0-1b, CAU, Bending fracture test, Opalinus Clay
URL (Experiments)	https://nextcloud.ifg.uni-kiel.de/index.php/s/pJxp2eNEJb6PfiS
URL (Numerics)	https://nextcloud.ifg.uni-kiel.de/index.php/s/ZBFN2rSZ99kPY9M
Subject	Bending fracture test, Opalinus Claye
Type of data	Experimental data, executable MATLAB P-file, input parameters
Subject	Bending fracture test, Opalinus Clay
Type of data	Experimental data, executable MATLAB P-file, input parameters
Data quality	Quality assured data
Status of data	Unprocessed data
Data format	txt, xlsx, MATLAB executable P-file
Creators	Kiel University, Institute of Geomechanics and Geotechnics, Ludewig-Meyn-Straße 10, 24118, Kiel
Source/Origin	In-house code
Publisher	Kiel University, Institute of Geomechanics and Geotechnics, Ludewig-Meyn-Straße 10, 24118, Kiel
Rights holders	Kiel University, Institute of Geomechanics and Geotechnics, Ludewig-Meyn-Straße 10, 24118, Kiel
Contributors	Kiel University, Institute of Geomechanics and Geotechnics: Amir Shoarian Sattari, Frank Wuttke
Time/period of creation	2018–2019
Language of the content	English
Update policy	Stored data is final
Access permissions	Full access

Table 5.8 MEX 0-1b (UFZ)

Data label	MEX 0-1b (UFZ)
URL	https://www.opengeosys.org/docs/benchmarks/
Subject	Bending fracture test, Opalinus Clay
Type of data	Data set (structured data in a defined format)
Data quality	Quality assured data by benchmarking
Status of data	Processed data
Data format	OGS files
Creators	Yoshioka, Keita
Source/Origin	Open source
Publisher	Helmholtz Centre for Environmental Research UFZ
Rights holders	Helmholtz Centre for Environmental Research UFZ
Contributors	Yoshioka, Keita
Time/period of creation	2019–2020
Language of content	English
Update policy	To be merged to OGS benchmarks (see below)
Access permissions	Free access

5.3.3 MEX 1-1: Swelling Process, Red Salt Clay

Participating institutions of MEX 1-1 (see Sect. 4.2): CAU, IfG (Table 5.9).

CAU Kiel

The required LEM code and the input variables for simulating the swelling process of the salt clay are uploaded to the IfG (Kiel) NextCloud server. The data is accessible through the following link: https://nextcloud.ifg.uni-kiel.de/index.php/s/JmZseQqrsbgWNqC.

The uploaded protected MATLAB file in a *.p format requires a MATLAB version with a built-in Voronoi Tessellation and Delaunay Triangulation functions.Fig. 5.8 shows the change of hydraulic conductivity with applied linear strains as described in Sect. 4.4.

Meta Data Overview (According to Dublin Core)
See (Table 5.10).

Table 5.9 MEX 1-1: Data overview

Type	Spec.	Owner	Access	Comment
EXP		IfG		Available on demand
MOD	LEM	CAU	Open	Executable MATLAB P-file
			Open	Input files available
MOD	DEM	IfG	Restricted	Please contact IfG

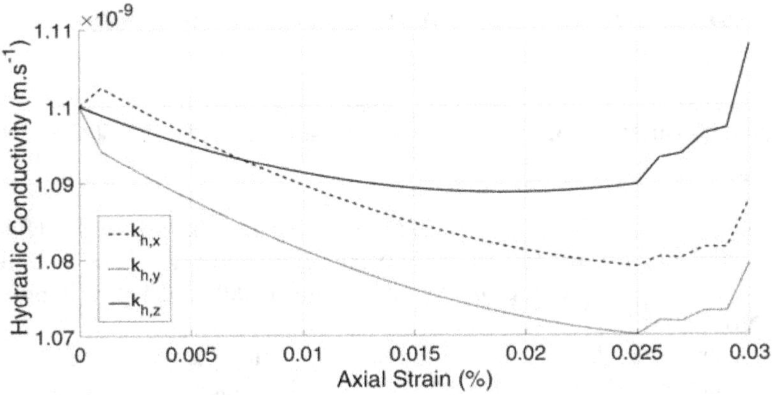

Fig. 5.8 The change of hydraulic conductivity with applied linear strains

Table 5.10 MEX 1-1 (CAU)

Data label	GeomInt—CAU—Swelling process, Red Salt Clay
URL (Numerics)	https://nextcloud.ifg.uni-kiel.de/index.php/s/JmZseQqrsbgWNqC
Subject	Swelling process (Red Salt Clay)
Type of data	Executable MATLAB P-file, input parameters
Data quality	Quality assured data
Status of data	Unprocessed data
Data format	txt, MATLAB executable P-file
Creators	Kiel University, Institute of Geomechanics and Geotechnics, Ludewig-Meyn-Straße 10, 24118, Kiel
Source/Origin	In-house code
Publisher	Kiel University, Institute of Geomechanics and Geotechnics, Ludewig-Meyn-Straße 10, 24118, Kiel
Rights holders	Kiel University, Institute of Geomechanics and Geotechnics, Ludewig-Meyn-Straße 10, 24118, Kiel
Contributors	Kiel University, Institute of Geomechanics and Geotechnics: Amir Shoarian Sattari, Frank Wuttke
Time/period of creation	2019–2020
Language of the content	English
Update policy	Stored data is final
Access permissions	Full access

5.3.4 *MEX 1-2: Drying and Wetting Paths of the Opalinus Clay*

Participating institutions of MEX 1-2 (see Sect. 4.5): CAU, UFZ (Table 5.11).

CAU Kiel

The experimental results of the drying and wetting paths of the sandy Opalinus Clay are uploaded to the IfG (Kiel) NextCloud server. The data is accessible through the following link: https://nextcloud.ifg.uni-kiel.de/index.php/s/q6g25nWyWJKqzNB.

The experimental data (*.xlsx) of drying and wetting paths are uploaded to the server. The data includes the reading number, time (day), stain values in perpendicular and parallel orientations, weight of the sample and measured water content values. Fig. 5.9 shows the change of the strains under the applied suction values.

The required LEM code and the input variables for simulating the drying and wetting paths of the sandy Opalinus Clay are uploaded to the IfG (Kiel) NextCloud server. The data is accessible through the following link: https://nextcloud.ifg.uni-kiel.de/index.php/s/fDNoPoXpXMqeAsK.

The uploaded protected MATLAB file in a *.p format requires a MATLAB version with a built-in Voronoi Tessellation and Delaunay Triangulation functions. Fig. 5.10 shows the change of hydraulic conductivity with applied linear strains as described in Sect. 4.5.

Meta Data Overview (According to Dublin Core)
See (Table 5.12).

Table 5.11 MEX 1-2: Data overview

Type	Spec.	Owner	Access	Comment
EXP	DRY	CAU	Open	Output files are uploaded
MOD	LEM	CAU	Open	Executable MATLAB P-file
			Open	Input files available
MOD	FEM	UFZ	Open	Branch to be merged into OGS
	VPF		Open	Input files available

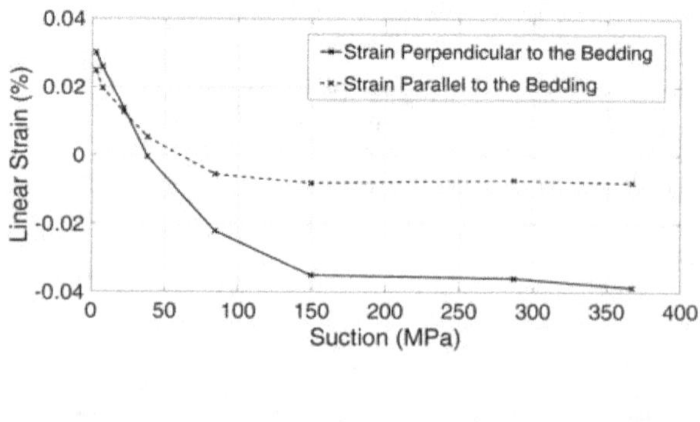

Fig. 5.9 The suction versus linear strains

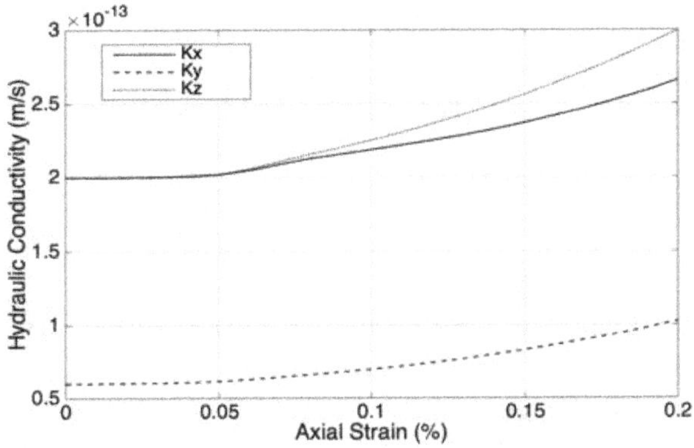

Fig. 5.10 The change of hydraulic conductivity with applied linear strains, Opalinus Clay

Table 5.12 MEX 1-2 (CAU)

Data label	GeomInt, MEX 1-2, CAU, drying and wetting paths, Opalinus Clay
URL (Experiments)	https://nextcloud.ifg.uni-kiel.de/index.php/s/q6g25nWyWJKqzNB
URL (Numerics)	https://nextcloud.ifg.uni-kiel.de/index.php/s/fDNoPoXpXMqeAsK
Subject	Drying and Wetting Paths (Opalinus Clay)
Type of data	Experimental data, executable MATLAB P-file, input parameters
Data quality	Quality assured data
Status of data	Unprocessed data
Data format	txt, xlsx, MATLAB executable P-file
Creators	Kiel University, Institute of Geomechanics and Geotechnics, Ludewig-Meyn-Straße 10, 24118, Kiel
Source/Origin	In-house code
Publisher	Kiel University, Institute of Geomechanics and Geotechnics, Ludewig-Meyn-Straße 10, 24118, Kiel
Rights holders	Kiel University, Institute of Geomechanics and Geotechnics, Ludewig-Meyn-Straße 10, 24118, Kiel
Contributors	Kiel University, Institute of Geomechanics and Geotechnics: Amir Shoarian Sattari, Frank Wuttke
Time/period of creation	2018–2020
Language of the content	English
Update policy	Stored data is final
Access permissions	Full access

Table 5.13 MEX 1-4: Data overview

Type	Spec.	Owner	Access	Comment
EXP	URL	BGR	Limited	UFZ DMP
MOD	FEM	UFZ	Open	OGS-5, OGS-6 versions
			Open	Input files available

5.3.5 MEX 1-4: CD/LP Experiment (BGR)

Participating institutions of MEX 1-4 (see Sect. 4.7): BGR (Table 5.13).

Link to the data set at UFZ data investigation portal (Download only for project members): www.ufz.de/record/dmp/archive/7471/.

The data set contains all project files that are relevant to reproduce the simulation results described in Sect. 4.7 and its corresponding subsections. For every simulation run with OGS-6, the data set comprises a project file (`*.prj`), a geometry file (`*.gml`) and the computational mesh (`*.vtu`). For simulations run with OGS-5, the project is split into files for boundary conditions (`*.bc`), initial conditions (`*.ic`), fluid properties (`*.mfp`), material properties (`*.mmp`), solid properties (`*.msp`), numerical parameters (`*.num`), output parameters (`*.out`), process definitions (`*.pcs`), tabulated functional dependencies (`*.rfd`), source terms (`*.sc`), and output parameters (`*.out`). In addition, separate files are provided for the geometry (`*.gli`) and the computational mesh (`*.msh`).

Meta Data Overview (According to Dublin Core)

See (Table 5.14).

5.3.6 MEX 2-1a: Pressure Driven Percolation in Salt

Participating institutions of MEX 2-1a (see Sect. 4.8): CAU, IfG, UFZ (Table 5.15).

UFZ

Link to the data set at UFZ data investigation portal (Download only for project members): https://www.ufz.de/record/dmp/archive/7706/.

The link contains two input decks for OGS-6 in which pressure driven percolation as described in MEX2 is simulated under different configurations of boundary loading. The first case applies the boundary loading of 12 MPa, 21 MPa, and 8 MPa in x-, y-, and z-direction respectively. It is called "case 1" and the corresponding OGS-6 input file is "me2_insitu_case1.prj". The second case is loaded with 4 MPa, 15 MPa, and 19 MPa in x-, y-, and z-direction respectively and the input file is named "me2_insitu_case2.prj". The remaining files are vtu files that describe the computing domain and the boundaries as shown in Fig. 4.58.

Table 5.14 MEX 1-4 (BGR)

Data label	GeomInt, MEX 1-4, BGR
URL	www.ufz.de/record/dmp/archive/7471
Subject	Clay stone, hydraulic-mechanical coupling
Type of data	Collection of various data
Data quality	Quality assured data
Status of data	Raw data
Data format	ASCII
Creators	Bundesanstalt für Geowissenschaften und Rohstoffe (BGR)
Source/Origin	Geological-Geotechnical Safety Analyses
Publisher	BGR, Stilleweg 2, 30655 Hannover
Rights holders	BGR, Stilleweg 2, 30655 Hannover
Contributors	Bernhard Vowinckel, Gesa Ziefle, Jobst Maßmann
Time/period of creation	01.12.2018–31.12.2019
Language of the content	English
Update policy	Stored data will not be extended
Access permissions	Limited access

MEX 2-1a (UFZ) will be also provided as an OGS benchmark case at: https://www.opengeosys.org/docs/benchmarks/phase-field/pf_perc/.

CAU Kiel

The required LEM code and the input variables of the percolation test on saltstone samples are uploaded to the IfG (Kiel) NextCloud server. The data is accessible through the following link: https://nextcloud.ifg.uni-kiel.de/index.php/s/9JZZcpS4S3JJT9S.

The uploaded protected MATLAB file in a *.p format requires a MATLAB version with a built-in Voronoi Tessellation and Delaunay Triangulation functions. The input variables are prepared in two files for two different stress configurations. Fig. 5.11 shows the frack surfaces under the percolation test as described in Sect. 4.8.

Meta Data Overview (According to Dublin Core)

See (Tables 5.16 and 5.17).

Table 5.15 MEX 2-1a: Data overview

Type	Spec.	Owner	Access	Comment
EXP	LIT	[2]	Restricted	Literature available online
MOD	LEM	CAU	License	Executable MATLAB P-file
			Free	Input files available
MOD	FEM	UFZ	Open source	via OpenGeoSys portal
	VFP		Free	I/O available

Fig. 5.11 The frack surfaces (red) under the percolation test, Saltstone

Table 5.16 MEX 2-1a (UFZ)

Data label	MEX 0-1a (UFZ)
URL	https://www.ufz.de/record/dmp/archive/7706/
Subject	Bending fracture test
Type of data	Data set (structured data in a defined format)
Data quality	Quality assured data by benchmarking
Status of data	Processed data
Data format	OGS files
Creators	Yoshioka, Keita
Source/Origin	Open source
Publisher	Helmholtz Centre for Environmental Research UFZ
Rights holders	Helmholtz Centre for Environmental Research UFZ
Contributors	Yoshioka, Keita; Wang Wenqing
Time/period of creation	2019–2020
Language of content	English
Update policy	To be merged to OGS benchmarks (see below)
Access permissions	Free access

Table 5.17 MEX 2-1a (CAU)

Data label	GeomInt I CAU I Percolation test, Saltstone
URI	https://nextcloud.ifg.uni-kiel.de/index.php/s/9JZZcpS4S3JJT9S (Numerics)
Subject	Percolation test (Saltstone)
Type of data	Executable MATLAB P-file, input parameters
Dataquality	Quality assured data
Status of data	Unprocessed data
Dataformat	txt, MATLAB executable P-file
Creators	Kiel University, Institute of Geomechanics and Geotechnics, Ludewig-Meyn-Straße 10, 24118, Kiel
Source/Origin	In-house code
Publisher	Kiel University, Institute of Geomechanics and Geotechnics, Ludewig-Meyn-Straße 10, 24118, Kiel
Rights holders	Kiel University, Institute of Geomechanics and Geotechnics, Ludewig-Meyn-Straße 10, 24118, Kiel
Contributors	Kiel University, Institute of Geomechanics and Geotechnics: Amir Shoarian Sattari, Frank Wuttke
Time/period of creation	2018–2020
Language of the content	English
Update policy	Stored data is final
Access permissions	Full access

5.3.7 MEX 2-1b: Pressure Driven Percolation, Opalinus Claystone

Participating institutions of MEX 2-1b (see Sect. 4.9): CAU, UFZ (Table 5.18).

CAU (LEM)

The experimental results of the pressure driven percolation test on the cubic Opalinus Clay samples are uploaded to the IfG (Kiel) NextCloud server. The data is accessible through the following link: https://nextcloud.ifg.uni-kiel.de/index.php/s/EMdNkdF4PRKWCqa.

Table 5.18 MEX 2-1b: Data overview

Type	Spec.	Owner	Access	Comment
EXP		CAU	Free	CAU-Cloud
MOD	LEM	CAU	Open	Executable MATLAB P-file
			Free	Input files available
MOD	FEM	UFZ	Open source	via OpenGeoSys portal
	VPF		Free	Input files available

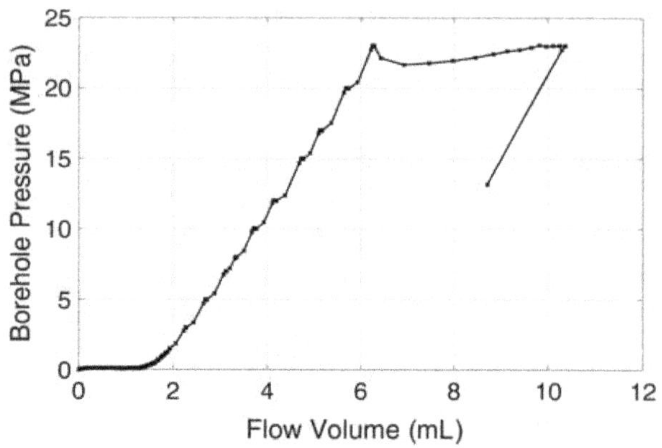

Fig. 5.12 The recorded results depicting the evolution of the borehole pressure versus flow volume for the 1st stress configuration

The experimental data (*.ASCII) of two different stress configurations (Sect. 2.4.2) are uploaded to the server. The data includes the time ($\Delta T = 1s$), pump volume (mL), given oil pressure (Bar) and actual oil pressure in the system (Bar). Fig. 5.12 illustrates an example of the plotted borehole pressure versus flow volume for the 1st stress configuration discussed in the Sect. 4.9.

The required LEM code and the input variables of the percolation test on Opalinus Clay samples are uploaded to the IfG (Kiel) NextCloud server. The data is accessible through the following link: https://nextcloud.ifg.uni-kiel.de/index.php/s/tFKKjxnSpgNG25b.

The uploaded protected MATLAB file in a *.p format requires a MATLAB version with a built-in Voronoi Tessellation and Delaunay Triangulation functions. The input variables are prepared in two files for two different stress configurations. Fig. 5.13 illustrates an example of the evolved frack surfaces for the 2nd stress configuration discussed in the Sect. 4.9).

UFZ (FEM-VPF)
MEX 2-1b (UFZ) will be also provided as an OGS benchmark case.

Meta Data Overview (According to Dublin Core)
See (Tables 5.19 and 5.20).

5.3.8 MEX 2-2: Closure and Healing of Cracks (IfG)

Participating institutions of MEX 2-2 (see Sect. 4.10): IfG, UFZ (Table 5.21).

The measured gas flow is converted into permeabilities, which are stored as time series in an Excel file. For each of the three experiment there are two columns. The

Fig. 5.13 The frack surfaces (red) for the 2nd stress configuration

Table 5.19 MEX 2-1b (CAU)

Data label	GeomInt, MEX 2-1b, CAU, percolation test, Opalinus Clay
URL (Experiments)	https://nextcloud.ifg.uni-kiel.de/index.php/s/EMdNkdF4PRKWCqa
URL (Numerics)	https://nextcloud.ifg.uni-kiel.de/index.php/s/tFKKjxnSpgNG25b
Subject	Percolation test (Opalinus Clay)
Type of data	Experimental data, Executable MATLAB P-file, Input parameters
Data quality	Quality assured data
Status of data	Unprocessed data
Data format	txt, ASCII, MATLAB executable P-file
Creators	Kiel University, Institute of Geomechanics and Geotechnics, Ludewig-Meyn-Straße 10, 24118, Kiel
Source/Origin	In-house code
Publisher	Kiel University, Institute of Geomechanics and Geotechnics, Ludewig-Meyn-Straße 10, 24118, Kiel
Rights holders	Kiel University, Institute of Geomechanics and Geotechnics, Ludewig-Meyn-Straße 10, 24118, Kiel
Contributors	Kiel University, Institute of Geomechanics and Geotechnics: Amir Shoarian Sattari, Frank Wuttke
Time/period of creation	2018–2020
Language of the content	English
Update policy	Stored data is final
Access permissions	Full access

Table 5.20 MEX 2-1b (UFZ)

Data label	GeomInt, MEX 2-1b, UFZ, percolation test, Opalinus Clay)
URL	https://www.opengeosys.org/docs/benchmarks/
Subject	Bending fracture test
Type of data	Data set (structured data in a defined format)
Data quality	Quality assured data by benchmarking
Status of data	Processed data
Data format	OGS files
Creators	Yoshioka, Keita
Source/Origin	Open source
Publisher	Helmholtz Centre for Environmental Research UFZ
Rights holders	Helmholtz Centre for Environmental Research UFZ
Contributors	Yoshioka, Keita
Time/period of creation	2019–2020
Language of content	English
Update policy	To be merged to OGS benchmarks (see below)
Access permissions	Free access

Table 5.21 MEX 2-2: Data overview

Type	Spec.	Owner	Access	Comment
EXP	LAB	IfG	Limited	Available on demand
MOD	DEM	IfG	License	Commercial code
			Restricted	I/O: Available on demand
MOD	FEM	UFZ	Open source	via OpenGeoSys portal
	VPF		Free	I/O Available

first contains the time in hours since the start of the experiment, the second contains the permeability (Fig. 5.14).

Meta Data Overview (According to Dublin Core)

See (Table 5.22).

Table 5.22 MEX 2-2 (IfG)

Data label	GeomInt, MEX 2-2, IfG, pressure driven percolation (healing)
URL	http://www.ufz.de/record/dmp/archive/7588
Subject	Rissschließung Steinsalz
Type of data	Dataset (structured data in a defined format)
Data quality	
Status of data	
Data format	
Creators	IfG GmbH
Source/Origin	IfG GmbH
Publisher	IfG GmbH, Friederikenstraße 60, 04279 Leipzig
Rights holders	IfG GmbH, Friederikenstraße 60, 04279 Leipzig
Contributors	Mathias Nest,
Time/period of creation	2019
Language of the content	German
Update policy	
Access permissions	Limited access

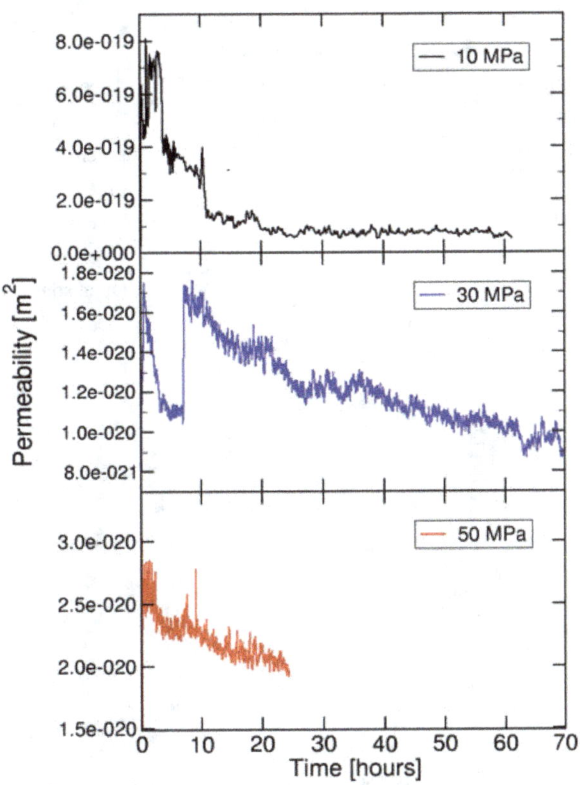

Fig. 5.14 Graphical representation of data

5.3.9 *MEX 2-3: Effect of Compressibility on Pressure Driven Percolation*

Participating institutions of MEX 2-3 (see Sect. 4.11): CAU, IfG, UFZ (Table 5.23).

CAU Kiel

The required LEM code and the input variables for simulating the effect of compressibility are uploaded to the IfG (Kiel) NextCloud server. The data is accessible through the following link: https://nextcloud.ifg.uni-kiel.de/index.php/s/6Mfg3P4PyKNN6By.

The uploaded protected MATLAB file in a *.p format requires a MATLAB version with a built-in Voronoi Tessellation and Delaunay Triangulation functions. The input variables are prepared in two different files for the simulation of the pressure drop in gas and brine reservoirs. Fig. 5.15 shows the differences between the pressure drop in gas and brine reservoirs.

Meta Data Overview (According to Dublin Core)
See (Table 5.24).

Table 5.23 MEX 2-3: Data overview

Type	Spec.	Owner	Access	Comment
EXP	LIT	[2]	Restricted	Literature available online
MOD	LEM	CAU	License	Executable MATLAB P-file
			Free	Input files available
MOD	DEM	IfG	License	Commercial software
			Free	I/O available
MOD	FEM	UFZ	Open source	via OpenGeoSys portal
	VFP		Free	I/O available

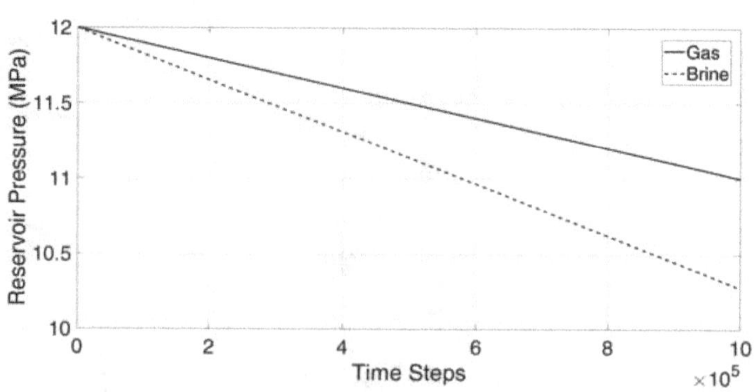

Fig. 5.15 The comparison between the pressure drop in gas and brine reservoirs

Table 5.24 MEX 2-3: Meta Data according to Dublin Core

Data label	GeomInt, MEX 2-3, CAU, effect of compressibility, salt
URL (Numerics)	https://nextcloud.ifg.uni-kiel.de/index.php/s/6Mfg3P4PyKNN6By
Subject	Effect of compressibility (Saltstone)
Type of data	Executable MATLAB P-file, Input parameters
Data quality	Quality assured data
Status of data	Unprocessed data
Data format	txt, MATLAB executable P-file
Creators	Kiel University, Institute of Geomechanics and Geotechnics, Ludewig-Meyn-Straße 10, 24118, Kiel
Source/Origin	In-house code
Publisher	Kiel University, Institute of Geomechanics and Geotechnics, Ludewig-Meyn-Straße 10, 24118, Kiel
Rights holders	Kiel University, Institute of Geomechanics and Geotechnics, Ludewig-Meyn-Straße 10, 24118, Kiel
Contributors	Kiel University, Institute of Geomechanics and Geotechnics: Amir Shoarian Sattari, Frank Wuttke
Time/period of creation	2019–2020
Language of the content	English
Update policy	Stored data is final
Access permissions	Full access

5.3.10 MEX 2-4: Large Wellbore Test (Springen)

Participating institutions of MEX 2-4 (see Sect. 4.12): IfG (Table 5.25).

Meta Data Overview (According to Dublin Core)
See (Table 5.26).

The data set contains a measured pressure curve over time (50 days) during an injection test in a large wellbore (Springen URL). The injection regime was conducted in 5 bar increasing pressure steps followed by shut-in periods for observing the related pressure diffusion. Corresponding pumping rates were between 1–5.5 liters

Table 5.25 MEX 2-4: Data overview

Type	Spec.	Owner	Access	Comment
EXP	URL	IfG	Restricted	Available on demand
MOD	DEM	IfG	License	Commercial software
			Restricted	I/O available on demand

Table 5.26 MEX 2-4 (IfG)

Data label	GeomInt, MEX 2-4, IfG, large wellbore test, URL Springen
URL	In-house data storage
Subject	Integrity, tidiness of salt rocks for different fluids (gases and brines)
Type of data	Dataset (structured data in a defined format)
Data quality	Quality assured data
Status of data	
Data format	ASCII
Creators	IfG GmbH
Source/Origin	IfG GmbH
Publisher	IfG GmbH, Friederikenstraße 60, 04279 Leipzig
Rights holders	IfG GmbH, Friederikenstraße 60, 04279 Leipzig
Contributors	Mathias Nest
Time/period of creation	2018–2019
Language of the content	German
Update policy	Ongoing updates
Access permissions	Limited access

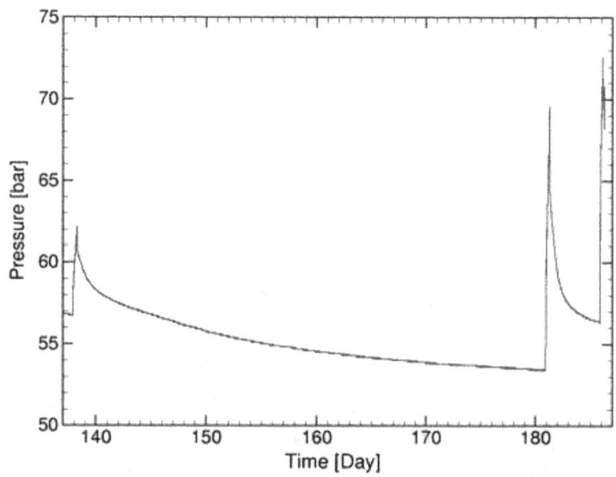

Fig. 5.16 Measured pressure curve data: $p(t)$

per second. Peak pressures of more than 60 bars were required to create in-situ discontinuities by overcoming the tensile/adhesive rock strength of salt. Additionally, acoustic emission (AE) data were recorded (Fig. 5.16).

5.3.11 *MEX 3-1: CNL Direct Shear Test Data (TUBAF)*

Participating institutions of MEX 3-1 (see Sect. 4.13): TUBAF (Table 5.27).

Link to the data set at UFZ data portal (DMP): www.ufz.de/record/dmp/archive/7925/.

The CNL data set contains four text files. One text file with the rock properties of the used granite (see Table 2.1). Two files with the scan data of the two surfaces. One point cloud can be seen in Fig. 5.17. The last file contains the laboratory data. In Fig. 5.18 the results for the four shear stress levels can be seen.

Meta Data Overview (According to Dublin Core) See (Table 5.28).

Table 5.27 MEX 3-1: Data overview

Type	Spec.	Owner	Access	Comment
EXP	LAB	TUBAF	Limited	Available, UFZ-DMP
MOD	FFS	TUBAF	Open source	Available via GitHub
			Free	I/O available

Fig. 5.17 Point cloud representing the surface of a granite sample from Saxony. The size is 65 mm by 170 mm and the cloud contains approx. 98000 points

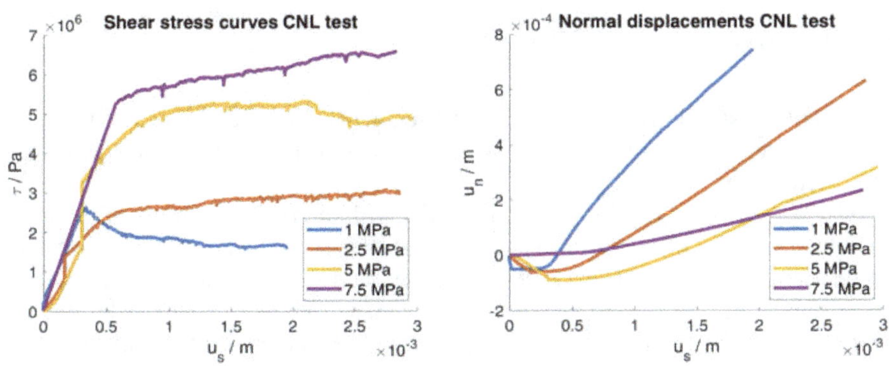

Fig. 5.18 CNL test results

Table 5.28 MEX 3-1 (TUBAF)

Data label	GeomInt I TUBAF I Data Set CNL
URL	http://www.ufz.de/record/dmp/archive/7925
Subject	Crystalline rock, direct shear test
Type of data	collection of various data
Data quality	quality assured data
Status of data	processed data
Data format	txt, jpg, png
Creators	TU Freiberg, Institut fr Geotechnik, Gustav-Zeuner-Str. 1, 09599 Freiberg
Source/Origin	Rock mechanical laboratory
Publisher	TU Freiberg, Institut fr Geotechnik, Gustav-Zeuner-Str. 1, 09599 Freiberg
Rights holders	TU Freiberg, Institut fr Geotechnik, Gustav-Zeuner-Str. 1, 09599 Freiberg
Contributors	TU Freiberg, Institut fr Geotechnik, Thomas Frühwirt and Daniel Pötschke
Time/period of creation	2018–2019
Language of the content	English
Update policy	Stored data will not be extended
Access permissions	Limited access

Table 5.29 MEX 3-2: Data overview

Type	Spec.	Owner	Access	Comment
EXP	LAB	TUBAF	Limited	Available, UFZ-DMP
MOD	FFS	TUBAF	Open source	Available via GitHub
			Free	I/O available

5.3.12 MEX 3-2: CNS Test

Participating institutions of MEX 3-2 (see Sect. 4.14): TUBAF (Table 5.29).

Link to the data set at UFZ data management portal (DMP): https://www.ufz.de/record/dmp/archive/7924/.

The data set of the CNS test contains a file with the rock properties of the used basalt (see Table 2.1), two files with the scan data of the two surfaces. One point cloud can be seen in Fig. 5.19. The results of the laboratory tests are available as ASCII files and the shear curves and the dilatation are visualized in Fig. 5.20. Additionally three photos of the basalt surface before, after the first and after the fourth shear test are included.

Meta Data Overview (According to Dublin Core)
See (Table 5.30).

Fig. 5.19 Point cloud representing the surface of a basalt sample from Thuringia. The size is 196 mm by 149 mm and the cloud contains approx. 252000 points

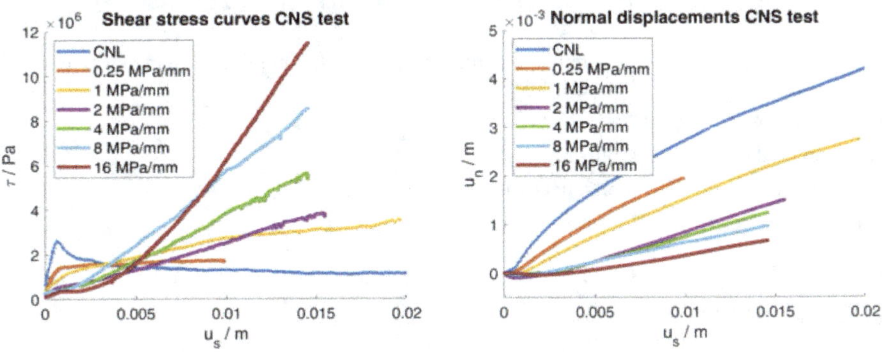

Fig. 5.20 CNS test results

5.3.13 *MEX 3-3: Inverse Analysis of Reiche Zeche Data and Harmonic Testing of a Single Fracture*

Participating institutions of MEX 3-3 (see Sect. 4.15): UoS, UFZ (Table 5.31).

Link to the data set stored at the Gitlab server located at the Institute of Applied Mechanics, University of Stuttgart:
https://galilei.isr.uni-stuttgart.de/gitlab/schmidt/geomint (Fig. 5.21).

The uploaded data set contains two compiled executables for simulations to fit experimental data for the Reiche Zeche fracture characterization tests and to reproduce the non-linear flow response throughout harmonic testing of a single fracture

Table 5.30 MEX 3-2 (TUBAF)

Data label	GeomInt, MEX 3-2, TUBAF, Data Set CNS
URI	http://www.ufz.de/record/dmp/archive/7924
Subject	Crystalline rock, direct shear test
Type of data	collection of various data
Dataquality	quality assured data
Status of data	processed data
Dataformat	txt, jpg, png
Creators	TU Freiberg, Institut fr Geotechnik, Gustav-Zeuner-Str. 1, 09599 Freiberg
Source/Origin	Rock mechanical laboratory
Publisher	TU Freiberg, Institut fr Geotechnik, Gustav-Zeuner-Str. 1, 09599 Freiberg
Rights holders	TU Freiberg, Institut fr Geotechnik, Gustav-Zeuner-Str. 1, 09599 Freiberg
Contributors	TU Freiberg, Institut fr Geotechnik, Thomas Frühwirt and Daniel Pötschke
Time/period of creation	2018–2019
Language of the content	English
Update policy	Stored data will not be extended
Access permissions	Limited access

Table 5.31 MEX 3-3: Data overview

Type	Spec.	Owner	Access	Comment
EXP	LAB	UoS	Free	UoS Cloud / GitLab
MOD	HDF	UoS	Open source	Executable available
			Free	I/O available
MOD	FEM	UFZ	Open source	via OpenGeoSys portal
	VFP		Free	I/O available

on the laboratory scale like described in Sect. 4.15. The folder includes executables, input files and the used discretization in terms of meshing files, required to perform the simulations. README.txt files provide further information how to start the simulations.

Meta Data Overview (According to Dublin Core)
See (Table 5.32).

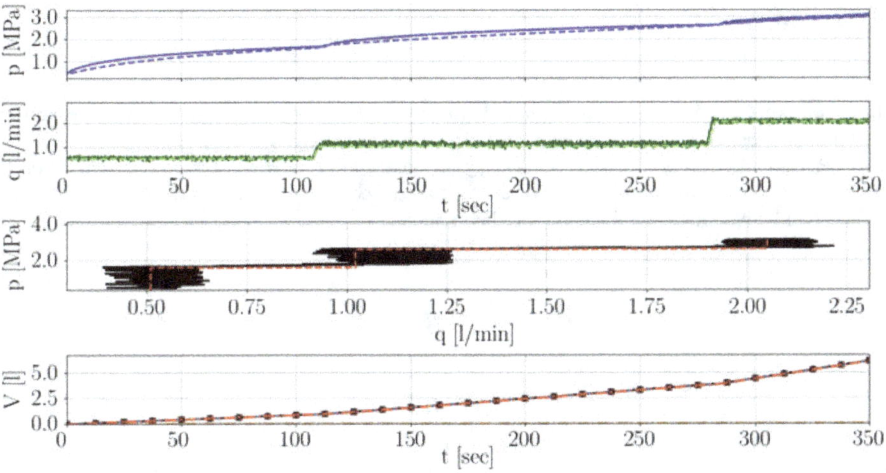

Fig. 5.21 Visualisation of data obtained from the pre-compiled executable used for inverse analysis computations of pumping tests performed at Reiche Zeche at a depth of 40.6 m

Table 5.32 MEX 3-3: Reiche Zeche Data

Data label	GeomInt, MEX 3-3, UoS, Data Set Reiche Zeche
URL	https://galilei.isr.uni-stuttgart.de/gitlab/schmidt/geomint
Subject	Non-linear hydro-mechanics / fracture flow
Type of data	Executable, mesh input, parameter input
Data quality	Quality assured data
Status of data	Unprocessed data
Data format	txt, msh, linux executable
Creators	University of Stuttgart, Institute of Applied Mechancis—Continuum Mechanics, Pfaffenwaldring 7, 70569 Stuttgart
Source/Origin	In-house code
Publisher	University of Stuttgart, Institute of Applied Mechancis—Continuum Mechanics, Pfaffenwaldring 7, 70569 Stuttgart
Rights holders	University of Stuttgart, Institute of Applied Mechancis—Continuum Mechanics, Pfaffenwaldring 7, 70569 Stuttgart
Contributors	University of Stuttgart, Institute of Applied Mechancis—Continuum Mechanics, Patrick Schmidt and Holger Steeb
Time/period of creation	2018–2019
Language of the content	English
Update policy	Stored data is final
Access permissions	Full access

References

1. A. Tarokh, C.-S. Kao, A. Fakhimi, and J.F. Labuz. Spalling and brittleness in surface instability failure of rock. *Geotechnique*, 66(2):161–166, 2016. cited By 5.
2. W.P. Kamlot. *Habilitationsschrift: Gebirgsmechanische Bewertung der geologischen Barrierefunktion des Hauptanhydrits in einem Salzbergwerk*. des Instituts für Geotechnik der Technischen Universität Bergakademie Freiberg, 2009.

6

GeomInt Research Project: Synthesis and Future Perspectives

Olaf Kolditz, Uwe-Jens Görke, Heinz Konietzky, Jobst Maßmann, Mathias Nest, Holger Steeb, Frank Wuttke, and Thomas Nagel

6.1 Synthesis—Directions

As a result of the GeomInt research project (Chap. 1) a broad combined experimental and numerical platform for the investigation of discontinuities due to swelling and shrinking processes (WP1, Sect. 2.3), pressure-driven percolation (WP2, Sect. 2.4) and stress redistribution (WP3, Sect. 2.5) for important reservoir and barrier rocks (clay, salt, crystalline) has been developed. Model comparisons for damage and fracture processes driven by different processes provide information on the optimal areas of application of the numerical methods (Sect. 6.1.1).

A comprehensive validation of the platforms ("Proof-of-Concept") was carried out by "Model-Experiment-Exercises" experiments (MEX) for the damage and fracture processes driven by different processes such as swelling and shrinking, pressure-driven percolation and stress redistribution (see Chap. 4). The MEX concept is the central synthesis element of GeomInt as it is directly linking models (MOD) with lab experiments (LAB) and paving the way towards the analysis of in-situ experiments

O. Kolditz (✉)
UFZ, Helmholtz Centre for Environmental Research, Technische Universität Dresden, Dresden, Germany
e-mail: olaf.kolditz@ufz.de

U.-J. Görke
UFZ, Helmholtz Centre for Environmental Research, Leipzig, Germany

H. Konietzky · T. Nagel
TUBAF, Technische Universität Bergakademie Freiberg, Freiberg, Germany

J. Maßmann
BGR, Federal Institute for Geosciences and Natural Resources, Hannover, Germany

M. Nest
IfG, Institut für Gebirgsmechanik Leipzig, Leipzig, Germany

H. Steeb
UoS, University of Stuttgart, Stuttgart, Germany

F. Wuttke
CAU, Christian-Albrechts-Universität zu Kiel, Kiel, Germany

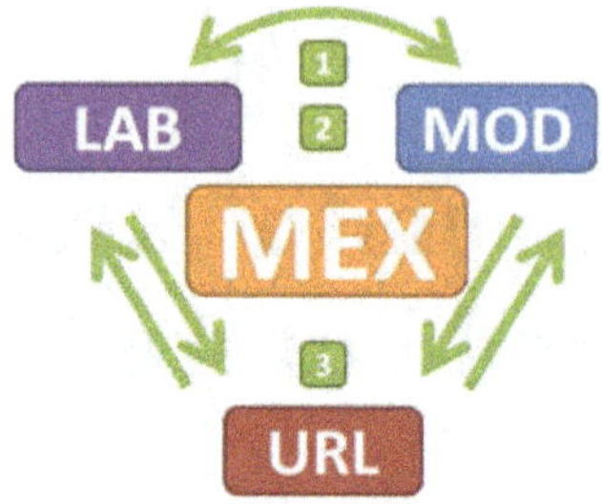

Fig. 6.1 Central MEX concept

(URL) which has been started in the current project phase and will be continued in future activities though (Fig. 6.1, see also Sect. 6.1.2).

The project results allow an improved understanding of the processes, the methods used and the application-oriented systems for realistic time and length scales in order to make the planning and implementation of geotechnical uses of the underground safer, more reliable and efficient. An important part of future work is the transferability of the experimental-numerical concepts and methods to other geotechnological applications (e.g. deep geothermal energy, energy storage, repository problems, methods for hydraulic stimulation, conventional and unconventional resource extraction or tunnel construction). Ongoing activities also aimed to intensify the internationalization started in the previous project in cooperation with complementary research projects (e.g. EURAD, DECOVALEX 2023, Mont Terri project). In future directions—the basic idea concerning the investigation of the three relevant process types for the formation of pathways in different rock types—shall be preserved. The proven project structure (WP1-3) is also to be retained (Fig. 6.8).

6.1.1 Numerical Methods Competencies

Significant emphasis in GeomInt was directed on the development of numerical methods for modelling of discontinuities in various rock types as well as the analysis of advantages/disadvantages of particular methods for specific fracture pattern at various scales (Fig. 3.1). The numerical methods have been described in detail in Chap. 3. The "competencies" of the numerical methods under use will be compactly summarized in the following set of tables.

Table 6.1 provides an overview of the capabilities of the numerical methods for simulation of fracturing processes, i.e. (i) crack representation by elements explicitly/separative or by damage variables, (ii) failure criteria, (iii) used mesh types and (iv) fracture aperture calculation.

The second Table 6.2 recalls fracturing processes within the adjacent rock mass (i.e. fractured porous medium) for the various rock types (brittle/ductile) investigated in the related work packages (WPs). In the context of hydro-mechanical (HM) coupled processes fluid migration from the fracture into the rock matrix (and vice versa) must be considered for both pre-defined and undefined cracks.

Table 6.1 Numerical methods comparison

Methods	Crack representation	Failure criteria	Mesh	Fracture apert.
LEM	Element separation	Element strength	Lattice	Direct
DEM	Element separation	Element strength	Discrete	Direct
SPH	Particle based	NA	NA	Particle based
FEM-LIE	Element explicit	Cohesive law	Conforming	Direct
FEM-VPF	Damage variable	Fracture mechanics	Non-conf.	Indirect
FEM-NLD	Damage variable	Stress based	Non-conf.	Indirect
FEM-HDF	Element explicit	Cohesive law	Non- & conf.	Direct

Legende: NA—not applicable

Table 6.2 Hydro-Mechanics (fracture mechanics) processes

Rock type	Crack PD	Crack UD	Leak-off	Non-brittle	Visc. diss.
Clay (WP1)	Yes	Yes	Maybe	Yes	Yes
Salt/Clay (WP2)	Yes	Yes	Yes	Yes	Yes
Crystalline (WP3)	Yes	Yes	Maybe	No	Maybe

Legende: PD—pre-defined, UD—undefined, Visc.—viscous, diss.—dissipation

Finally an evaluation of the competencies of all used methods is provided in Table 6.3 based mainly on the results of the MEX studies. An important part of the MEX concept was the application of multiple (at least) methods to these exercises (e.g. for MEX 0-1A three methods, LEM, DEM and FEM-VPF have been applied). For the analysis of lab scale experiments all present methods (discontinuous and continuous) could be applied successfully. The advantages of the discontinuous methods (LEM, DEM) lies in the simulation of small scale processes, e.g. fracture initiation and propagation, and because they rely on the fundamental fracture mechanical phenomena. Discontinuous methods exhibit some difficulties when it comes to coupled processes and at larger scales where continuum methods (FEM#) are rather strong. We also added 3D capabilities and HPC implementation status to the table.

The application of a specific numerical methods is driven by the specific purpose:

- fracturing process details, i.e. initiation and propagation
- process coupling, i.e. THMC
- scale of application

An important result of GeomInt is the significant further development of individual methods and their extension of application areas [1–3], respectively. Therefore, the GeomInt numerical platform as a whole advanced significantly.

Table 6.3 Methods competencies

Method	Crack PD	Crack UD	Leak-off	Non-brittle	Visc. diss.	3D	HPC
LEM	✓	✓	✓	✓	✓	✓	✓
DEM	✓	✓	✓	✓	✓	✓	✓
SPH	✓	✓	✓	✓	✓	✓	✓
FEM-LIE	✓	×	✓	✓	✓	×	×
FEM-VPF	✓	✓	✓	✓	✓	✓	✓
FEM-NLD	✓	✓	✓	✓	✓	×	✓
FEM-HDF	✓	×	✓	✓	✓	✓	✓

Legende: PD—pre-defined, UD–undefined, Visc.—viscous, diss.—dissipation, 3D—three-dimensional simulation, HPC—High-Performance-Computing

6.1.2 Proof-of-Concepts

Even though the focus of GeomInt was directed on further improvement of modeling and experimental methods, numerical simulation of real-world geotechnical applications is sort of standard meanwhile. The Proof-of-Concept (PoC) for modelling consists of multiple steps (calibration, validation, verification). Without doubt calibrated models can be provided for individual, site-specific applications—also thanks to the increasing computational power available. The challenge is still, can we reliably transfer models over multiple scales or for various rock types. The latter challenge is subject to GeomInt.

Figure 6.2 depicts a number of geotechnical applications in the North-German-Basin (NGB) including shallow and deep geothermal systems, energy storage in aquifers and caverns as well as geological sequestration of carbon dioxide in the subsurface. All applications studies have been conducted with the OpenGeoSys platform using various T-H-M-C coupled models, i.e. using weak or strong coupling of related thermo-hydro-mechanical-chemical processes. This comprehensive numerical modeling of various geoenergy applications in the NGB may serve as a validation of the THMC concept.

The second Proof-of-Concept study is related to the Underground Research Laboratory (URL) in Mont Terri where numerous (MT) experiments are conducted simultaneously [4, 5]. Figure 6.3 shows the tunnels system of the extended URL and a number of ongoing and planned experiments. Numerical simulation is used for both analysis of experimental results and design of new experiments, i.e. for their optimal planning. Again OGS is used for the indicated experiments and therefore providing a unified context for numerical simulation. Moreover, all models are included into a Virtual Reality context (see next section).

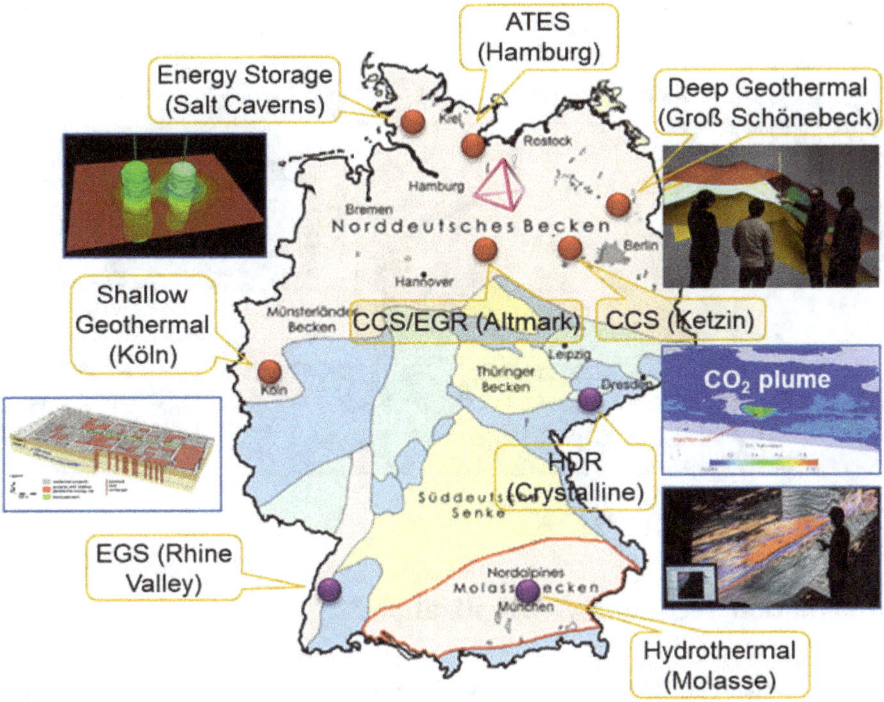

Fig. 6.2 OGS applications in the North German Basin

Fig. 6.3 Numerical simulation of various MT experiments [12]

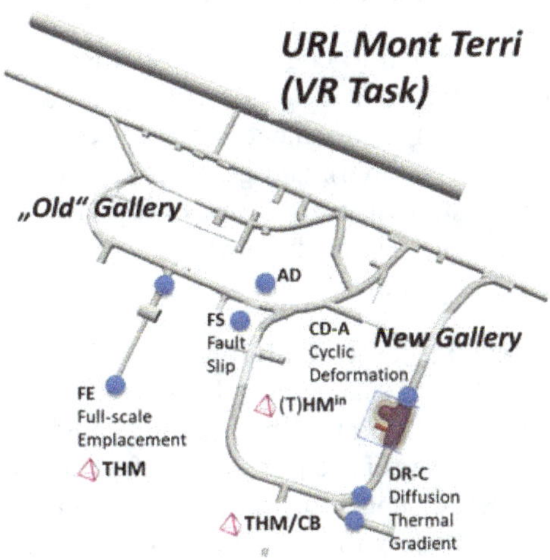

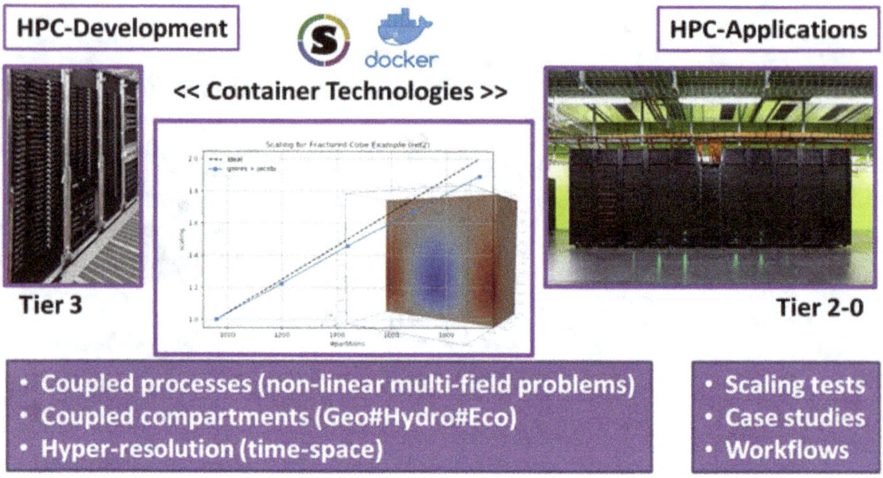

Fig. 6.4 High-Performance-Computing

Information Technology—Digitization

Many research areas largely benefit from new developments in information technology in many ways in particular data- and computational geosciences. Figure 6.4 shows the HPC concept starting with development and testing methods at smaller machines (Tier 3) and then implementation of high-end infrastructure (Tier 2-0). Container technologies (e.g. Docker, Singularity) are used for standardisation and portability of HPC solutions [6]. Scalability of numerical methods for coupled THMC simulations is a particular challenge.

In addition to computational efficiency, visual data analysis is promising field for many domain sciences as well. Figure 6.5 depicts several examples of visualisation from environmental sciences concerning terrestrial (hydrology), atmospheric (climate research), and geological systems. Virtual Reality concepts allow for the combination of various kind of data and models in a real geo-referenced context—making those integrated VR Labs a useful planning tool e.g. for experimental design in URLs or infrastructures for urban systems.

Intelligent use of Information Technology will foster the development and implementation of seamless analysis workflows in many research areas. Figure 6.6 illustrates the concept for geotechnical applications starting with data integration from URLs, set-up of adequate models for experimental analysis and design. HPC concepts for THMC simulations allow realistic analysis of process complexity and predictions. Virtual Reality makes outcomes accessible for specialists and a public audience as well.

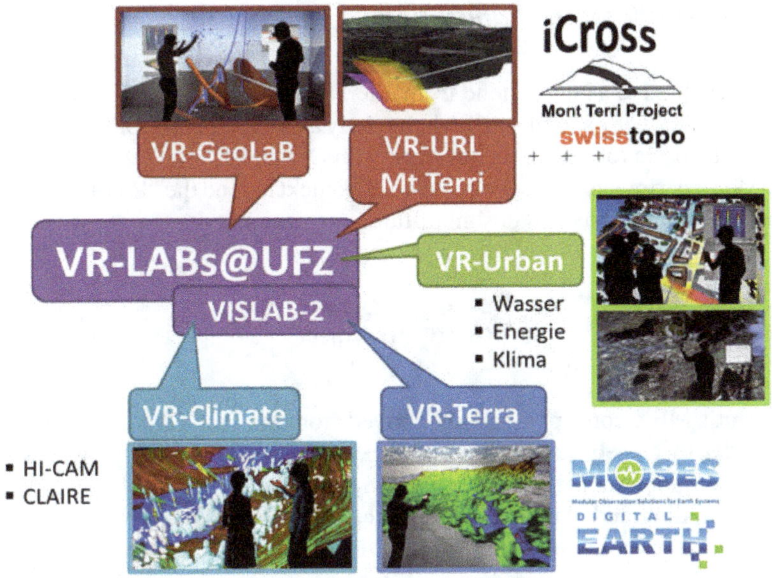

Fig. 6.5 Virtual Reality Labs

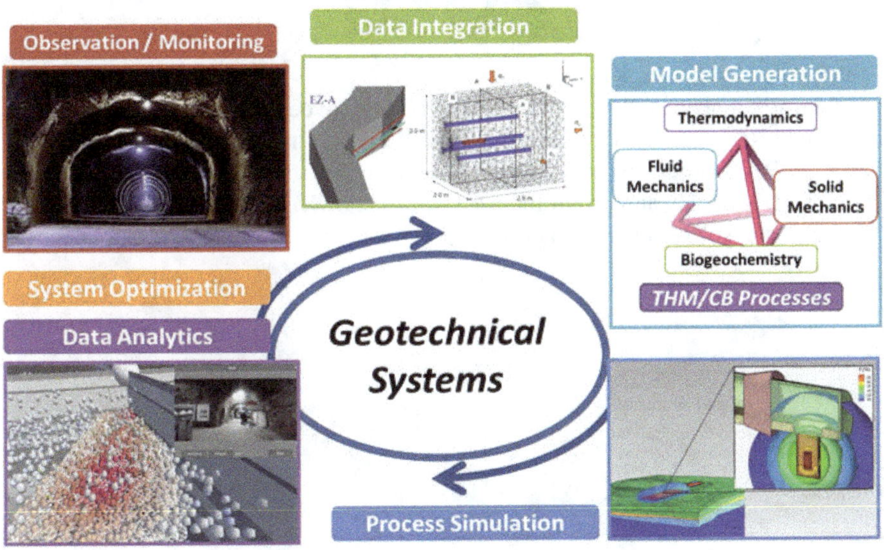

Fig. 6.6 Seamless analysis Workflows

6.1.3 International Collaboration

GeomInt benefited to a huge extend from international cooperation and produced an impact vice versa. Clay samples of both the shaly an sandy facies from the URL Mont Terri were the basis for experimental work with Opalinus Clay. GeomInt models for variuos MT experiments embedded into a VR context helped the "Proof-of-Concept" for experimental analysis and design of future campaigns.

Mont Terri Project

GeomInt's MEX concept is mainly derived from the DECOVALEX benchmarking tasks idea by closely combining own experimental with modelling works. Using different (continuous and discontinuous) numerical methods helped the general process understanding and elaboration of particular methods' dis-/advantages as well. GeomInt partners will participate the new DECOVALEX 2023 phase.

DECOVALEX 2019

GeomInt's concept and project results will be deployed for the new European Joint Program on Radioactive Waste Management EURAD. In addition to Opalinus Clay (OPA), EURAD is dealing with various European clays such as Callodo-Oxfordian (COx) and Boom clay and providing a European knowledge base for clays as barrier rock for waste isolation and energy storage.

eurad
European Joint Programme on Radioactive Waste Management

Webpage

The webpage of GeomInt (Fig. 6.7) is the entry point an reference of the project. It contains an overview of the project as well as quarterly updated news of ongoing activities. GeomInt publications are listed and an access point to the data management system is provided (see Chap. 5).

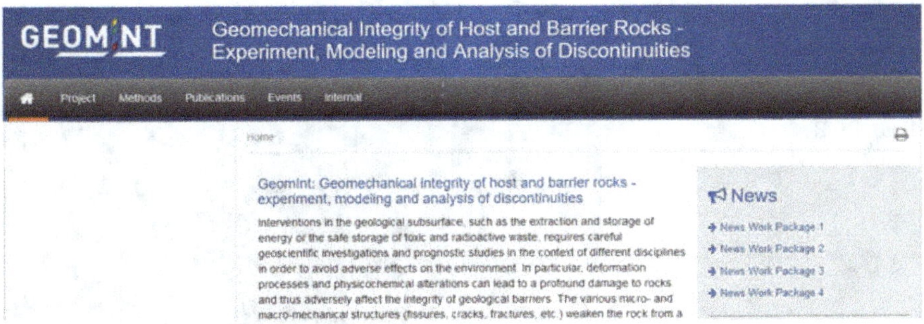

Fig. 6.7 Webpage of the GeomInt project ⇒ www.ufz.de/geomint

6.2 GeomInt Outlook—Future Work

After a very strongly methodically oriented first phase, the continuation project is intended to demonstrate practical applicability under real conditions (URLs). Furthermore, knowledge gaps in the experimental as well as numerical field shall be closed.

GeomInt is to be worked on by the interdisciplinary consortium of partners from universities, public and private research institutions with complementary, long-standing experience in the analysis of geosystems, which has proven itself from the previous project. New findings are expected, in particular regarding system understanding of the effects of discontinuities on underground geosystems. In this context, the focus of experimental investigations will shift significantly to the performance and evaluation of in-situ experiments in various underground laboratories (URLs). For the efficient numerical simulation of coupled mechanical, thermal and hydraulic processes in the formation and development of discontinuities on the scale under consideration, the adaptation of models and algorithms investigated and further developed in GeomInt to methods of high-performance computing (HPC) is a new aspect. The descriptive presentation of structural, experimental and model results in the real context (e.g.. URLs) is to be carried out within the framework of an integrated visual data analysis (virtualization). In the Fig. 6.8 shows a graphical illustration of the project continuation (Fig. 6.9).

Fig. 6.8 Graphical abstract of the GeomInt research concept

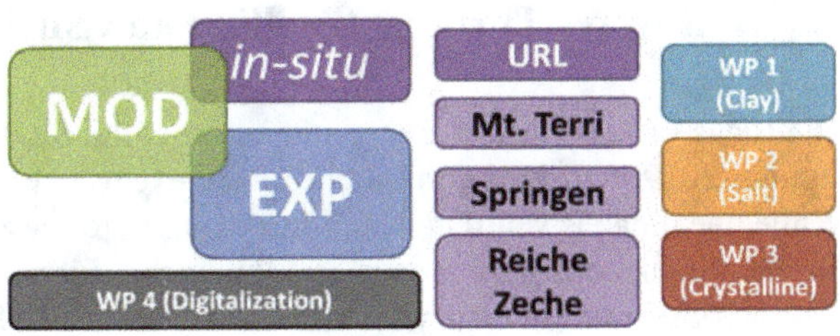

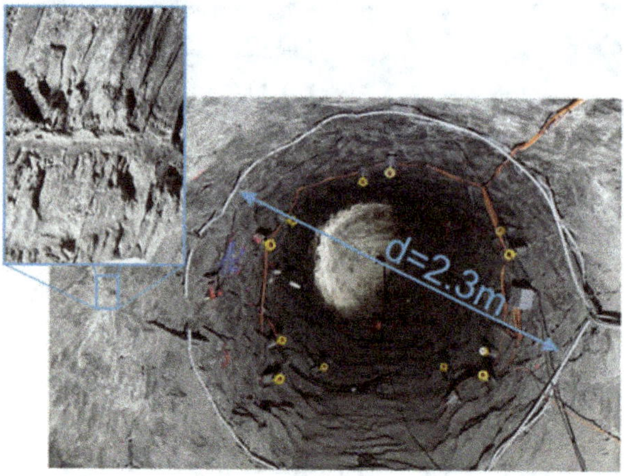

Fig. 6.9 URL Mont Terri: CD-A experiment: desaturation cracks at the ventilated open niche (Photo: BGR)

Future research will contain following aspects:

- experiments: The focus is on the evaluation of the latest in-situ experiments in the rocklaboratories Mont Terri (Opalinus Clay), Springen (salt) and Reiche Zeche (crystalline). Selected laboratory experiments are planned in order to specifically close still existing knowledge gaps (e.g. mechanical anisotropy of clay rocks).
- modelling: Numerical methods are to be further developed and applied for in-situ use in a focused manner. For the scaling from micro- to macroscale models, different model approaches are to be specifically linked together (e.g. integration of the LEM in the FE method).
- digitization: New developments in computer science, such as high-performance computing and visual data analysis, will be incorporated into the continuation of the project. This is necessary on the one hand to achieve the necessary scaling of the models (from micro to macro scale) and on the other hand to allow the representation of the model results in the real context of the underground laboratories.
- Internationalization: In addition, international cooperation with the Mont Terri project and DECOVALEX 2023 will be intensified, also to increase the visibility of the Geo:N project.

6.3 Pathways Through Swelling and Shrinking Processes

The focus of the work in the continuation is the Mont Terri CD-A experiment. In autumn 2019, two 11 m long niches were excavated and extensively instrumented. While one niche is ventilated and thus exposed to seasonal fluctuations in humidity, the second niche is kept closed by a bulkhead. First geophysical measurements as well as observations of drying cracks on the walls already indicate the influence of desaturation in the ventilated niche, while the closed niche, where a constantly

Fig. 6.10 Pathways through pressure-driven percolation, clay rock (Photo: CAU Kiel)

high humidity has been established, does not show any such cracks. Based on the experimental analyses, the anisotropic swelling and shrinking behaviour under these boundary conditions is analysed and characterised in the laboratory. The analysis is the basis for further modelling on larger scales. The numerical methods [7, 8] developed in GeomInt are to be used and, if necessary, further refined in order to model and analyse the observed differences between the two niches. In a first step, the two-dimensional HM calculations of the CD/LP experiment (in [9]) performed within GeomInt will be extended by the newly developed non-linear transversal-isotropic approaches in mechanics and in the swelling and shrinking model. Furthermore, the possibility of modelling shrinkage cracks in the HM context is investigated on the mechanical side by the phase field method and plastic models and on the hydraulic side with multicontinua models. In a second step, this approach will be applied to the CD-A experiment and quantitatively analyzed by comparison with measured data, so that a basic validation of the model approach on field scale is possible (Fig. 6.10).

6.4 Displacements Due to Pressure-Driven Percolation

6.4.1 Pressure-Driven Percolation in Clay Rock Under In-Situ Conditions

In the GeomInt continuation, the laboratory-experimental focus is on the investigation of anisotropy effects. This material state determines all physical properties, especially for claystone. For the characterization of the anisotropic material behavior, in particular fracture behavior and pathability under temperature, saturation and pressure changes are to be analyzed. Selected investigations, e.g. fracture toughness analyses, fluid-driven percolation tests or high-pressure cube pressure tests are to be carried out with special consideration of the material anisotropy. Any additional material samples that may be required could be provided by ongoing drilling activities in the new galleries in the Mont Terri rock laboratory. The aim is to determine the anisotropic fracture and strength parameters, conductivities or fracture paths required for further simulations during fluid percolation under in-situ conditions.

The already developed thermo-hydro-break-mechanical lattice element method (LEM) is used with a vectorized meshing for anisotropic boundary conditions and verified by the above mentioned specific laboratory tests. For the transition from micro- or mesoscale LEM to macroscale FEM simulations, the "Mesh Fragmentation Technique (MFT)" is to be extended in such a way that vectorized Voronoi LEM grids are directly transferred to the finite element meshes according to the MFT. The material properties of the MFT interface elements are then replaced by the physical fracture properties of the lattice beams. Since a pre-definition of the crack path is not necessary with this approach even in large scale FE models, this approach is extremely attractive for future realistic simulations. The planned combination of LEM with FEM and thus a consideration of micro-properties for larger scale simulations corresponds to the concept of the GeomInt continuation regarding scaling from laboratory to in-situ scale and thus the direct use of the GeomInt2 modeling platform for URL experiments.

The LIE method [10], which has been further developed in GeomInt, will also be used to simulate pressure-driven crevice reactivation in another Mont Terri experiment (Fault-Slip FS experiment). The hydromechanical coupling is investigated in the case of an inelastic change in the properties of an existing fissure or crack network in the claystone as a result of fluid injection. Due to the special boundary conditions, modeling is of particular importance for the interpretation of this complex in-situ experiment, since the observed behavior in laboratory experiments can only be simulated in a very simplified way (Fig. 6.11).

6.4.2 Pressure-Driven Percolation in Salt Rock Under In-Situ Conditions

In the in-situ laboratory of the Springen mine, a large bore hole of 50 m^3 is used to investigate the pressure-driven percolation of both liquids and gases. This was carried out with compressed air, representative of natural gas, and the rupture of the rock was monitored by an array of sensors for acoustic emissions. This large-scale experiment will be evaluated in two ways in the continuation. On the one hand, the influence of the local stress field and the rupture zone on pressure development and propagation direction shall be investigated by means of a discontinuum mechanical method. On the other hand, it shall be determined whether the same process can be reproduced with a continuum mechanical method. The model application is again driven by the idea of scaling from the smaller to the larger in-situ scale. As an extended continuum mechanical approach, the phase field method is used to simulate the large-scale Springen experiment. It is planned to compare the two methods (FLAC3D and

Fig. 6.11 Underground laboratory Springen

OpenGeoSys phase field method) with regard to their suitability for scalability to larger systems. This will also involve computational efficiency using HPC methods (parallelisation).

6.5 Displacements Due to Stress Redistribution

In the continuation of the project, there is the chance to observe on a smaller scale selected phenomena of the hydro-mechanically coupled behaviour of discontinuities, which were integrally investigated in the previous work package, i.e. averaged over the size of the sample on the cm to dm scale. For this purpose, test paths for mechanical and hydraulic scenarios can be built up on the experimental basis already developed with regard to the boundary conditions. In particular, the discretization of the shear surfaces with the aid of the 3D scanner represents the link between the measured hydro-mechanically coupled fracture behaviour and local phenomena in the mm range. Methodologically, a further development towards observational procedures, which allow an appropriate spatial resolution (at least in the mm range) with simultaneous mechanical or coupled loading of the fissure body sample, is being carried out. Specifically, the shear box of the shear apparatus in the Freiberg laboratory is modified in such a way that the arrangement of piezoceramic sensors for recording acoustic emissions (AE) and the fixing of micro-accelerometers directly next to, or above and below the fissure surface is possible. By operating the AE sensors in active mode, i.e. as ultrasonic actuators for pulse generation, damage parallel to the fissure surface and dislocation on the fissure surface can be quantified. The combined use of the AE system and the microseismic network significantly improves the possibilities for locating the damage processes and for generating the hearth surface solution, since shear waves can now also be reliably detected in their 3D spatial position. In contrast to AE sensors, the micro-accelerometers also provide measurement data on the absolute energy of the recorded seismic events, thus enabling quantitative analysis of the crack initiation process in its spatio-temporal evolution.

The gneiss from Freiberg (URL Reiche Zeche) is selected as sample material, with the focus primarily on fissures and fault zones in the otherwise massive gneiss body. Since preliminary tests showed that the foliation of the gneiss is mechanically visible but hydraulically only of limited relevance, crevasses and fault zones were identified as potential pathways in the gneiss. The Freiberg data from the experimental tests are directly used by the Stuttgart team for numerical simulation to characterize the hydro-mechanical coupling of fluid-saturated cracks or fracture networks (using hybrid FE methods). Transient stimulation experiments will be simulated. The combination of e.g. harmonic pump experiments and the numerical simulation tools developed here allows a realistic description of the pressure-dependent effective permeability of the crack networks as well as an estimation of the associated effective crack storage capacity.

This research focus will also strengthen the international networking of GeomInt2 with the DECOVALEX 2023 project. There it is planned to work on complementary objectives to the work package described here—for example, on fissure behavior under thermal or polyaxial boundary conditions.

Both projects GeomInt2 and STIMTEC2 want to strengthen their cooperation in the continuation. Of great interest for GeomInt2 are the experimental stimulation experiments of the STIMTEC project in the rich coal mine [11] for testing the modeling platform under in-situ conditions. For STIMTEC2 the determination of mechanical and hydraulic properties for the crystalline (characterization of fault zones, fractures and rock matrix) from the GeomInt2 concept is of particular interest. Thus both projects can create significant added value in the cooperation.

6.6 Data and Model Integration Using Virtualization and High-Performance Computing

In the context of the follow-up project, the synthesis activities are to be strengthened. In particular, modern methods of digitization will be further developed and applied. This includes scientific software development, virtualization and high-performance computing—which will be applied specifically to selected GeomInt studies in the sense of demonstrating the GeomInt platform.

The procedure will be explained using the example of the CD-A experiment in the underground laboratory at Mont Terri (see Sect. 6.3). In the first project phase, the constitutive model was developed and implemented [10]. In order to use the models as realistically as possible, they are built into the underground laboratory

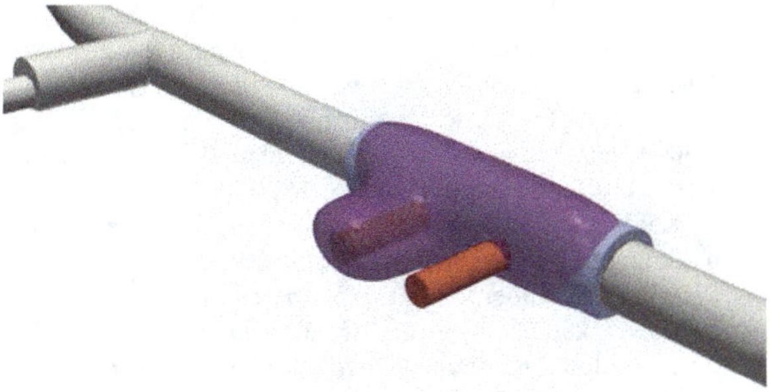

Fig. 6.12 Virtualization of the CD-A experiment (*Source* BGR/UFZ)

using virtualization (see Fig. 6.12). The model has already been used for the planning of the CD-A experiment to determine the optimal distance between the niches (no significant interaction within 20 years). A further increase of the process complexity (nonlinear mechanics in the HM context) requires the unconditional use of HPC methods to simulate a large number of variants. The model results should be continuously compared with the running experiments and also integrated into the existing virtualization context. This is the basic idea of continuous data and model integration.

References

1. A. S. Sattari, H. B. Motra, Z. H. Rizvi, and F. Wuttke. A new lattice element method (lem) with integrated interface elements for determining the effective thermal conductivity of rock solids under thermo-mechanical processes. *International Symposium on Energy Geotechnics (SEG), Energy Geotechnics*, pages 266–275, 2019.
2. P. Schmidt and H. Steeb. Numerical aspects of hydro-mechanical coupling of fluid-filled fractures using hybrid-dimensional element formulations and non-conformal meshes. *Int J Geomath*, 10(14), 2019.
3. Keita Yoshioka, Dmitri Naumov, and Olaf Kolditz. On Crack Opening Computation in Variational Phase-Field Models for Fracture. *Computer Methods in Applied Mechanics and Engineering*, under review.
4. P. Bossart. Twenty years of research at the mont terri rock laboratory: what we have learnt. *Swiss Journal of Geosciences*, 110(1):405–411, 2017. cited By 0.
5. P. Bossart, F. Bernier, J. Birkholzer, C. Bruggeman, P. Connolly, S. Dewonck, M. Fukaya, M. Herfort, M. Jensen, J.-M. Matray, J.C. Mayor, A. Moeri, T. Oyama, K. Schuster, N. Shigeta, T. Vietor, and K. Wieczorek. Mont terri rock laboratory, 20 years of research: introduction, site characteristics and overview of experiments. *Swiss Journal of Geosciences*, 110(1):3–22, 2017. cited By 44.
6. Lars Bilke, Bernd Flemisch, Olaf Kolditz, Rainer Helmig, and Thomas Nagel. Development of open-source porous-media simulators: principles and experiences. *Transport in Porous Media*, 2019. under review.

7. K. Yoshioka, F. Parisio, D. Naumov, R. Lu, O. Kolditz, and T. Nagel. Comparative verification of discrete and smeared numerical approaches for the simulation of hydraulic fracturing. *GEM-International Journal on Geomathematics*, 10(1), 2019.
8. F. Parisio, A. Tarokh, R. Makhnenko, D. Naumov, X.-Y. Miao, O. Kolditz, and T. Nagel. Experimental characterization and numerical modelling of fracture processes in granite. *International Journal of Solids and Structures*, 163:102–116, 2019. cited By 4.
9. O. Kolditz and et al. GeomInt-Geomechanical integrity of host and barrier rocks-experiment, modeling and analysis of discontinuities. In *Terrestrial Environmental Sciences*, page ca. 250. Springer, Heidelberg, 2020. in Vorbereitung.
10. B. Vowinckel, A. Alfarra, J. Maßmann, D. Naumov, G. Zielfe, O. Kolditz, and T. Nagel. Computational modeling of hydraulically induced fault slip in argillaceous rock using lower-dimensional interface elements with local enrichment of the displacement field. *preparation for the Geo:N Topical Collection in Environmental Earth Sciences*, 2020.
11. P. Schmidt, J. Renner, and H. Steeb. Numerical characterization of hydraulically tested fractures at reiche zeche. *preparation for the Geo:N Topical Collection in Environmental Earth Sciences*, 2020.
12. K. Rink, L. Bilke, and O. Kolditz. Visualisation strategies for environmental modelling data. *Environmental Earth Sciences*, 72(10):3857–3868, 2014. cited By 22.

7

Explanation of Code

Lars Bilke, Thomas Fischer, Dmitri Naumov, Daniel Pötschke, Karsten Rink, Amir Shoarian Sattari, Patrick Schmidt, Wenqing Wang, and Keita Yoshioka

7.1 FFS—Forces on Fracture Surfaces

The FFS method (see Sect. 3.2.1) was developed to simulate direct shear tests. To provide a tool for the project work and get things easier done a graphical user interface (GUI) was also created. The GUI simply calls all necessary functions by letting the user either fill form fields or choose input files from the working folder. The rock parameters and the conditions of the direct shear test with the normal stress levels and shear displacements have to be selected. If an experiment is simulated the lab results can be selected as a text file so a visual comparison is possible. The geometry has to be loaded as a point cloud or an artificial surface can be generated. With small modifications the code can do multiple executions using artificial surfaces.

The GUI can be found at www.github.com/Poetschke/Ecodist/. At github an executable is available which allows (after some installation) to test it without needing a Matlab licence. The scheme of the FFS algorithm is illustrated in Fig. 7.1.

L. Bilke (✉) · T. Fischer · D. Naumov · K. Rink · W. Wang · K. Yoshioka
UFZ, Helmholtz Centre for Environmental Research, OGS Core Team, Leipzig, Germany
e-mail: lars.bilke@ufz.de

D. Naumov · D. Pötschke
TUBAF, Technische Universität Bergakademie Freiberg, Freiberg, Germany

A. S. Sattari
CAU, Christian-Albrechts-Universität zu Kiel, Kiel, Germany

P. Schmidt
UoS, University of Stuttgart, Stuttgart, Germany

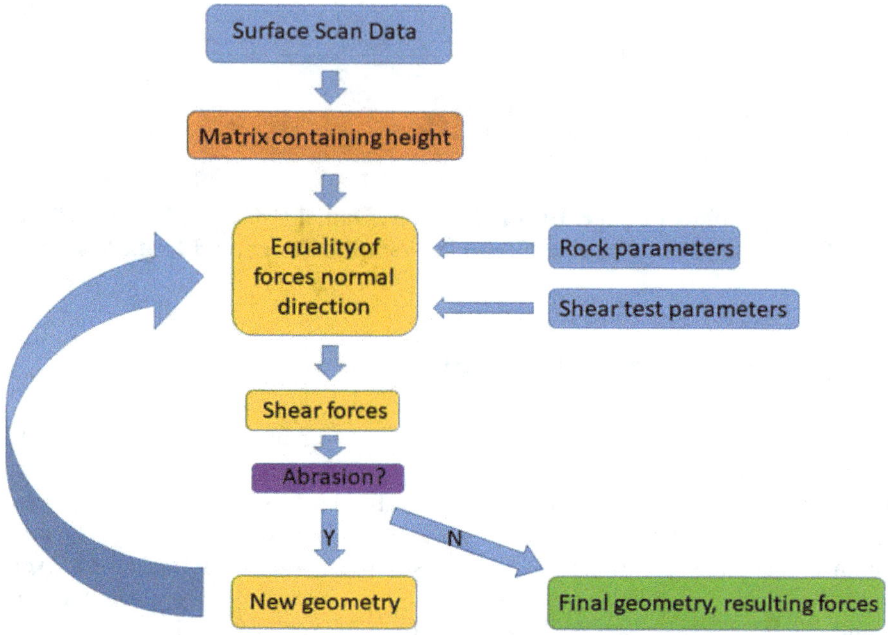

Fig. 7.1 Scheme of the FFS code

7.2 LEM—Lattice-Element-Method

The lattice element method (LEM) is a well-known model for the simulation of the fracture in the cemented geomaterial and concrete. In comparison to the discrete element method (DEM), where the contact search and contact mechanics are implemented, the LEM represents the medium with a series of spring or beam elements to simulate the fracking process. The considered LEM in this study is fully developed in Kiel University (CAU Kiel) and is implemented in various engineering applications. In earlier studies, the application of LEM was restricted to fracture simulation in concrete, where the heterogeneity was introduced with defining the aggregates, mortar and interface bond zone [1–3]. With the development of LEM its application is extended to failure behavior of cemented geomaterials such as bio-cemented granular material [4]. The LEM is also used to simulate the fracture under dynamic loading for the foam concrete [5], masonry walls [6] and cemented geomaterial [7, 8]. Figure 7.2 depicts the coupled THM processes and affected geomaterial parameters, which are implemented in LEM algorithm.

In its recent application, the evaluation of effective properties in shallow crustal rock is investigated [9]. The developed in-house LEM model is applied for the simulation of the heat transfer in modified granular material and assessment of effective thermal conductivity [10–13] as well as the Nano geocomposites [14]. The thermo-mechanical LEM model is implemented to simulate the change of the thermal conduc-

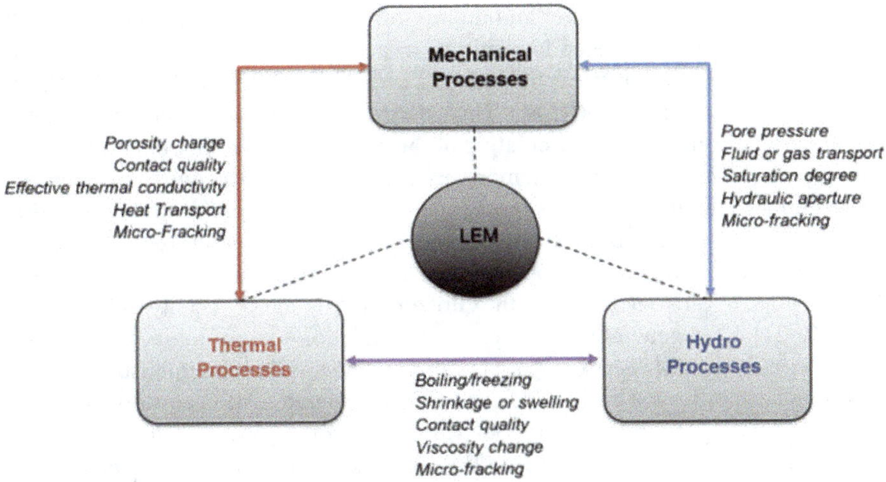

Fig. 7.2 The simulation of the coupled THM processes with LEM

tivity of the rocks under mechanical and thermal loadings [15]. With an integration of the interface element, the LEM is able to simulate the fully coupled TM processes in cemented geomaterial [16]. The application of LEM is extended to model the hydro-mechanical processes [17, 18]. In these models, the dual lattice setup is considered, where lattice elements transfer the mechanical loads between the nodes and conduct elements only carry the fluid flow. Similar to DEM models [19], the LEM is extended to simulate the shrinkage and swelling processes in rock material. In the scope of this study, the LEM is also used for the simulation of pressurized percolation tests in rock material. In this model, the mass conservation law is implemented and artificial cavities for fluid or gas transport are defined. In CAU Kiel, we are devoted to continue the development of the LEM and overcome its application limitations. In this sense, the parallel computing for computing efficiency is under process and development. The ongoing work incorporate the plasticity, visco-plasticity, flow, hardening, fatigue and creep rules to establish a constitutive lattice model, which can be implemented in the practical applications to simulate the geomaterial response under the coupled THM processes.

7.3 SPH—Smoothed-Particle-Hydrodynamics

The (explicit) discrete nodal formulation of the Navier-Stokes equations basically results in computations of loops over all considered particles and for each additional nested loops over neighbouring particles. On the one hand, this circumstance renders SPH a computationally demanding method, on the other hand, the parallelization of this structure assembled from subroutines is quite generic on CPUs and even GPUs.

Despite the Lagrangian character and the meshfree formulation, SPH codes can be compared to collocation methods resulting in particle-particle interactions (linear algebra operations) similar to explicit particle codes like Molecular Dynamics (MD) or Discrete Element Methods (DEM). Thus, they exhibit the same challenges as these explicit particle codes: for the calculating of the particle interactions, data from neigbouring particles is needed, and memory access and load balancing is unstructured. The neighbour search algorithm is most expensive and considerable communication as well as data migration between processors is necessary. Therefore the presented SPH formulation is implemented on top of the highly optimized and MPI-parallelized HOOMD-blue library developed by the Glotzer group at the University of Michigan, USA [20, 21]. This general purpose particle simulation toolkit, initially developed for MD, comes with MPI-based spatial domain decomposition, demonstrated weak and strong scalability for both, GPU- and CPU-accelerated HPC clusters, heuristic load balancing, algorithms for neigbour search and sorting methods to ensure optimal memory access patterns. The HOOMD-Blue software package is employed in a large selection of research areas, cf. the mentioned homepage. It is open-source, published under a BSD 3-clause license and a and comprehensive documentation is available. Recently, weak and strong scaling tests of fluid flow through porous media has been investigated on CPU- and GPU-HPC platforms, [22]. The implemented SPH model [23] includes both, CUDA and MPI features and uses the above mentioned advantages. Setup of the boundary value problem and initialisation of the geometry and particle data is implemented as user-friendly Python scripts. The main implementation are programmed in C++ and CUDA. This comprises among other things the evaluation of the kernel, the computation of density rate, pressure fields and particle accelerations as well as the time integration. Besides single-phase flow models based on the Navier-Stokes equations [24], multi-phase flow models of two immiscible fluids including surface tension has been investigated [25] as well as suspension-flow of a Newtonian/non-Newtonian carrier fluid and solid non-colloidal particles [26].

7.4 OpenGeoSys—Finite-Element-Method

OpenGeoSys (OGS) is a scientific open-source initiative for the numerical simulation of thermo-hydro-mechanical/chemical (THMC) processes in porous and fractured media, inspired by FEFLOW [27] and ROCKFLOW concepts and continuously developed since the mid-eighties (Fig. 7.3), see e.g. [28–31]. Meanwhile, more than 50 PhD projects have been dedicated to the OGS development since the merger in the nineties.

The OGS framework is targeting applications of various disciplines in environmental geoscience, e.g., in the fields of regional [32], contaminant [33] and coastal hydrology [34], fundamental geothermal processes [35] and geothermal energy systems [36, 37]. OGS is applied for energy storage applications in technical systems

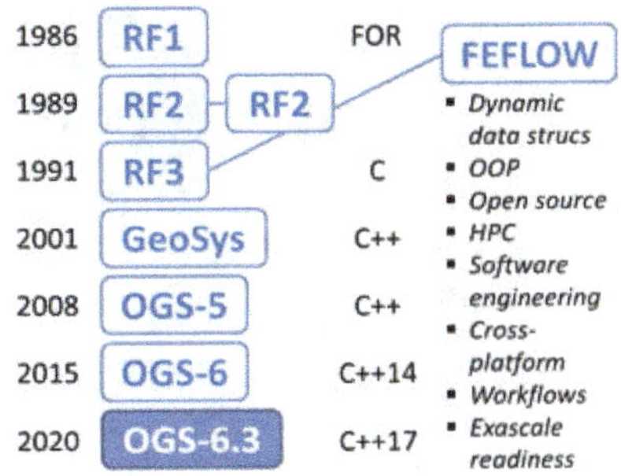

Fig. 7.3 OpenGeoSys development history

such as concrete [38] or zeolite-based heat storage [39] and natural systems such as salt caverns [40, 41]. OGS is also used in fundamental studies for nuclear waste management [42].

The most recent version, OpenGeoSys-6 (OGS-6) [43, 44], is a complete re-implementation of the multi-physics code OpenGeoSys-4/5 [45, 46] using advanced methods in software engineering and architecture with a focus on code quality, modularity, performance and comprehensive documentation. The current release version OpenGeoSys 6.2.0 [47] will be dedicated to analyze and predict the behaviour of geosystems becoming more and more relevant in future like nuclear waste deposition, geothermal use of subsurface resources for power and heat production, and geological storage of various energy carriers. Particular emphasis is put on the implementation of advanced numerical methods for the propagation of discontinuities, such as enriched finite element function spaces [48], non-local formulations [49] and phase-field models for fracture [50] with the ability to utilize HPC platforms [51, 52].

OpenGeoSys is participating in several international model development, validation and benchmarking initiatives, e.g., DEVOVALEX (with applications mainly in the assessment of waste repositories, see [53]), CO2BENCH (geological CO_2 storage, see [54]), SeS Bench (reactive transport processes, see [55]) and HM-Intercomp (coupled hydrosystems, see [56]). The OGS community provides an ongoing series of benchmark books [57] and tutorials [58]. For more information please refer to the OpenGeoSys webpage www.opengeosys.org.

VPF—Variational Phase-Field model

The variational phase-field model (V-pf) is increasingly becoming a popular numerical method for fracture computation because of its ability to account for arbitrary numbers of pre-existing or propagating cracks in terms of energy minimization, without any a priori assumption on their geometry or restriction on the growth to specific grid directions. The variational phase-field model applied in this study has been based on the model proposed by [59, 60] where each process (e.g. mechanical or hydraulic) is solved in a staggered manner as in Fig. 7.4 and has been implemented in OGS utilizing its linear algebraic and finite element method platform.

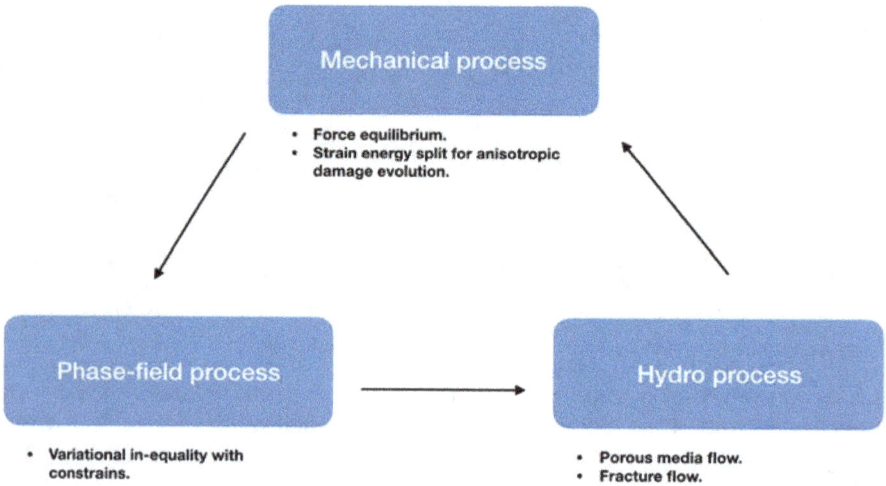

Fig. 7.4 Computation scheme with the variational phase-field model

The mechanical process solves the force equilibrium under the presence of the crack (damage) field in which the damage is accounted differently depending on the state of the load (e.g. compression or tension) in order to distinguish the material's response under different types of loading. Various approaches have been proposed for the energy split strategy and the three of the most established approaches [61–63] have been implemented. Though the process for the phase-field is an elliptic problem, the solution space is bounded in [0, 1] and is constrained by the irreversibility (i.e. fracture is not allowed to heal). Therefore, its solution requires a variational inequality solver and it is achieved through PETSc [64, 65]. Once the displacement and the phase-field are solved, the crack opening displacement will be reconstructed following an approximation proposed by [66] and the computation result will be passed onto the hydro process where fluid flows both in porous medium and fracture will be solved. These processes will be repeated in a staggered manner until the convergence is met (currently its judgement is based on the phase-field process).

7.5 HDF—Hybrid-Dimensional-Formulation

The Hybrid-Dimensional-Formulation results in a numerically strongly coupled system of governing equations. Different numerical strategies, namely the weak/staggered and strong/monolithic coupling schemes have been implemented in course of this project to solve the interaction between fluid flow and deformation of the surrounding porous matrix. Dependent on the method different technical requirements are demanded from the numerical framework. Hence for each one of the two cou-

Fig. 7.5 Packages used to develop Hybrid-Dimensional Framework

pling strategies an individual numerical framework has been chosen to guarantee numerical efficiency (Fig. 7.5).

For the strongly coupled scheme high performance has been ensured by choosing the Distributed and Unified Numerics Environment (DUNE) [67] to monolithically build and solve the global system of governing equations. The Dune implementation is based on modern C++ programming techniques to provide a unique combination of highly efficient and flexible code by providing a common interface at a very low overhead for various mesh based methods. Combined with the generalized discretization module PDELab [68] the basis for Finite Element calculations of the implemented solver has been built. Nevertheless, extensive work on the existing framework has been performed in order to allow the integration of zero-thickness elements; a feature which is not provided by default.

The staggered algorithm of the weak coupling scheme allows for calculations on different numerical domains. Efficient numerical implementation combined with a versatile way to handle continuously varying boundary conditions provided by the FEniCS computing platform [69] form the basis of the developed solver. Since the coupling between both domains is numerically strong an implicit coupling iteration is required. The Precise Code Interaction Coupling Environment (preCICE) [70] provides an easily accessible interface for parallel communication between existing solvers allowing for non-conformal discretization of computational domains in combination with a highly developed Quasi-Newton method to guarantee numerical stability.

Discretization for both methods, namely the construction of interface elements and separation of fracture surfaces is a challenging task especially in three dimensions. This challenge has been overcome by an in house meshing tool based on the Gmsh meshing facility [71].

References

1. J.X. Liu, S.C. Deng, J. Zhang, and N.G Liang. Lattice type of fracture model for concrete. *Theoretical and Applied Fracture Mechanics*, 48:269–284, 2007.
2. E.P. Prado and J.G.M. van Mier. Effect of particle structure on mode i fracture process in concrete. *Engineering Fracture Mechanics*, 70:1793–1807, 2003.
3. J.G.M. van Mier, M.R.A. van Vliet, and K. Wang Tai. Fracture mechanisms in particle composites: statistical aspects in lattice type analysis. *Mechanics of Materials*, 34, 2002.

4. Z. H. Rizvi, M. Nikolic, and F. Wuttke. Lattice element method for simulations of failure in bio-cemented sands. *Granular Matter*, 21(18), 2019.
5. Z. H. Rizvi, A. S. Sattari, and F. Wuttke. Meso scale modelling of infill foam concrete wall for earthquake loads. *16th European Conference on Earthquake Engineering (16ECEE), Thessaloniki, Greece*, 2018.
6. A. S. Sattari, Z. H. Rizvi, and F. Wuttke. The reinforcement of masonry walls in a mortar-brick interface: The experimental and numerical approach using the lattice element method (lem). *12th international congress on mechanics (12HSTAM), Thessaloniki, Greece*, 2019.
7. Z.H. Rizvi, F. Wuttke, and A.S. Sattari. Dynamic analysis by lattice element method simulation. *Springer Series in Geomechanics and Geoengineering*, (216849):405–409, 2018. cited By 2.
8. Z.H. Rizvi, S.H. Mustafa, A.S. Sattari, S. Ahmad, P. Furtner, and F. Wuttke. Dynamic lattice element modelling of cemented geomaterials. In Amit Prashant, Ajanta Sachan, and Chandrakant S. Desai, editors, *Advances in Computer Methods and Geomechanics*, pages 655–665, Singapore, 2020. Springer Singapore.
9. Z.H. Rizvi, M.A. Khan, H.B. Motra, F. Wuttke, and J. Ahmad. Effective physical parameter evaluation of shallow crustal rocks by lattice element method. *Materials Today: Proceedings*, 2019.
10. Z.H. Rizvi, H.H. Zaidi, S.J Akhtar, A.S. Sattari, and F. Wuttke. Soft and hard computation methods for estimation of the effective thermal conductivity of sands. *Heat and Mass Transfer*, Feb 2020.
11. Z.H. Rizvi, D. Shrestha, A.S. Sattari, and F. Wuttke. Numerical modelling of effective thermal conductivity for modified geomaterial using lattice element method. *Heat Mass Transf*, 54(2):483–499, 2018.
12. Z.H. Rizvi, A.S. Sattari, and F. Wuttke. Numerical analysis of heat conduction in granular geo-material using lattice element method. *Energy Geotechnics—Proceedings of the 1st International Conference on Energy Geotechnics, ICEGT*, pages 367–372, 2016.
13. D. Shrestha, Z.H. Rizvi, and F. Wuttke. Effective thermal conductivity of unsaturated granular geocomposite using lattice element method. *Heat and Mass Transfer*, 55(6):1671–1683, 2019.
14. Z.H. Rizvi, K. Sembdner, A. Suman, M.J. Giri Prasad, and F. Wuttke. Experimental and numerical investigation of thermo-mechanical properties for nano-geocomposite. *International Journal of Thermophysics*, 40(5), 2019. cited By 1.
15. A.S. Sattari, Z.H. Rizvi, H.B. Motra, and F. Wuttke. Meso-scale modeling of heat transport in a heterogeneous cemented geomaterial by lattice element method. *Granular Matter*, 19(66), 2017.
16. A. S. Sattari, H. B. Motra, Z. H. Rizvi, and F. Wuttke. A new lattice element method (lem) with integrated interface elements for determining the effective thermal conductivity of rock solids under thermo-mechanical processes. *International Symposium on Energy Geotechnics (SEG), Energy Geotechnics*, pages 266–275, 2019.
17. P. Grassl. A lattice approach to model flow in cracked concrete. *Cement & Concrete Composites*, 31:454–460, 2009.
18. P. Grassl, C. Fahy, D. Gallipoli, and J. Bolander. A lattice model for liquid transport in cracked unsaturated heterogeneous porous materials. *VIII International Conference on Fracture Mechanics of Concrete and Concrete Structures*, 2013.
19. J. Sima, M. Jiang, and C. Zhou. Modelling desiccation cracking in thin clay layer using three-dimensional discrete element method. *AIP Conference Proceedings*, 1542(1), 2013.
20. Joshua A. Anderson, Chris D. Lorenz, and Alex Travesset. General purpose molecular dynamics simulations fully implemented on graphics processing units. *Journal of Computational Physics*, 227(10):5342–5359, 2008.
21. Jens Glaser, Trung Dac Nguyen, Joshua A Anderson, Pak Lui, Filippo Spiga, Jaime A Millan, David C Morse, and Sharon C Glotzer. Strong scaling of general-purpose molecular dynamics simulations on GPUs. *Computer Physics Communications*, 192:97–107, 2015.
22. M. Osorno, M. Schirwon, N. Kijanski, R. Sivanesapillai, H. Steeb, and D. Göddeke. A cross-platform, high-performance SPH implementation of flow in μCT imaged porous media for digital rock physics. *Computer Physics Communications*, 2019. submitted for publication.

23. Rakulan Sivanesapillai. Pore-scale study of non-darcian fluid flow in porous media using smoothed-particle hydrodynamics. 2016.
24. R Sivanesapillai, H Steeb, and A Hartmaier. Transition of effective hydraulic properties from low to high reynolds number flow in porous media. *Geophysical Research Letters*, 41(14):4920–4928, 2014.
25. Rakulan Sivanesapillai, Nadine Falkner, Alexander Hartmaier, and Holger Steeb. A csf-sph method for simulating drainage and imbibition at pore-scale resolution while tracking interfacial areas. *Advances in water resources*, 95:212–234, 2016.
26. D. Markauskas, H. Kruggel-Emden, R. Sivanesapillai, and H. Steeb. Comparative study on mesh-based and mesh-less coupled cfd-dem methods to model particle-laden flow. *Powder Technology*, 305:78–88, 2017.
27. H.-J.G. Diersch. *FEFLOW: Finite element modeling of flow, mass and heat transport in porous and fractured media.* Springer, 2014. cited By 156.
28. Olaf Kolditz. *Zur Modellierung und Simulation geothermischer Transportprozesse in untertägigen Zirkulationssystemen.* PhD thesis, Akademie der Wissenschaften der DDR, Berlin, 1990.
29. J. Wollrath. *Ein Strömungs- und Transportmodell für klüftiges Gestein und Untersuchungen zu homogenen Ersatzsystemen.* PhD thesis, Institut für Strömungsmechanik und Elektronisches Rechnen im Bauwesen, Universität Hannover, 1990.
30. K.P. Kroehn. *Simulation von Transportvorgängen im klüftigen Gestein mit der Methode der Finiten Elemente.* PhD thesis, Institut für Strömungsmechanik und Elektronisches Rechnen im Bauwesen, Universität Hannover, 1991.
31. Reiner Helmig. *Theorie und Numerik der Mehrphasenströmungen in geklüftet-porösen Medien.* PhD thesis, Institut für Strömungsmechanik und Elektronisches Rechnen im Bauwesen, Universität Hannover, 1993.
32. M. Jing, F. Heße, R. Kumar, W. Wang, T. Fischer, M. Walther, M. Zink, A. Zech, L. Samaniego, O. Kolditz, and S. Attinger. Improved regional-scale groundwater representation by the coupling of the mesoscale hydrologic model (mhm v5.7) to the groundwater model opengeosys (ogs). *Geoscientific Model Development*, 11(5):1989–2007, 2018.
33. E. Nixdorf, Y. Sun, M. Lin, and O. Kolditz. Development and application of a novel method for regional assessment of groundwater contamination risk in the songhua river basin. *Science of the Total Environment*, 605-606:598–609, 2017.
34. M. Walther, T. Graf, O. Kolditz, R. Liedl, and V. Post. How significant is the slope of the sea-side boundary for modelling seawater intrusion in coastal aquifers? *Journal of Hydrology*, 551:648–659, 2017.
35. F. Parisio, S. Vinciguerra, O. Kolditz, and T. Nagel. The brittle-ductile transition in active volcanoes. *Scientific Reports*, 9(1), 2019.
36. B. Meng, T. Vienken, O. Kolditz, and H. Shao. Modeling the groundwater temperature response to extensive operation of ground source heat pump systems: A case study in germany. *Energy Procedia*, 152:971–977, 2018.
37. Philipp Hein, Ke Zhu, Anke Bucher, Olaf Kolditz, Zhonghe Pang, and Haibing Shao. Quantification of exploitable shallow geothermal energy by using borehole heat exchanger coupled ground source heat pump systems. *Energy Conversion and Management*, 127:80 – 89, 2016.
38. X.-Y. Miao, O. Kolditz, and T. Nagel. Modelling thermal performance degradation of high and low-temperature solid thermal energy storage due to cracking processes using a phase-field approach. *Energy Conversion and Management*, pages 977–989, 2019.
39. C. Lehmann, O. Kolditz, and T. Nagel. Modelling sorption equilibria and kinetics in numerical simulations of dynamic sorption experiments in packed beds of salt/zeolite composites for thermochemical energy storage. *International Journal of Heat and Mass Transfer*, 128:1102–1113, 2019.
40. Norbert Böttcher, Uwe-Jens Görke, Olaf Kolditz, and Thomas Nagel. Thermo-mechanical investigation of salt caverns for short-term hydrogen storage. *Environmental Earth Sciences*, 76(3):98, 2017.
41. Thomas Nagel, Wolfgang Minkley, Norbert Böttcher, Dmitri Yu. Naumov, Uwe-Jens Görke, and Olaf Kolditz. Implicit numerical integration and consistent linearization of inelastic constitutive models of rock salt. *Computers & Structures*, 182:87–103, apr 2017.

42. H. Shao, J. Hesser, O. Kolditz, and W. Wang. Hydraulic characterisation of clay rock under consideration of coupled thm properties. *Environmental Science and Engineering*, pages 33–40, 2019.
43. Dmitri Yu. Naumov, Lars Bilke, Thomas Fischer, Yonghui Huang, Christoph Lehmann, Xing-Yuan Miao, Thomas Nagel, Francesco Parisio, Karsten Rink, Haibing Shao, Wenqing Wang, Norihiro Watanabe, Tianyuan Zheng, and Olaf Kolditz. Appendix a: Opengeosys-6. In Olaf Kolditz, Thomas Nagel, Hua Shao, Wenqing Wang, and Sebastian Bauer, editors, *Thermo-Hydro-Mechanical-Chemical Processes in Fractured Porous Media: Modelling and Benchmarking*, pages 271–277. Springer International Publishing, 2018.
44. Lars Bilke, Bernd Flemisch, Olaf Kolditz, Rainer Helmig, and Thomas Nagel. Development of open-source porous-media simulators: principles and experiences. *Transport in Porous Media*, 2019. under review.
45. O. Kolditz and S. Bauer. A process-oriented approach to computing multi-field problems in porous media. *Journal of Hydroinformatics*, 6(3):225–244, 2004.
46. Wenqing Wang and Olaf Kolditz. Objectoriented finite element analysis of thermohydromechanical (thm) problems in porous media. *International Journal for Numerical Methods in Engineering*, 69(1):162–201, 2006.
47. Dmitry Yu. Naumov, Tom Fischer, Lars Bilke, Karsten Rink, Christoph Lehmann, Norihiro Watanabe, Wenqing Wang, Yonghui Huang, Renchao Lu, Chaofan Chen, Jasper Bathmann, Xingyuan Miao, Keita Yoshioka, Haibing Shao, Marc Walther, Tianyuan Zheng, Francesco Parisio, Jan Thiedau, Carolin Helbig, Jörg Buchwald, and Thomas Nagel. ufz/ogs: 6.1.0, January 2018.
48. N Watanabe, W Wang, J Taron, UJ Görke, and O Kolditz. Lower-dimensional interface elements with local enrichment: application to coupled hydro-mechanical problems in discretely fractured porous media. *International Journal for Numerical Methods in Engineering*, 90(8):1010–1034, 2012.
49. F. Parisio, D. Yu. Naumov, O. Kolditz, and T. Nagel. Material forces: An insight into configurational mechanics. *Mechanics Research Communications*, 93:114–118, 2018.
50. K. Yoshioka, F. Parisio, D. Naumov, R. Lu, O. Kolditz, and T. Nagel. Comparative verification of discrete and smeared numerical approaches for the simulation of hydraulic fracturing. *GEM—International Journal on Geomathematics*, 10(1), 2019.
51. Wenqing Wang, Thomas Fischer, Björn Zehner, Norbert Böttcher, Uwe-Jens Görke, and Olaf Kolditz. A parallel finite element method for two-phase flow processes in porous media: OpenGeoSys with PETSc. *Environmental Earth Sciences*, 73(5):2269–2285, 2015.
52. Wenqing Wang, Olaf Kolditz, and Thomas Nagel. Parallel finite element modelling of multiphysical processes in thermochemical energy storage devices. *Applied Energy*, 185(P2):1954–1964, 2017.
53. J.T. Birkholzer, A.E. Bond, J.A. Hudson, L. Jing, C.-F. Tsang, H. Shao, and O. Kolditz. Decovalex-2015: an international collaboration for advancing the understanding and modeling of coupled thermo-hydro-mechanical-chemical (thmc) processes in geological systems. *Environmental Earth Sciences*, 77(14), 2018. cited By 1.
54. O. Kolditz, S. Bauer, C. Beyer, N. Böttcher, P. Dietrich, U.-J. Görke, T. Kalbacher, C.-H. Park, U. Sauer, C. Schütze, H. Shao, A. Singh, J. Taron, W. Wang, and N. Watanabe. A systematic benchmarking approach for geologic CO2 injection and storage. *Environmental Earth Sciences*, 67(2):613–632, 2012.
55. C.I. Steefel, C.A.J. Appelo, B. Arora, D. Jacques, T. Kalbacher, O. Kolditz, V. Lagneau, P.C. Lichtner, K.U. Mayer, J.C.L. Meeussen, S. Molins, D. Moulton, H. Shao, J. Šimůnek, N. Spycher, S.B. Yabusaki, and G.T. Yeh. Reactive transport codes for subsurface environmental simulation. *Computational Geosciences*, 19(3):445–478, 2015.
56. Reed M. Maxwell, Mario Putti, Steven Meyerhoff, JensOlaf Delfs, Ian M. Ferguson, Valeriy Ivanov, Jongho Kim, Olaf Kolditz, Stefan J. Kollet, Mukesh Kumar, Sonya Lopez, Jie Niu, Claudio Paniconi, YoungJin Park, Mantha S. Phanikumar, Chaopeng Shen, Edward A. Sudicky, and Mauro Sulis. Surfacesubsurface model intercomparison: A first set of benchmark results to diagnose integrated hydrology and feedbacks. *Water Resources Research*, 50(2):1531–1549, 2016.

57. Olaf Kolditz, Thomas Nagel, Hua Shao, Wenqing Wang, and Sebastian Bauer, editors. *Thermo-Hydro-Mechanical-Chemical Processes in Fractured Porous Media: Modelling and Benchmarking*. Terrestrial Environmental Sciences. Springer International Publishing, Cham, 2018.
58. Christoph Lehmann, Olaf Kolditz, and Thomas Nagel. *Models of Thermochemical Heat Storage*. Computational Modeling of Energy Systems. Springer International Publishing, 2018.
59. Blaise Bourdin, Chukwudi P. Chukwudozie, and Keita Yoshioka. A Variational Approach to the Numerical Simulation of Hydraulic Fracturing. In *the 2012 SPE Annual Technical Conference and Exhibition*, 2012.
60. C. Chukwudozie, B. Bourdin, and K. Yoshioka. A variational phase-field model for hydraulic fracturing in porous media. *Computer Methods in Applied Mechanics and Engineering*, 347:957–982, 2019.
61. Hanen Amor, Jean-jacques Marigo, and Corrado Maurini. Regularized formulation of the variational brittle fracture with unilateral contact: numerical experiments. *Journal of Mechanics and Physics of Solids*, 57(8):1209–1229, 2009.
62. C. Miehe, F. Welschinger, and M. Hofacker. Thermodynamically consistent phase-field models of fracture: Variational principles and multi-field fe implementations. *International Journal for Numerical Methods in Engineering*, 83(10):1273–1311, 2010.
63. Francesco Freddi and Gianni Royer-Carfagni. Regularized variational theories of fracture: A unified approach. *Journal of the Mechanics and Physics of Solids*, 58(8):1154–1174, 2010.
64. S. Balay, S. Abhyankar, M.F. Adams, J. Brown, P. Brune, K. Buschelman, L Dalcin, A. Dener, V. Eijkhout, W.D. Gropp, D. Karpeyev, D. Kaushik, M.G. Knepley, D.A. May, L.C. McInnes, R.T. Mills, T. Munson, K. Rupp, P. Sanan, B.F. Smith, S. Zampini, H. Zhang, and H. Zhang. PETSc users manual. Technical Report ANL-95/11—Revision 3.11, Argonne National Laboratory, 2019.
65. S. Balay, S. Abhyankar, M.F. Adams, J. Brown, P. Brune, K. Buschelman, L Dalcin, A. Dener, V. Eijkhout, W. D. Gropp, D. Karpeyev, D. Kaushik, M.G. Knepley, D.A. May, L.C. McInnes, R.T. Mills, T. Munson, K. Rupp, P. Sanan, B.F. Smith, S. Zampini, H. Zhang, and H. Zhang. PETSc Web page, 2019.
66. Keita Yoshioka, Dmitri Naumov, and Olaf Kolditz. On Crack Opening Computation in Variational Phase-Field Models for Fracture. *Computer Methods in Applied Mechanics and Engineering*, under review.
67. M. Blatt, A. Burchardt, A. Dedner, Ch. Engwer, J. Fahlke, B. Flemisch, Ch. Gersbacher, C. Gräser, F. Gruber, Ch. Grüninger, D. Kempf, R. Klöfkorn, T. Malkmus, S. Müthing, M. Nolte, M. Piatkowski, and O. Sander. The Distributed and Unified Numerics Environment, Version 2.4. *Archive of Numerical Software*, 4(100):13–29, 2016.
68. P. Bastian, F. Heimann, and S. Marnach. Generic implementation of finite element methods in the Distributed and Unified Numerics Environment (DUNE). *Kybernetika*, 46:294–315, 2010.
69. Martin S. Alnæs, Jan Blechta, Johan Hake, August Johansson, Benjamin Kehlet, Anders Logg, Chris Richardson, Johannes Ring, Marie E. Rognes, and Garth N. Wells. The fenics project version 1.5. *Archive of Numerical Software*, 3(100), 2015.
70. Hans-Joachim Bungartz, Florian Lindner, Bernhard Gatzhammer, Miriam Mehl, Klaudius Scheufele, Alexander Shukaev, and Benjamin Uekermann. preCICE—a fully parallel library for multi-physics surface coupling. *Computers and Fluids*, 141:250–258, 2016. Advances in Fluid-Structure Interaction.
71. Christophe Geuzaine and Jean-François Remacle. Gmsh: A 3-d finite element mesh generator with built-in pre- and post-processing facilities. *International Journal for Numerical Methods in Engineering*, 79(11):1309–1331, 2009.

Appendix A

Ergebnisse des GeomInt-Vorhabens

Die wichtigsten erreichten Ergebnisse des GeomInt-Vorhabens werden entsprechend der Arbeitspakete (APs) im Folgenden zusammengestellt (Abschn. A.1–A.3). Eine geografische Verortung der Aktivitäten ist in der Abb. 1.4 dargestellt. Die APs sind eng verknüpft mit den Untertagelaboren (URLs) Mt. Terri (AP1/AP2), Springen (AP2) und Reiche Zeche (AP3). Aus den URLs stammt das Probenmaterial für die geotechnischen Labore in Freiberg (TUBAF), Kiel (CAU) und Leipzig (IfG) (Abschn. A.4.1). Daten aus In-situ-Versuchen in Mt. Terri (BGR) und der Grube Springen (IfG) fanden Eingang in die Modellierungsplattform des GeomInt-Vorhabens (Abschn. A.4.2) unter Koordination des UFZ an der sich alle Projektpartner beteiligt haben. Als Synthese-Instrument wurden sogenannte Modell-EXperimente (MEX) definiert, in denen experimentelle und Modellierungsarbeiten von Beginn an systematisch zusammengeführt wurden (Abschn. A.4).

Bisher wurden bereits 12 GeomInt-Publikationen in internationalen Fachzeitschriften und 3 in Proccedings veröffentlicht, 3 Manuskripte sind zur Begutachtung eingereicht und 6 Manuskripte sind in Vorbereitung für die Geo:N Topical Collection in Environmental Earth Sciences.[1] Eine umfassende Darstellung der Ergebnisse des Gesamtvorhabens befindet sich im vorliegenden GeomInt-Buch.

A.1 AP1: Wegsamkeiten Durch Quell- und Schrumpfungsprozesse

Zur Gewinnung von Probenmaterial wurden im Felslabor Mt. Terri von der BGR zwei Bohrkampagnen durchgeführt. Neben der geologischen Charakterisierung der Kerne wurde das Gestein vom Bohrloch aus durch elektrische Resistivitätstomo-

[1] https://link.springer.com/journal/12665/topicalCollection/AC_efd2f04ddbcca9be8213d04bf1e1fd42.

Abbildung A.1 Messung von Quell-Dehnungen in und senkrecht zur Schichtungsebene (CAU)

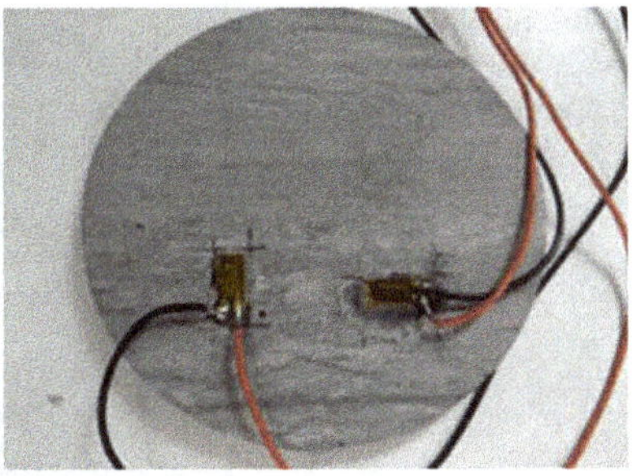

graphie und Miniseismik untersucht. Wissenschaftler der BGR und der CAU Kiel führten die Probenahme durch, bei der mehr als 20 m Kernproben von Opalinuston gewonnen wurden. Diese Kerne sind in Laborversuchen weiter analysiert worden, um Diskontinuitäten zu untersuchen (Abbildung A.1).

Zu diesem Zweck wurden an der CAU Kiel und am IfG Leipzig das Schrumpf- und Quellverhalten von Tonstein unter freien Randbedingungen der Saugspannung und Temperatureinfluss sowie das Verschließen von Rissen im Tonstein durch Aufquellen untersucht. Die Erkenntnisse aus diesen Laborexperimenten wurden erfolgreich auf numerische Modellierungsansätze, wie die Lattice-Element-Methode auf einer Mikro- und Mesoskala der Materialabbildung, übertragen [1–3]. Ferner wurden erste hydraulisch-mechanisch gekoppelte Modellierungen von saisonalen Sättigungs-/Entsättigungsprozessen im In-situ Maßstab für das CD/LP-Experiment im Felslabor Mt. Terri mittels FEM (OpenGeoSys) durchgeführt.

A.2 AP2: Wegsamkeiten Durch Druckgetriebene Perkolation

Die druckgetriebene Perkolation von Fluiden in einer geologischen Verwerfung im Tonstein wurde auf In-situ-Skala an Hand des Fault-Slip (FS) Experimentes im Felslabor Mt. Terri analysiert und mit Modellergebnissen, die durch der Lower-dimensional Interface Element (LIE) Methode gewonnen wurden, verglichen [5]. Darüber hinaus wurden Methoden zur Beschreibung der Ausbreitung von Fluiden auf den Grenzflächen polykristallinen Gesteins verfeinert und erweitert. Dies geschah in enger Wechselwirkung mit Laborversuchen, in denen Permeabilitäten, Rissschließung und Verheilung unter unterschiedlichen Bedingungen mit Gas und Sole an Steinsalz und Tonstein gemessen wurden. Im Ergebnis konnten quantitative Voraussagen getroffen und das Spektrum an beschreibbaren Prozessen erweitert werden (Abbildung A.2).

Ein systematischer Vergleich von verschiedenen Berechnungsmethoden (LIE, Phasenfeld-Methode, Non-Local Deformation) für druckgetriebene Perkolation wurde in [6] beschrieben. Da sich Kriech- und Schädigungsprozesse im Steinsalz auf

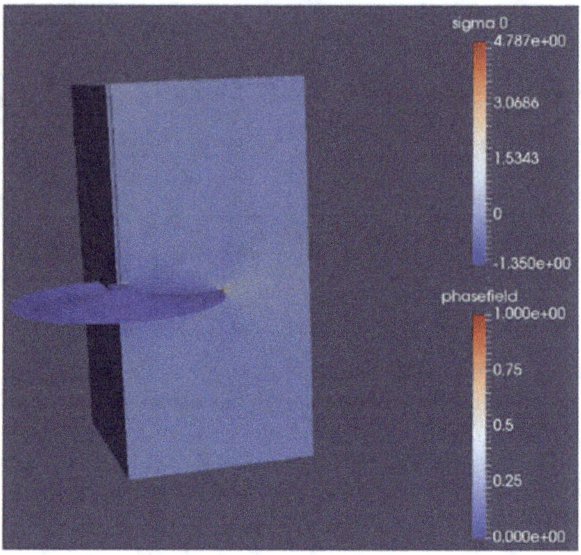

Abbildung A.2 Simulation von Bruchvorgängen mit der Phasenfeld-Methode [4]

sehr unterschiedlichen Zeitskalen entwickeln und die Prozessbeschreibung auf stark nichtlinearen Gleichungssystemen beruht, wurde ein adaptives Zeitschrittverfahren für implizite FEM-Simulationen implementiert und anhand monotoner sowie zyklischer Belastungsversuche an Steinsalz getestet [7]. Mit dem Verfahren steigen sowohl die Effizienz als auch die Genauigkeit der Simulationen, wobei letztere durch den Nutzer sehr konkret mittels einer definierten oberen Fehlerschranke kontrolliert werden kann. Alle o.g. numerischen Methoden wurden in die wissenschaftliche Open-Source-Software OpenGeoSys implementiert und stehen damit über die GeomInt-Modellierungsplattform zur Verfügung.

A.3 AP3: Wegsamkeiten Durch Spannungsumlagerungen

Im Freiberger Geotechnik-Labor wurden die experimentellen Grundlagen zum Verständnis der Grundlagen für die Migration von Fluiden entlang von Diskontinuitäten in kristallinem Gestein mit unterschiedlichem Gefüge gelegt. Um das Systemverständnis weiter zu verbessern, wurden Gesteine mit Variation der Korngröße (Kristallgröße) über mehrere Größenordnungen untersucht—sehr grob-kristalliner Granit mit porphyrischem Gefüge bis zu fein-körnigem, dichtem Basalt. Die geometrischen Charakteristika sämtlicher Kluftflächen vor deren Belastung wurden mit einem 3D-Scanner flächenhaft erfasst und ausgewertet (Abbildung A.3).

Den Extremfall stellte jeweils eine makroskopisch glatte Kluft dar, die durch einen Sägeschnitt mit der Diamantkreissäge hergestellt wurde. Neben mechanischen Versuchsrandbedingungen mit konstanter Auflastspannung bei quasi-statischer

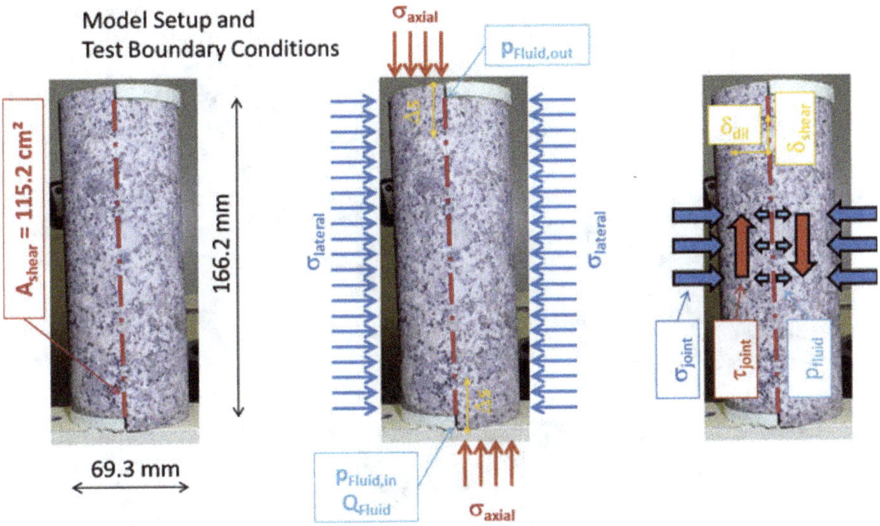

Abbildung A.3 Vorbereitung des Kernmaterials

Belastung und dynamisch-zyklischer Anregung wurden definierte Steifigkeiten des umgebenden Gebirges im Labor vorgegeben. Durch die Realisierung hydromechanisch gekoppelter Szenarien konnte die hydraulische Wirksamkeit der Diskontinuitäten unter den sich ändernden Spannungsrandbedingungen experimentell beobachtet werden. Der lokale Schädigungszustand der Probe wurde durch Scannen der Kluftflächen nach jedem Belastungsschritt dokumentiert. Im Ergebnis dieses klar definierten, mehrstufigen Vorgehens war es möglich, die experimentellen Grundlagen zur Abbildung typischer Belastungssituationen von potentiellen Wegsamkeiten im Kristallingestein zu schaffen. Durch den Fokus auf gekoppelte mechanische und hydraulische Phänomene, die zum Teil aus dem Probematerial selbst (z.B. Korngröße, Kluftgeometrie, lokal unterschiedliche Festigkeiten, lokale Schädigung) zum Teil auch aus den Belastungsrandbedingungen (z.B. quasi-statisch vs. dynamisch, Steifigkeit des umgebenden Gebirges) herrühren, konnte das integrale Verhalten von Diskontinuitäten im kristallinen Gestein dokumentiert werden. Ein weiterer Schwerpunkt lag in der Entwicklung eines numerischen Algorithmus zur Berechnung von scherbeanspruchten Klüften im kristallinen Gestein. Da als wesentliche Modelleingangsparameter neben den makroskopischen Gesteinseigenschaften primär die geometrischen Eigenschaften der Kluftufer gewählt wurden, können sowohl konkrete Einzelsituationen nachgerechnet al.s auch das Verhalten von Diskontinuitäten prädiktiv abgeschätzt werden [8]. Der erstgenannte Fall dient zur Kalibrierung des numerischen Modells mit Hilfe der durchgeführten Laborversuche und zur Berechnung ausgewählter, gut dokumentierter Einzelfälle. Im zweitgenannten Fall können Sets von Klüften mit typischen Eigenschaften auf Basis geologischer Beobachtungen anhand statistisch generierter Parametersätze erstellt werden. Durch den sehr geringen Zeitaufwand zur Lösung einer Einzelberechnung ist das Mod-

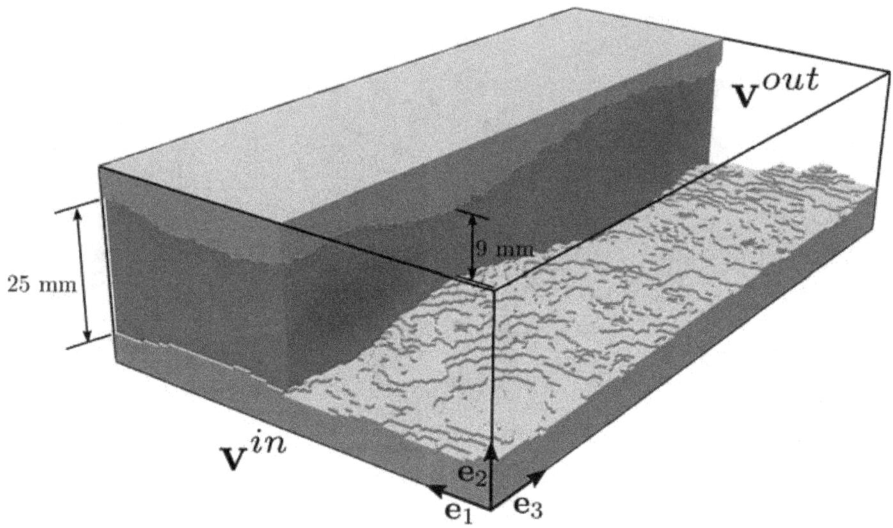

Abbildung A.4 Strömungssimulation in Klüften mit der FEM [9]

ell gut geeignet das erwartbare Verhalten der Diskontinuitäten im Kristallin als Ergebnis einer Vielzahl von Einzelberechnungen zu ermitteln und (als Einhüllende) darzustellen. Mit diesem Ansatz können trotz unvermeidbarer statistischer Unsicherheiten hinsichtlich der konkreten Ausbildung der geometrischen und mechanischen Eigenschaften von geologischen Strukturen Aussagen zum Verhalten von Diskontinuitäten getroffen und in geotechnischen Studien berücksichtigt werden (Abbildung A.4).

Bezüglich der Modellierung wurde die hydro-mechanische Interaktion im kristallinen Gebirge (Granit, Gneis) mittels gekoppelten mehrskaligen numerischen Methoden (Smoothed Particle Hydrodynamics—SPH und Finite Elemente—FEM) analysiert. In kleinskaligen Untersuchungen einzelner Risse wurde gezeigt, dass der Effekt von Rissrauigkeiten Auswirkungen auf Transport- und Druckdiffusionsprozesse hat [10]. Sind Rissgeometrie und Oberflächenrauigkeit bekannt, kann numerisch die effektive hydraulische Leitfähigkeit mit effizienten direkten numerischen Verfahren (SPH) mit hoher Genauigkeit berechnet werden [10]. Im Rahmen der hybrid-dimensionalen FEM wurden die Rissrauigkeiten mittels eines neuen effektiven Spannungskonzept-Ansatzes integriert [11]. Die numerische Effizienz und Genauigkeit wird dabei durch eine starke oder eine schwache Kopplung mittels nichtkonformen Netzen realisiert. In hochauflösenden zeit-harmonischen Laborexperimenten konnte eine gute Übereinstimmung des Modells für nichtlineare Prozesse gezeigt werden, bei denen die hydraulische Leitfähigkeit stark von der Deformation des Risses einschließlich der Deformation von Kontaktflächen teilweise geschlossener Risse abhängt [12]. Schädigungs- und Bruchprozesse im Granit wurden mit nichtlokalen Ansätzen modelliert und konnten sowohl lokal aufgelöste als auch integrale Messgrößen aus Dreipunktbiegeversuchen erfolgreich nachbilden, ohne dass ein gesondertes Anpassen der Materialparameter an diese Versuche nötig gewesen wäre [13].

Erste Übertragungen der Modelle auf die Feldskala, und damit die numerische Interpretation von transienten Pumpversuchen zur Charakterisierung der hydromechanischen Eigenschaften geklüfteter kristalliner Gesteinsformationen wurden in enger Kooperation mit den experimentellen Daten aus Pumpversuchen in Untertagelabors durchgeführt [14] (In-situ-Stimulation and Circulation experiment—ISC, Grimsel Test Site) und [9] (STIMTEC, Reiche Zeche, Freiberg).

A.4 Synthese

Auf die Synthese von Experiment und Modell wurde von Anfang an großer Wert gelegt, daher wurden zu Beginn des Vorhabens sogenannte Modell-EXperimente (MEX) definiert. Dabei handelt es sich um eine systematische, umfassende Untersuchung der zu Diskontinuitäten führenden Prozesse (Quellen/Schrumfen, Perkolation/Heilung, Spannungsumlagerung) durch die Kombination von Experiment und Modell.

A.4.1 Experimentelle Plattform

Die experimentellen Analysen wurden bei den beteiligten Projektpartnern an unterschiedlichen Materialien durchgeführt. Die Untersuchungen am Tonstein, Steinsalz und dem kristallinen Gestein erforderten spezialisierte Laborexperimente. Standard-Triaxialversuche, um die Parameter im Versagenszustand zu bestimmen, können in allen Laboren in Freiberg, Leipzig und Kiel zu Vergleichszwecken durchgeführt werden. Weitergehende Untersuchungen wurden in enger Absprache in den spezialisierten Labors durchgeführt. Perkolationsversuche an zylindrischen Salzproben unter hydrostatischen und deviatorischen Bedingungen können im Leipziger sowie an Tonstein unter triaxialen Bedingungen in der Würfeldruckpresse im Kieler Labor durchgeführt werden. Dort erfolgten ebenfalls die Analyse des Quell- und Schrumpfverhaltens des Tonsteins unter verschiedenen Matrixspannungen sowie die Bestimmung der Bruchzähigkeit mittels Dreipunktbiegeversuch sowie der Zugfestigkeit aus Splitversuchen bei unterschiedlichen thermischen und hydraulischen Randbedingungen. Die Analyse des Scherverhaltens von kristallinen Gesteinen in Fels-Rahmenscherversuchen erfolgte im Freiberger Labor. Für die kleinskalige Charakterisierung von Rissen in Gesteinsproben (typische Größen: $D = 30\,\mathrm{mm}$) wurde ein hochauflösendes Röntgentomographiegerät (XRCT) im Stuttgarter Labor verwendet. Die Oberflächen- und XRCT-Scans können für weiterführende Untersuchungen in der Fortsetzung des Vorhabens weiter verwendet werden. Die Zusammenarbeit der Projektpartner im experimentellen Bereich ist in der Abb. 1.4 grafisch dargestellt—insbesondere die gegenseitige Bereitstellung und Aufbereitung des

Probenmaterials aus den verschiedenen URLs für die geotechnischen Labore war ein erheblicher Mehrgewinn für die experimentelle Plattform in GeomInt.

A.4.2 Modellierungs-Plattform

Ein wesentliches Ergebnis des GeomInt-Vorhabens ist die Entwicklung einer Modellierungsplattform. Dabei wurden verschiedene Modellierungskonzepte weiterentwickelt und verifiziert, die insbesondere die Evolution von Diskontinuitäten auf verschiedenen Raum-Zeit-Skalen abbilden können. In einer Reihe von Modell-EXperiment-Studien (MEX) wurden diese Konzepte für die verschiedenen Prozesstypen (AP1-AP3) miteinander verglichen (proof-of-concept), um die Vor- und Nachteile der einzelnen Methoden herauszuarbeiten [6]. Ein besonderer Schwerpunkt lag in der Entwicklung von Open-Source-Software [15], um eine kontinuierliche Entwicklung der Plattform auch in der Zukunft zu gewährleisten. Zur Software-Entwicklung und „best-practices“ wurde ein Workshop für alle GEO:N Verbundprojekte im April 2020 in Leipzig durchgeführt.

Appendix B
Symbols

This section provides a comprehensive compilation of used symbols throughout the book. As concepts and results are compiled from various disciplines (i.e. solid and fluid as well as porous media mechanics, geomechanics, geotechnics etc.) representing both theoretical and experimental work—a certain degree of redundancy is still necessary in this first book edition and will be homogenized further in subsequent work (Table B.1).

Table B.1 Table of symbols

Symbol	Parameter	Unit
Greek Symbols		
α	Biot coefficient	–
α	Intergranular radius	–
α	Thermal expansion coefficient	K^{-1}
α	van Genuchten parameter	m^{-1}
α_k	Kinetic isotope fractionation factor	–
α_L	Longitudinal dispersion length	m
α_{Li}	Linear thermal expansion coefficient	K^{-1}
α_R	Randomness factor	–
α_s	Shrinkage and swelling coefficient	–
α_T	Transversal dispersion length	m
α_V	Volumetric thermal expansion coefficient	K^{-1}
β	Cubic thermal expansion coefficient	K^{-1}
β	fitting parameter	–
β_c	Burial constant	–
$\beta^{\mathfrak{f}}$	Fluid compressibility	Pa^{-1}
γ	Activity coefficient for dissolved species	–
γ	Dimensionless temperature	–
γ	Angle of inclination	–
γ_l	First-order degradation rate	day^{-1}
Γ	Domain boundary	–
Γ_-^{Fr}	Negative fracture surface domain	m^2
Γ_+^{Fr}	Positive fracture surface domain	m^2
δ	Dirac delta function	–
δ	Fracture aperture	m
2δ	Fracture width	m
Δ	Half of aspect ratio	–
ΔX	Change of quantity X	[X]
ϵ	Error tolerance	–
ϵ	Strain	–
$\dot{\epsilon}$	Strain rate	s^{-1}
$\boldsymbol{\epsilon}$	Strain tensor	–
$\boldsymbol{\epsilon}_{\mathrm{el}}$	Elastic train tensor	–
$\boldsymbol{\epsilon}_{\mathrm{pl}}$	Plastic train tensor	–

(continued)

Table B.1 (continued)

Symbol	Parameter	Unit
ϵ_b	Element strain	–
$\epsilon_{i,j,k,m}$	Element strain alignment	–
ϵ_f	Failure strain	–
ϵ_{flex}	Flexural strain	–
ϵ_p	Peak strain	–
$\epsilon_{\mathbb{R}}$	Continuum strain	–
ϵ_v	Volume plastic strain	–
ϵ_V	Volume strain	–
η	Porosity	–
η^{fR}	Effective dynamic viscosity	Pa s
η_{M}	Maxwell viscosity	Pa d
η_{K}	Kelvin viscosity	Pa d
θ^*	Apparent dip angle	deg
θ^*_{max}	Maximum apparent dip angle	deg
κ	Thermal conductivity	$\text{W m}^{-1}\text{ K}^{-1}$
$\kappa_{i,j}$	Curvature strain	–
λ	Lamé coefficients	GPa
λ	Plastic multiplier	–
λ	Thermal conductivity	$\text{W m}^{-1}\text{ K}^{-1}$
λ_{arith}	Arithmetic effective thermal conductivity	$\text{W m}^{-1}\text{ K}^{-1}$
λ_b	Bulk thermal conductivity	$\text{W m}^{-1}\text{ K}^{-1}$
λ_c	Virgin compression index	–
λ_{eff}	Effective thermal conductivity	$\text{W m}^{-1}\text{ K}^{-1}$
λ_{geom}	Geometric effective thermal conductivity	$\text{W m}^{-1}\text{ K}^{-1}$
λ_{harm}	Harmonic effective thermal conductivity	$\text{W m}^{-1}\text{ K}^{-1}$
λ_p	Hardening parameter	–
λ_{pm}	Thermal conductivity of porous medium	$\text{W m}^{-1}\text{ K}^{-1}$
λ_s	Solid thermal conductivity	$\text{W m}^{-1}\text{ K}^{-1}$
μ	Lamé coefficient	GPa
μ	Dynamic viscosity	Pa s
μ_0	Reference dynamic viscosity	Pa s
ν	Poisson's ratio (Poisson number)	–
ν_{dyn}	Dynamic Poisson's ratio	–
ν^k_f	Kinematic fluid viscosity	$\text{m}^2\text{ s}^{-1}$

(continued)

Table B.1 (continued)

Symbol	Parameter	Unit
$\boldsymbol{\nu}_{\mathrm{M}}$	Maxwell viscosity tensor	Pa d
$\boldsymbol{\nu}_{\mathrm{K}}$	Kelvin viscosity tensor	Pa d
ρ	Electrical resistivity	Ω m
ρ_a	Apparent electrical resistivity	Ω m
ρ (ϱ)	Phase density	kg m^{-3}
$\rho^{\{}$	Fluid density	kg m^{-3}
ρ_f	Fluid density	kg m^{-3}
$\rho^{\mathfrak{f}R}$	Effective fluid density	kg m^{-3}
ρ_s	Density of solid	kg m^{-3}
ρ_w	Density of water	kg m^{-3}
ρ_d^s	Density of bentonite bulk	kg m^{-3}
ρ_0	Reference fluid density	kg m^{-3}
ϱ_{SR}	Real density of solid	kg m^{-3}
ϱ_{LR}	Real density of liquid	kg m^{-3}
ϱ_{IR}	Real density of ice	kg m^{-3}
$\boldsymbol{\sigma}$	Cauchy stress tensor	Pa
σ'	Effective stress	Pa
σ_a	Contact stress	Pa
σ_c	Critical stress	Pa
σ_c	Compressive strength	Pa
σ_{con}	Confining stress	Pa
σ_{dev}	Deviatoric stress	Pa
σ_{eff}	Effective stress	Pa
σ_{flex}	Flexural stress	Pa
σ_{iso}	Isotropic stress	Pa
σ_{loc}	Locally acting stress	Pa
σ_n	Normal stress	Pa
$\sigma_{\mathbb{R}}$	Continuum stress	Pa
$\sigma_{\mathrm{max}}^{sw}$	Tested maximum swelling stress	Pa
σ_{sp}	Splitting stress	Pa
σ_t	Tensile strength	Pa
σ_V	von Mises equivalent stress	Pa
$\boldsymbol{\tau}$	Shear stress tensor	Pa
τ	Shear stress	Pa
τ_p	Peak shear stress	Pa
τ_r	Residual shear stress	Pa
ϕ	Porosity	–
ϕ_0	Phase shift angle	deg

(continued)

Table B.1 (continued)

Symbol	Parameter	Unit
ϕ, (Φ)	Friction angle	deg
φ_b	Basic friction angle	deg
χ	Bishop coefficient	–
ψ	Dilatancy angle	deg
ψ	Strain energy density	$\mathrm{J\,m^{-3}}$
ψ_{total}	Total suction	Pa
ω	Intergranular thickness	m
ω	Saturation index	–
ω	Water content	–
Roman Symbols		
a	Specific surface area	$\mathrm{m^2\,m^{-3}}$
a	Grid constant	m
a_{flex}	Flexural notch length	m
a_h	Hydraulic aperture	m
$\dot{a}$	Effective diameter of ion	m
a^{σ}	Activity of stressed solid	–
A	Surface area	$\mathrm{m^2}$
A_0	Active surface area after Grasselli	–
A_b	Area of element	$\mathrm{m^2}$
b	Fracture aperture	m
$\mathbf{b}$	Body force vector	N
b_0	Initial fracture aperture	m
b_h	Fracture hydraulic aperture	m
b_m	Fracture mechanical aperture	m
B_{flex}	Flexural sample thickness	m
c	Cohesion	–
c	Normalized concentration	–
c	Specific heat capacity of a cell	$\mathrm{J\,K^{-1}}$
c_f	Specific fluid heat capacity	$\mathrm{J\,kg^{-1}\,K^{-1}}$
$c_{\{}$	Speed of sound	$\mathrm{m\,s^{-1}}$
c_p	Heat capacity	$\mathrm{J\,kg^{-1}\,K^{-1}}$
C	Concentration	$\mathrm{kg\,m^{-3}}$
C	Roughness after Grasselli	–
C^h_{eq}	Solubility under hydrostatic pressure	$\mathrm{mol\,m^{-3}}$
C_i	Intergranular concentration	$\mathrm{mol\,m^{-3}}$
$C_{i,j,k,m}$	Element stiffness matrix	Pa

(continued)

Table B.1 (continued)

Symbol	Parameter	Unit
C_p	Pore-space concentration	$\mathrm{mol\,m^{-3}}$
Cp	Heat capacity	$\mathrm{J\,kg^{-1}\,K^{-1}}$
Cr	Courant number	–
$C_{\mathbb{R}}$	Continuum stiffness	Pa
$\mathbb{C}$	Elasticity tensor	Pa
d	Order parameter	–
$d_{i,j}$	Curvature stiffness	Pa
D	Diffusivity coefficient	$\mathrm{m^2\,s^{-1}}$
D_{cyl}	Diameter of cylindrical sample	m
D_f	Intergranular diffusion coefficient	$\mathrm{m^2\,s^{-1}}$
e	Void ratio	–
e_0	Initial void ratio	–
E	Young's modulus	Pa
E	Energy total	J
E_b	Element Young's modulus	Pa
E_{dyn}	Dynamic Young's modulus	Pa
E_{flex}	Flexural Young's modulus	Pa
f_{flex}	Flexural force	N
f_p	Peak axial load	N
f_{sp}	Splitting force	N
f_x	Axial force	N
f_y	Shear force	N
F_b	Element forces	N
F_n	Normal force	N
F_{shear}	Shear-off force	N
F_{slide}	Sliding force	N
g	Gravitational coefficient	$\mathrm{m\,s^{-2}}$
g	Plastic potential	J
$\mathbf{g}$	Gravity vector	$\mathrm{m\,s^{-2}}$
$\mathbf{g}$	Gravitational acceleration	$\mathrm{m\,s^{-2}}$
g_c	Fracture toughness	$\mathrm{N\,m^{-1}}$
G	Gibbs energy	J
G	Gibbs energy	$\mathrm{J\,mol^{-1}}$
G	Shear modulus	Pa
G_b	Element shear modulus	Pa
G_c	Fracture toughness	Pa m
G_{dyn}	Dynamic shear modulus	Pa
G_{F}	Plastic potential	J

(continued)

Table B.1 (continued)

Symbol	Parameter	Unit
G_{M}	Maxwell shear modulus	–
G_{K}	Kelvin shear modulus	–
G_s	Specific gravity	$\mathrm{kg\,m^{-3}}$
H	Fault height	m
H_{PW}	Pipe water level	m
h	Hardening parameter	–
h	Height of sample	m
h_b	Thermal conductance	$\mathrm{W\,K^{-1}}$
h_c	Thickness of colmation layer	m
h_f	Freshwater hydraulic head	m
h_s	Saltwater hydraulic head	m
Δh_{I}	Specific enthalpy of fusion	J
I	Ionic strength	–
I	Identity tensor	–
$I(L)$	intensity values of X-rays	–
I_1	First principal invariant of the stress tensor	Pa
I_b	Element moment of inertia	$\mathrm{kg\,m^2}$
$\mathbf{J}_f$	Fluid mass flux	$\mathrm{kg\,m^{-3}\,s{-}1}$
J_2	Second principal invariant of deviatoric stress tensor	$\mathrm{Pa^2}$
J_3	Third principal invariant of deviatoric stress tensor	$\mathrm{Pa^3}$
JCS	Joint wall compressive strength	Pa
JRC	Joint roughness coefficient	–
k	Residual stiffness parameter	–
$\mathbf{k}$	Permeability tensor	$\mathrm{m^2}$
k_b	Thermal conductivity	$\mathrm{W\,K^{-1}\,m^{-1}}$
k_c	Swelling/recompression index	–
k_h	Hydraulic conductivity	$\mathrm{m^1\,s^{-1}}$
$k^{\mathfrak{s}}_{Fr}$	Local fracture permeability	$\mathrm{m^2}$
$k^{\mathfrak{s}}_{Fr,t}$	Local transversal fracture permeability	$\mathrm{m^2}$
k_{perp}	Permeability perpendicular to the bedding plane	$\mathrm{m^2}$
k_{parallel}	Permeability parallel to the bedding plane	$\mathrm{m^2}$
k_{rel}	Relative permeability	–

(continued)

Table B.1 (continued)

Symbol	Parameter	Unit
k^+	Dissolution rate constant	$\mathrm{mol\,m^{-2}\,s^{-1}}$
k°	Reaction rate constant	$\mathrm{mol\,m^{-2}\,s^{-1}}$
K_{eff}	Effective thermal conductivity	$\mathrm{W\,K^{-1}\,m^{-1}}$
K_{eq}	Equilibrium constant	–
K_f	Fluids bulk modulus	Pa
K^{f}	Fluid bulk modulus	Pa
$K^{\{}$	Fluid bulk modulus	Pa
K_I	Fracture toughness	$\mathrm{Pa\,m^{0.5}}$
K_I	Stress intensity factor	$\mathrm{Pa\,m^{0.5}}$
K_{Ic}	Fracture toughness	$\mathrm{Pa\,m^{0.5}}$
K_{M}	Maxwell bulk modulus	–
K_{n}	Normal stiffness	$\mathrm{Pa\,m^{-1}}$
K_{s}	Shear stiffness	$\mathrm{Pa\,m^{-1}}$
K_s	Saturated hydraulic conductivity	$\mathrm{m\,d^{-1}}$
K_s	Solids bulk modulus	Pa
ℓ	Regularization length	m
ℓ_c	Internal length of continuum	m
L_b	Element length	m
L'_b	Element contact length	m
L_{cyl}	Length of cylindrical sample	m
m	Mass	kg
m	Fitting parameter	–
m	Van Genuchten parameter	–
M	Kinetic coefficient	$\mathrm{mm^2\,N^{-1}\,s^{-1}}$
M	Slope of critical state line	–
M_b	Element moment	N m
M_w	Molecular mass of water vapor	$18.016\,\mathrm{g\,mol^{-1}}$
n	Porosity	$\mathrm{m^3\,m^{-3}}$
n	van Genuchten parameter	–
$n_{i,j,k,m}$	Element alignment	–
$\mathbf{n}$	Normal of fracture surface	–
$\mathbf{n}^+$	Normal of positive fracture surface	–
$\mathbf{n}^-$	Normal of negative fracture surface	–
N_b	Element's total number of bonds	–
N_c	Number of Voronoi cells in the boundary	–
p, P	Pressure [Pa]	$\mathrm{kg\,s^{-1}\,m^{-1}}$

(continued)

Table B.1 (continued)

Symbol	Parameter	Unit
p_s	Mean stress	Pa
p_{scn}	Isotropic pre-consolidation pressure	Pa
p^+	Fluid pressure at positive fracture surface	Pa
p^-	Fluid pressure at negative fracture surface	Pa
P	Load	Pa
P_c	Capillary pressure	Pa
P_d	air entry pressure	Pa
$P_{\mathfrak{f}}$	Averaged fluid pressure	Pa
P_{vap}	Vapor pressure	Pa
Pe	Péclet number	–
q	Source/sink term	–
q	Shear stress	Pa
q	Heat source	W
q	Darcy velocity	m s^{-1}
$\mathbf{q}$	Darcy velocity vector	m s^{-1}
$\dot{q}$	Heat density	W kg^{-1}
q_{ave}	Average Heat flux	W m^{-2}
q_b	Heat flow	W
q^b_{cell}	Heat flow from boundary Voronoi cells	W
q_{lk}	Leak-off	Pa s^{-1}
q_{T}	Heat flux through unit area	W m^{-2}
Q	Ion activity product	–
Q_{leak}	Leakage flow	$\text{m}^3\,\text{s}^{-1}$
Q_ω	Source/sink term	$\text{kg m}^{-3}\,\text{s}^{-1}$
Q_f	Source/sink term	$\text{kg m}^{-3}\,\text{s}^{-1}$
R	Universal gas constant (8.31432)	$\text{J mol}^{-1}\,\text{K}^{-1}$
R	Transformation matrix	–
R_c	Contact area ratio	–
R^b_h	Hydraulic resistance	s^{-1}
$(R)'$	First stiffness coefficient	Pa
$(R)''$	Second stiffness coefficient	Pa
Ra	Rayleigh number	–
Ra_{crit}	Critical Rayleigh number	–
s	Soil suction	kPa
S	Saturation	–
S	Storage	Pa^{-1}

(continued)

Table B.1 (continued)

Symbol	Parameter	Unit
S_e	Effective saturation	–
S_{eff}	Effective saturation	–
S_{flex}	Flexural sample span	m
$S_{\max}$	Maximum water saturation	–
Sr	Saturation degree	–
S_r	Residual saturation	–
S_r	Residual water saturation	–
SA	Reactive surface area	m^2
t	Time	s
$\mathbf{t}$	Traction force vector	N
$\mathbf{t}_c$	Traction force vector at interface	N
T	Temperature	K
T	Period	d
T_{tot}	total simulation time	d
T_c	Top temperature (cold)	K
T_h	Bottom temperature (hot)	K
T_{iso}	Isotropic thermal loading	K
T_{init}	Initial temperature	K
T_v	Approximation temperature	K
u	Displacement	m
$\mathbf{u}$	Displacement vector	m
u_b	Element displacement	m
u_n	Normal displacement	m
u_s	Shear displacement	m
U_a^b	Axial strain energy	N m
U_{cell}	Cell strain energy	N m
U_m^b	Moment strain energy	N m
$U_{\mathbb{R}}$	Continuum strain energy	N m
U_s^b	Shear strain energy	N m
U_t^b	Total stored strain energy	N m
v	Volume of a Voronoi cell	m^3
$\mathbf{v}$	Velocity vector	$m\,s^{-1}$
v_p, v_s	Ultrasonic velocities (p-s wave related)	$m\,s^{-1}$
V	Volume	m^3
V_{cav}	Volume of the cavity	m^3
V_m	Molar volume	$m^3\,mol^{-1}$
V_{mol}	Molecular volume	m^3
$V_{\mathbb{R}}$	Continuum volume	m^3

(continued)

Table B.1 (continued)

Symbol	Parameter	Unit
V_u	Cell volume	m^3
w	Margules parameter	$J\,mol^{-1}$
w_a	Test function of auxiliary element	–
$\mathbf{w}_\mathfrak{f}$	Relative fluid velocity	$m\,s^{-1}$
$\mathbf{w}_\mathfrak{f}^-$	Relative fluid velocity positive fracture surface	$m\,s^{-1}$
$\mathbf{w}_\mathfrak{f}^+$	Relative fluid velocity negative fracture surface	$m\,s^{-1}$
$\mathbf{w}_\mathfrak{f}^l$	Relative longitudinal fluid velocity	$m\,s^{-1}$
$\mathbf{w}_\mathfrak{f}^t$	Relative transversal fluid velocity	$m\,s^{-1}$
W	SPH kernel function	
$\mathbf{W}_\mathfrak{f}$	Averaged relative fluid velocity	$m\,s^{-1}$
W_{flex}	Flexural sample width	m
$\mathbf{x}$	Position vector	m
$\hat{x}_{\text{cell}}$	Relative coordinates	m
X	Molar fraction	–
Z	Ionic charge	–
Indices		
$\parallel$	Co-linear direction	
$\perp$	Orthogonal direction	
$+$	Positive fracture surface	
$-$	Negative fracture surface	
e	Efficient value	
f	Fluid phase	
$\mathfrak{f}$	Fracture	
l	Longitudinal fracture direction	
R	Effective value	
s	Solid phase	
t	Transversal fracture direction	
w	Water	
0	Reference value	
Operators		
div, $\nabla\cdot$	Divergence operator	
grad, ∇	Nabla, gradient operator	
tr	Trace	

References

1. A.S. Sattari, H.B. Motra, Z.H. Rizvi, and F. Wuttke. A new lattice element method (lem) with integrated interface elements to determine the effective thermal conductivity of rock solids under thermo-mechanical processes. *Springer Series in Geomechanics and Geoengineering*, 0(217729):266–275, 2019. cited By 0.
2. D. Shrestha, Z.H. Rizvi, and F. Wuttke. Effective thermal conductivity of unsaturated granular geocomposite using lattice element method. *Heat and Mass Transfer/Waerme- und Stoffuebertragung*, 55(6):1671–1683, 2019. cited By 4.
3. Z.H. Rizvi, H.H. Zaidi, S.J. Akhtar, A.S. Sattari, and F. Wuttke. Soft and hard computation methods for estimation of the effective thermal conductivity of sands. *Heat and Mass Transfer*, Feb 2020.
4. K. Yoshioka, D. Naumov, and O. Kolditz. On Crack Opening Computation in Variational Phase-Field Models for Fracture. *Computer Methods in Applied Mechanics and Engineering*, under review.
5. B. Vowinckel, A. Alfarra, J. Maßmann, D. Naumov, G. Zielfe, O. Kolditz, and T. Nagel. Computational modeling of hydraulically induced fault slip in argillaceous rock using lower-dimensional interface elements with local enrichment of the displacement field. *preparation for the Geo:N Topical Collection in Environmental Earth Sciences*, 2020.
6. K. Yoshioka, F. Parisio, D. Naumov, R. Lu, O. Kolditz, and T. Nagel. Comparative verification of discrete and smeared numerical approaches for the simulation of hydraulic fracturing. *GEM - International Journal on Geomathematics*, 10(1), 2019.
7. N. Zhang and T. Nagel. Error-controlled implicit time integration of elasto-visco-plastic constitutive models for rock salt. *International Journal for Numerical and Analytical Methods in Geomechanics*, in press:1–18, 2020. cited By 0.
8. D. Poetschke and et al. Development of a matlab-code to simulate direct shear tests. *preparation for the Geo:N Topical Collection in Environmental Earth Sciences*, 2020.
9. P. Schmidt, J. Renner, and H. Steeb. Numerical characterization of hydraulically tested fractures at reiche zeche. *preparation for the Geo:N Topical Collection in Environmental Earth Sciences*, 2020.
10. H. Steeb. Hydro-mechanics of porous and granular material – Poro-elasticity and beyond. In *Proceedings of Alert Workshop and School*. 2019.
11. P. Schmidt and H. Steeb. Numerical aspects of hydro-mechanical coupling of fluid-filled fractures using hybrid-dimensional element formulations and non-conformal meshes. *Int J Geomath*, 10(14), 2019.
12. P. Schmidt, J. Renner, and H. Steeb. Non-linear hydromechanics of fractures throughout harmonic testing. *Geophysical Research Letters, submitted*, 2020. to be submitted.
13. F. Parisio, A. Tarokh, R. Makhnenko, D. Naumov, X.-Y. Miao, O. Kolditz, and T. Nagel. Experimental characterization and numerical modelling of fracture processes in granite. *International Journal of Solids and Structures*, 163:102–116, 2019. cited By 4.
14. P. Schmidt, N. Dutler, J. Renner, and H. Steeb. Hydro-mechanical analysis of fracture flow and deformation. *Journal of Geophysical Research: Solid Earth, submitted*, 2020. to be submitted.
15. L. Bilke, B. Flemisch, T. Kalbacher, O. Kolditz, R. Helmig, and T. Nagel. Development of open-source porous media simulators: Principles and experiences. *Transport in Porous Media*, 130(1):337–361, 2019. cited By 3.

Permissions

All chapters in this book were first published by Springer; hereby published with permission under the Creative Commons Attribution License or equivalent. Every chapter published in this book has been scrutinized by our experts. Their significance has been extensively debated. The topics covered herein carry significant information for a comprehensive understanding. They may even be implemented as practical applications or may be referred to as a beginning point for further studies.

The contributors of this book come from diverse backgrounds, making this book a truly international effort. We would like to thank all the contributing authors for lending their expertise to make the book truly unique. They have played a crucial role in the development of this book. Without their invaluable contributions this book wouldn't have been possible. They have made vital efforts to compile up to date information on the varied aspects of this subject to make this book a valuable addition to the collection of many professionals and students.

This book was conceptualized with the vision of imparting up-to-date and integrated information in this field. To ensure the same, a matchless editorial board was set up. Every individual on the board went through rigorous rounds of assessment to prove their worth. After which they invested a large part of their time researching and compiling the most relevant data for our readers.

The editorial board has been involved in producing this book since its inception. They have spent rigorous hours researching and exploring the diverse topics which have resulted in the successful publishing of this book. They have passed on their knowledge of decades through this book. To expedite this challenging task, the publisher supported the team at every step. A small team of assistant editors was also appointed to further simplify the editing procedure and attain best results for the readers.

Apart from the editorial board, the designing team has also invested a significant amount of their time in understanding the subject and creating the most relevant covers. They scrutinized every image to scout for the most suitable representation of the subject and create an appropriate cover for the book.

The publishing team has been an ardent support to the editorial, designing and production team. Their endless efforts to recruit the best for this project, has resulted in the accomplishment of this book. They are a veteran in the field of academics and their pool of knowledge is as vast as their experience in printing. Their expertise and guidance has proved useful at every step. Their uncompromising quality standards have made this book an exceptional effort. Their encouragement from time to time has been an inspiration for everyone.

The publisher and the editorial board hope that this book will prove to be a valuable piece of knowledge for students, practitioners and scholars across the globe.

Index

V

W

www.ingramcontent.com/pod-product-compliance
Lightning Source LLC
LaVergne TN
LVHW011516200926
848484LV00001B/47

* 9 7 9 8 8 8 8 3 6 0 5 8 3 *